Key Technologies on New Energy Vehicles

Key Technologies on New Energy Vehicles publishes the latest developments in new energy vehicles - quickly, informally and with high quality. The intent is to cover all the main branches of new energy vehicles, both theoretical and applied, including but not limited:

- Control Technology of Driving System
- Hybrid Electric Vehicle Coupling Technology
- Cross Disciplinary Design Optimization Technology
- Single and Group Battery Technology
- Energy Management Technology
- Lightweight Technology
- New Energy Materials and Device
- Internet of Things (IoT)
- Cloud Computing
- 3D Printing
- Virtual Reality Technologies

Within the scopes of the series are monographs, professional books or graduate text-books, edited volumes, and reference works purposely devoted to support education in related areas at the graduate and post-graduate levels.

Jichao Hong

Big Data Analysis and Application of Power Battery Operation Safety

机械工业出版社
CHINA MACHINE PRESS

Springer

Jichao Hong
School of Mechanical Engineering
University of Science and Technology
Beijing
Beijing, China

ISSN 2662-2920 ISSN 2662-2939 (electronic)
Key Technologies on New Energy Vehicles
ISBN 978-981-96-4398-1 ISBN 978-981-96-4399-8 (eBook)
https://doi.org/10.1007/978-981-96-4399-8

Jointly published with China Machine Press Co., Ltd.
The print edition is not for sale in China (Mainland). Customers from China (Mainland) please order the print book from: China Machine Press Co., Ltd.
ISBN of the Co-Publisher's edition: 978-7-111-72607-4

The original submitted manuscript has been translated into English. The translation was done using artificial intelligence. A subsequent revision was performed by the author(s) to further refine the work and to ensure that the translation is appropriate concerning content and scientific correctness. It may, however, read stylistically different from a conventional translation.

This Springer imprint is published by the registered company Springer Nature Singapore Pte Ltd.
The registered company address is: 152 Beach Road, #21-01/04 Gateway East, Singapore 189721, Singapore

Contents

Chapter 1
Overview

1.1 Background of New Energy Vehicle Development

1.1.1 Necessity of New Energy Vehicle Development

1. **Developing New Energy Vehicles is the Only Way for Our Country to Transition from a Major Automobile Country to a Powerful Automobile Country**

   To effectively alleviate energy and environmental pressures, promote the healthy and sustainable development of China's automobile industry, and accelerate the transformation and upgrading of the automobile industry, the State Council issued the "Energy Saving and New Energy Vehicle Industry Development Plan (2012–2020)" (State Council [2012] No. 22, hereinafter referred to as the "Plan") in June 2012. The "Plan" established "pure electric drive as the main strategic direction for the development of new energy vehicles and the transformation of the automobile industry in China", and proposed the goals of "by 2020, the production capacity of pure electric vehicles and plug-in hybrid vehicles will reach 2 million units, and the cumulative production and sales volume will exceed 5 million units" and "by 2020, the average fuel consumption of passenger vehicles produced in that year will be reduced to 5.0 L/100 km".

   In May 2014, General Secretary Xi Jinping pointed out during his inspection in Shanghai that "the development of new energy vehicles is the only way for China to transform from a big automobile country to a strong automobile country", which pointed out the direction for the transformation and upgrading of China's automobile industry.

   In order to implement the strategic deployment of the Party Central Committee and the State Council, scientifically and effectively implement the "Plan", the leaders of the State Council have conducted several special investigations and held meetings to solve major problems encountered in industrial development. The General Office of the State Council has issued the "Guiding Opinions on

J. Hong, *Big Data Analysis and Application of Power Battery Operation Safety*, Key Technologies on New Energy Vehicles, https://doi.org/10.1007/978-981-96-4399-8_1

Accelerating the Promotion and Application of New Energy Vehicles" (Guo Ban Fa [2014] No. 35) and the "Guiding Opinions on Accelerating the Construction of Electric Vehicle Charging Infrastructure" (Guo Ban Fa [2015] No. 73), further promoting the establishment of a long-term stable new energy vehicle policy system. In November 2013, with the approval of the State Council, an inter-ministerial joint meeting system for the development of the energy-saving and new energy vehicle industry was established, led by the Ministry of Industry and Information Technology, with the participation of 18 departments including the Development and Reform Commission, the Ministry of Science and Technology, the Ministry of Finance, and the Ministry of Public Security; 31 provinces (regions, cities) have also established coordination mechanisms for the development of the energy-saving and new energy vehicle industry, building a horizontal coordination and vertical penetration promotion system.

As of 2021, the total number of vehicles in China has reached 395 million, with 7.84 million new energy vehicles, accounting for 2.60% of the total number of vehicles.

2. **Low-carbonisation of Vehicles is an Important Support for Achieving Peak Carbon and Carbon Neutrality in the Transport Sector**

In September 2020, President Xi Jinping solemnly announced at the 75th United Nations General Assembly: "China will enhance its national contributions, adopt more powerful policies and measures, strive to peak carbon dioxide emissions by 2030, and strive to achieve carbon neutrality by 2060." The Central Economic Work Conference's communique clearly stated that doing a good job in peaking carbon and achieving carbon neutrality is one of this year's key tasks, requiring the urgent formulation of an action plan for peaking carbon emissions by 2030while supporting regions with conditions to peak first. The "30·60" target proposed by China will bring profound changes to sustainable economic and social development. Currently, China is still the country with the largest total carbon emissions in the world, with carbon emissions in the transport sector accounting for more than 8% of the total carbon emissions in the whole society, and road transport, mainly automobiles, accounting for more than 82% in the transport sector. The automobile industry, as one of the important supporting industries for China to achieve peak carbon and carbon neutrality, needs to reduce energy consumption and emissions across the industry. According to the development goals of the "Energy Saving and New Energy Vehicle Technology Roadmap 2.0", traditional fuel vehicles are still an important part of the automobile industry, accounting for 80% in 2025, 60% in 2030, and 50% in 2035. Hybrid vehicles are the main direction of energy-saving development. At present, the fuel-saving rate of advanced hybrid technology is about 30%, and the future fuel-saving rate of hybrid power will reach 40%. Hybrid power is an important technical route for the transformation, upgrading, and low-carbon development of automobiles.

3. **Governments Around the World Have Put Forward Strict Regulatory Requirements for Vehicle Emissions**

In 2014, the European Parliament passed a new emission standard bill, requiring that from 2021, the average CO_2 emissions of new cars should not exceed 95 g/km, which is equivalent to a fuel consumption of about 4 L/100 km. In 2018, representatives of EU governments and the European Parliament set the average CO_2 emission target for 2030 at: 59.4 g/km (2.5 L/100 km), a reduction of 37.5% from the 95 g/km in 2021. In 2012, the US government announced the Corporate Average Fuel Economy (CAFE) standards for 2017 to 2025, requiring a start from 35.5 mpg (about 6.63 L/100 km) in 2016, and gradually increasing fuel efficiency by about 5% per year, reaching 54.5 mpg (about 4.3 L/100 km) by 2025. Japan, South Korea and other developed countries have also set vehicle fuel consumption/emission standards.

In order to cope with increasingly stringent fuel consumption and emission regulations, for traditional energy vehicles, it is difficult to achieve lower fuel consumption targets simply by improving engine efficiency, and it is necessary to continuously increase the application of hybrid power technology. All major car companies around the world have significantly reduced the research and development efforts of traditional power systems in the next five years, while increasing the research and development ratio of hybrid power systems. Foreign hybrid power system research and development started early, and the products have gone through multiple rounds of iteration and optimization. In the past 7 years, the global production of hybrid vehicles has increased nearly threefold.

1.1.2 Current Status of New Energy Vehicle Development

The strategic orientation towards pure electric drive is an important judgment and decision made by our country based on a systematic analysis of the dual pressures of energy security and environmental protection, the latest technological trends in the international automotive industry, and the foundation of more than a decade of the early development of electric vehicles. It can be said to have consolidated the development consensus of the entire industry and pointed out the direction for the transformation of China's automotive power systems. After nearly a decade of implementation, China's new energy vehicles have achieved production and sales of millions of units, energy-saving vehicles represented by hybrid vehicles have covered all levels of models, the average fuel consumption of passenger cars has significantly decreased, some key component products have entered the international supporting ranks, and the overall technical level and industrial competitiveness of China's automotive industry have been greatly improved, achieving huge energy-saving and emission-reduction benefits. At present, the global automotive industry has taken electrification as an important direction for a century of major changes. Major countries and multinational car companies are constantly releasing new strategies and product plans for car electrification. The development of new energy vehicles has become a global consensus. China was the first to propose this strategic orientation, fully verifying the foresight and accuracy of the decision, and it is an important measure for China to seize the opportunity of automotive industry transformation and achieve leapfrog development. Since 2012, China's energy-saving and new energy vehicle industry has made significant progress, with continuous improvement in independent research and development capabilities, continuous improvement in product technology levels, and gradual improvement in the industrial chain. The production capacity and cumulative production and sales of new energy vehicles, the technical level of whole vehicles, power batteries, and drive motors, the industrial chain of power batteries and drive motors, and backbone enterprises have all been completed, with some indicators exceeding the target; however, there is still a certain gap from the planning target for the average fuel economy of passenger cars.

1.1.3 Safety of Power Battery Systems

The technology of power batteries and key components has reached the international advanced level as a whole, and certain progress has also been made in production equipment, key common technologies, standardization, and major basic and frontier technology research. In terms of power battery matching, since 2013, with the gradual expansion of China's new energy vehicle market, the matching volume of power batteries has been increasing year by year, as shown in Fig. 1.1. Figure 1.1 Production and power battery matching volume of new energy vehicles in China from 2013 to 2021 New energy vehicles have a high risk in the electrical system, and power batteries pursue high performance, posing certain safety risks. Compared with traditional vehicles, new energy vehicles contain complex electrical systems such as power batteries, drive motors, and high and low-voltage wiring harnesses, and the probability of fire accidents caused by electrical system failures has significantly increased. At the same time, power batteries pursue high specific energy, choose high-nickel ternary materials with higher gram capacity but poorer thermal stability, and pose certain safety risks. When new battery technologies are just developed, they are difficult to fully verify and evaluate without extensive testing, and it is difficult to push them to the market on a large scale; at the same time, some new technologies, such as power battery thermal prevention and control technology, currently have no mandatory standard requirements, and the application cost is relatively high, currently only installed in high-end models. New energy vehicle fire accidents are still in a high incidence state. In 2021, China implemented 59 recalls of new energy vehicles, involving 830,000 vehicles, an increase of 31.1% and 75.9% respectively from the previous year, and 52.5% of new energy vehicle defect clues reflected power battery, motor, and electronic control system problems. According to statistics from the Ministry of Emergency Management, more than 3000 new energy vehicle fires occurred throughout the year in 2021. The reasons include not only external factors such as unqualified charging facilities, improper user use, and irregular maintenance, but also internal factors such as aging of power batteries and high-voltage components of early products, vehicle, and battery quality, and also face the problem of battery safety after traffic accident collisions. In terms of power battery safety performance: At present, key technologies have been developed around power battery thermal runaway safety, battery system high voltage insulation safety and collision safety, the thermal stability of ternary material power batteries, charging safety protection, etc., and the goal of no thermal diffusion of the battery system has been achieved through heat source isolation, bidirectional conversion, high-temperature insulation, intelligent cooling, automatic fire extinguishing, and other technologies. At the same time, efforts have been made to strengthen the key material technology research on solid electrolytes, flame-retardant electrolytes, etc., develop all-solid-state batteries, lithium metal batteries, and other new system batteries with high safety, and ensure the safety of power batteries through product updates and iterations. In terms of safety monitoring and early warning and fire safety technology: At present, relevant companies have carried out retrospective analysis research on safety monitoring data and historical data of accident vehicles, carried out analysis of power battery hidden dangers and faults, collected abnormal phenomena corresponding to different faults through a large number of monitoring cases, identified key features that can locate fault points, and provided appropriate response measures for different risk types. At the same time, the automotive industry has strengthened joint research with the fire industry, studied the development mechanism and spread law of various battery fires, developed early fire detection technology, long-lasting fire extinguishing agents, new automatic fire extinguishing devices, and achieved early warning and rapid response to fire after safety accidents. The high-voltage platform and new technology applications of new energy vehicles bring safety risks. With the rapid development of new energy vehicles, to achieve shorter charging times and improve the overall output efficiency of the powertrain, many car companies are developing 800 V high-voltage platform technology, but the high-voltage platform brings serious heating, insulation design, and other problems. At the same time, the power battery system has developed heat prevention and control technologies, such as cloud-based real-time early warning, active cooling, low thermal conductivity structure, all-around thermal insulation design efficient heat dissipation channels, etc. The representative companies and products mainly include BYD's blade battery, GAC's magazine battery, Great Wall's Dayu battery, Jikrypton's Jixin battery Nezha's Tiangong battery, etc. However, new technologies often bring potential risks, and a large amount of testing and verification is needed to prevent safety accidents.

1.1.4 Power Battery Management System

The Battery Management System (BMS) is a comprehensive system for monitoring and managing high-voltage electrical energy of power batteries for electric vehicles (EV, PHEV, HEV). The battery management system monitors and controls the power battery in real time, providing the vehicle with information about the state of the power battery, such as voltage, current, temperature, state of charge (SOC), health

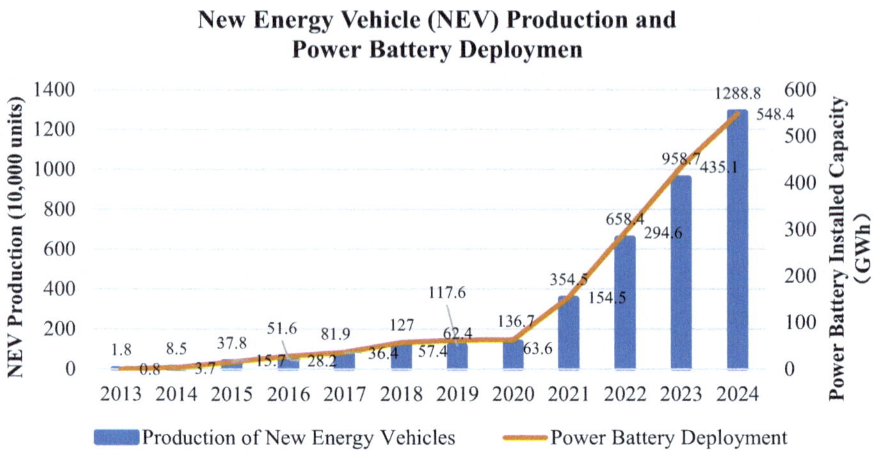

Fig. 1.1 Production and power battery matching volume of new energy vehicles in China from 2013 to 2024

status (SOH), insulation status, high voltage interlock status, etc., and at the same time, it judges in real time whether the operation of the power battery is normal. If a fault occurs, it will take corresponding measures, such as sending a fault signal to the vehicle controller and giving an alarm prompt, reducing power handling, etc.

The battery management system mainly has the following functions: voltage detection of the battery cell and the whole pack, detection of the charging and discharging current of the battery pack, control of the temperature field of the battery box, airtightness detection of the battery box, estimation of the SOC and SOH of the battery pack, communication with the vehicle controller and display system, charging control, real-time status judgment and fault control of the battery pack, energy management, high voltage safety management, thermal management, etc. Traditional battery management systems generally have voltage, current, temperature collection functions, SOC estimation functions, data communication, and fault management functions. The functions of the battery management system are shown in Fig. 1.2.

In addition to the above functions, the future BMS will have more in-depth development in several other aspects: more intelligent energy management, through modeling and self-learning functions of individual battery cells, to achieve a more accurate state of charge (SOC), health status (SOH), energy status (SOE) calculation: software layering and distributed development based on AUTOSAR, and rapid iteration and OTA remote update of the software: a combination of cloud services and big data, analysis of all data of the battery pack, realization of remote online diagnosis, timely discovery and repair of hidden: fully meet the functional safety requirements of ISO26262, further enhance the safety level of electric vehicles.

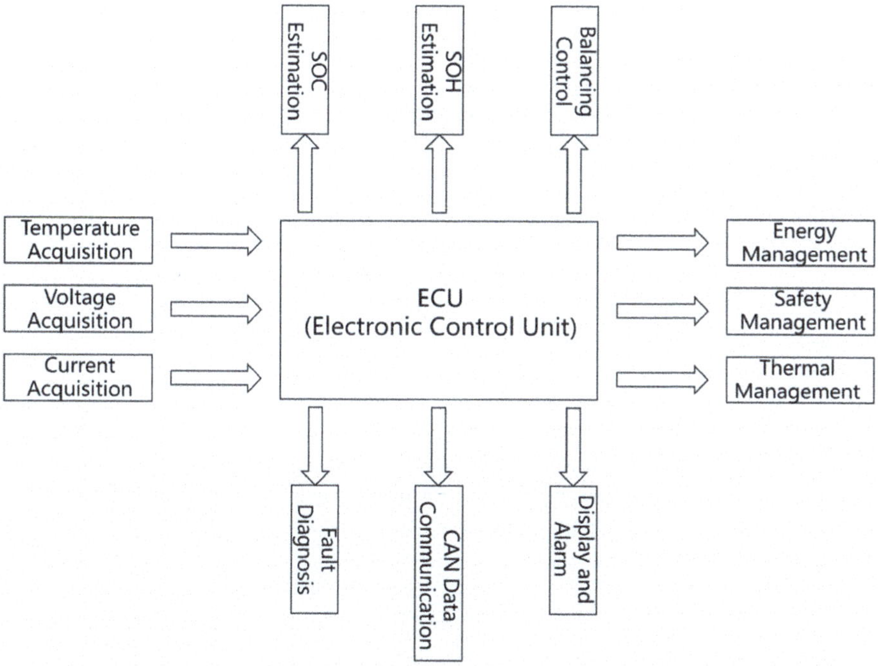

Fig. 1.2 Functions of the battery management system

1.2 Application Requirements of Power Battery and Management System

1.2.1 Pure Electric Vehicles

A Battery Electric Vehicle (BEV) is a vehicle that is completely powered by a rechargeable battery (such as a lead-acid battery, nickel–cadmium battery, nickel–hydrogen battery, or lithium-ion battery) and uses an electric motor as its drive system. Compared with traditional internal combustion engine vehicles, pure electric vehicles are also composed of a power system, body, chassis and electrical equipment. The main difference between them lies in the power system. Pure electric vehicles differ from traditional vehicles in that they have an electric drive system and a power management system. Pure electric vehicles are mainly composed of an electric drive system, a power system, and an auxiliary system.

The electric drive system is a system that converts the chemical energy in the power battery into the kinetic energy of the wheels. At the same time, the electric drive system of a pure electric vehicle can also convert the kinetic energy of the wheels into electrical energy to be charged into the power battery during vehicle deceleration or braking. The electric drive system, which is mainly composed of an electronic controller, a power converter, an electric motor, and a mechanical transmission system, directly affects the power performance and economic performance of the whole vehicle.

The power system is the system that provides driving energy to the electric motor of a pure electric vehicle, which mainly includes battery energy management systems and charging devices, etc. The power system should also have the function of monitoring battery status and charging management. Similar to the auxiliary systems on traditional vehicles, the auxiliary systems of pure electric vehicles consist of auxiliary power sources, air conditioning systems, power steering systems, lighting systems, auxiliary water devices, defrosting devices, etc. The overall composition of a pure electric vehicle is shown in Fig. 1.3. In addition to having a battery in the power system, pure electric vehicles are also equipped with a set of auxiliary power sources. The function of the auxiliary power source is to supply power to the power steering system, air conditioning system, and other body electrical equipment.

The structure of pure electric vehicles is quite flexible, currently mainly including central motor drive and wheel motor drive. Among them, the central motor drive also includes various types of front-wheel drive and rear-wheel drive with or without a drive shaft, while the wheel motor is divided into two-wheel and four-wheel drive types, including wheel-side drive and hub drive. The different motor drive forms of pure electric vehicles determine their power system layout forms, and the common layout forms can be divided into six types, as shown in Fig. 1.4.

(a) The traditional type of drive is similar to common internal combustion engine vehicles, retaining the clutch, transmission, and differential, and replacing the internal combustion engine with a drive motor, usually converted from traditional internal combustion vehicles.

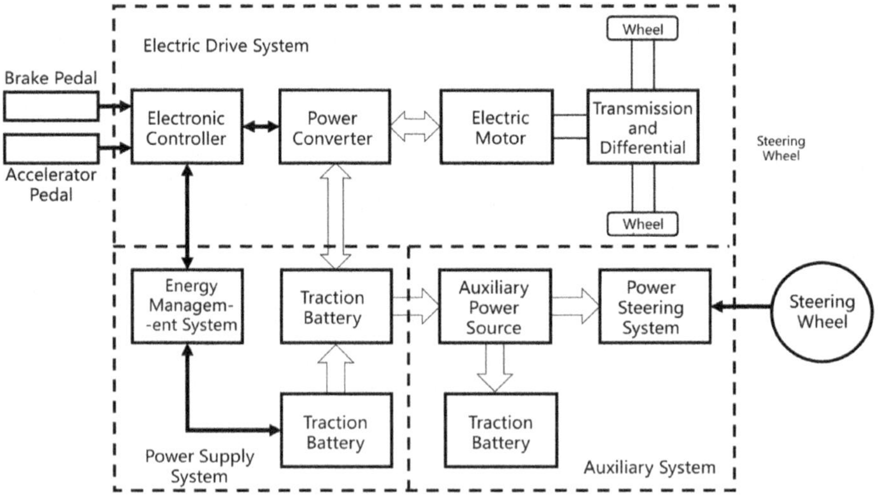

Fig. 1.3 Composition framework of pure electric vehicle

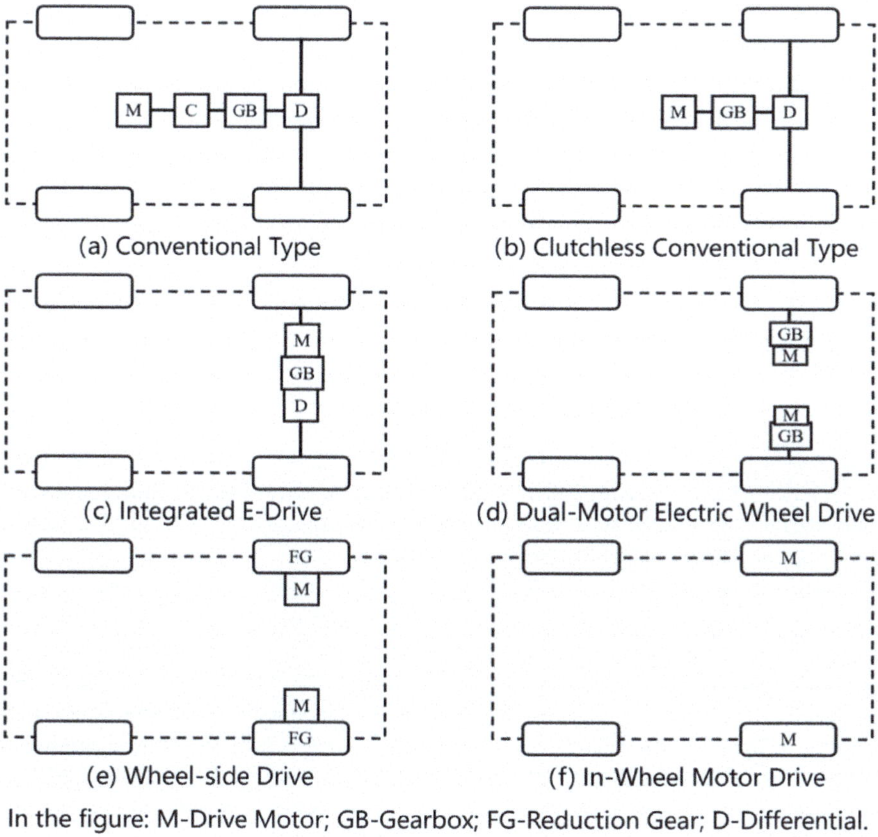

Fig. 1.4 Common power system layout forms

(b) The clutchless type of drive is similar to the traditional type, using the constant power characteristics of the motor, eliminating the clutch, and reducing the volume and weight of the mechanical transmission device. But at the same time, it has higher requirements for the speed regulation performance of the motor.

(c) The integrated type of drive integrates the drive motor, reducer, and differential, the structure is more compact, and such a structure is more common in micro pure electric vehicles.

(d) The characteristic of the dual motor wheel drive type is that two motors drive two wheels to travel through the reducer. This type of drive uses electronic differential control to achieve steering function instead of the mechanical structure of the traditional differential.

(e) The wheel hub wheel-side drive type installs the motor and the fixed-ratio planetary gear reducer inside the wheel, without a drive shaft and differential, the drive structure is relatively simple, but it has higher requirements for the characteristics of the motor and control strategy.
(f) For the wheel hub wheel-in drive type, similar to the (e) drive form, but driven by a low-speed outer rotor motor directly driving the wheel, eliminating the general mechanical transmission device, and has higher requirements for the performance of the motor.

Energy-type power batteries are mainly used in the field of pure electric vehicles (Electric Vehicle, EV). At present, international high-end cylindrical batteries use high-nickel low-cobalt nickel cobalt aluminum (NCA) cathode materials matched with a small amount of silicon monoxide graphite anode materials. The energy density of mass-produced 21,700 cylindrical batteries reaches 260 Wh/kg; the energy density of soft pack batteries mass-produced with high-nickel ternary cathode materials (NCM811, etc.) reaches 288 Wh/kg; the energy density of square aluminum shell batteries mass-produced with ternary cathode materials (NCM523, etc.) matched with graphite anode materials reaches 240 Wh/kg.

Chinese lithium iron phosphate batteries use lithium iron phosphate cathode materials matched with graphite anode materials, and the energy density generally reaches 140–180 Wh/kg, some products can reach 190 Wh/kg, and mass production has been achieved; using ternary cathode materials (NCM523, etc.) matched with graphite anode materials, the energy density of mass-produced square aluminum shell batteries is between 210 and 240 Wh/kg, and the energy density of mass-produced soft pack batteries is between 230 and 260 Wh/kg; the energy density of cylindrical batteries mass-produced with high-nickel ternary cathode materials is between 240 and 260 Wh/kg. Using high-nickel ternary (NCM811) cathode materials and silicon carbon anode materials, soft pack batteries with an energy density of 300 Wh/kg and square aluminum shell batteries with an energy density of 260 Wh/kg have been developed.

1.2.2 Hybrid Electric Vehicles

HEV is an abbreviation for Hybrid Electric Vehicle, that is, a hybrid electric vehicle. HEV is a compromise between traditional vehicles and fully electric vehicles: it uses both the internal combustion engine of a traditional vehicle (which can be designed smaller) and the motor of a fully electric vehicle (Purely Electric Vehicle) for hybrid driving (including battery and inverter links), reducing the demand for fossil fuels, improving fuel economy, thereby achieving energy saving and emission reduction and mitigating the greenhouse effect. Toyota Prius and Honda Insight are two major manufacturers of HEVs.

The widespread use of petrol engine cars has many drawbacks. Statistics show that under more than 80% of road conditions, a regular car only utilizes 40% of its power potential, which can drop to 25% in urban areas. More seriously, the emission of exhaust gases pollutes the environment. Since the 1990s, there has been a growing call for environmental improvements worldwide, and various electric cars have emerged. Although people generally believe that the future belongs to electric cars, battery technology issues have hindered the application of electric cars. As the energy density of batteries is hundreds of times less than that of petrol, it has not reached the values people require. Experts estimate that within the next 10 years, electric cars will not be able to replace petrol engine cars (unless there is a significant breakthrough in fuel cell technology). Reality has forced engineers to come up with a win–win solution, developing a hybrid power device (Hybrid-Electric Vehicle, abbreviated as HEV). The so-called hybrid power device combines an electric motor with an auxiliary power unit on a car for propulsion. The auxiliary power unit is a small fuel engine or power generator. In simpler terms, it minimizes the traditional engine as much as possible, allowing a portion of the power to be borne by the battery-electric motor system. This hybrid power device not only utilises the advantages of the engine's long continuous working time and good power performance but also the benefits of the electric motor's pollution-free, low-noise operation. The two "fight side by side", complementing each other's strengths, and the thermal efficiency of the car can be increased by more than 10%, and exhaust emissions can be improved by more than 30%. Hybrid power refers to a car using both petrol and electric drive, the advantage being that when the vehicle starts and stops, it is driven only by the electric motor. The engine does not work until a certain speed is reached, thus allowing the engine to always remain in the best working condition, with good power performance and very low emissions. Moreover, the source of electricity is the engine, so only refueling is required. The key to a hybrid car is the hybrid power system, its performance directly relates to the overall performance of the hybrid car. After more than ten years of development, the hybrid power system has evolved from the original discrete structure of the engine and motor to an integrated structure of the engine, motor, and gearbox, i.e., an integrated hybrid powertrain system. Hybrid powertrains can be classified into series, parallel, and mixed types based on power transmission routes. Series: The series power is composed of the engine, generator, and motor, which are connected in series to form the SHEV power unit system. The engine drives the generator to generate electricity, and the electric energy is delivered to the battery and motor through the controller. The motor drives the car through the transmission mechanism. When the load is small, the battery drives the motor to drive the wheels. When the load is large, the engine drives the generator to generate electricity to drive the motor. When the vehicle is in the start, acceleration, and uphill conditions, the engine, generator set, and battery pack provide electric energy to the motor together. The battery pack drives the motor when the electric vehicle is in low speed, coasting, and idle conditions. When the battery pack is out of power, the engine-generator set charges the battery pack. The series structure is suitable for frequent starts and low-speed operation in the city, which can adjust the engine to run stably near the optimal operating point, and adjust the vehicle speed by adjusting the output of the battery and motor. This avoids idle and low-speed operation of the engine, thereby improving the engine's efficiency and reducing exhaust emissions. However, its disadvantage is that the energy is converted several times, and the mechanical efficiency is low. Parallel: The parallel device's engine and motor drive the car together. The engine and motor belong to two separate systems, which can independently provide torque to the car's transmission system. They can drive together or separately on different roads. When the car accelerates uphill, the motor and engine can provide power to the transmission mechanism at the same time. Once the car speed reaches cruising speed, the car will rely solely on the engine to maintain that speed. The motor can be used as both a motor and a generator, also known as a motor-generator set. Since there is no separate generator, the engine can directly drive the wheels through the transmission mechanism. This device is closer to the traditional car drive system, the mechanical efficiency loss is almost the same as that of ordinary cars, and it has been widely used. Mixed: The mixed device includes the characteristics of series and parallel. The power system includes an engine, generator, and motor. Depending on the auxiliary device, it is divided into engine-main and motor-main. In the engine-main form, the engine is the primary power source, and the motor is the auxiliary power source. In the motor-main form, the engine is the auxiliary power source, and the motor is the primary power source. The advantage of this structure is easy to control, but the disadvantage is a more complex structure. Toyota's Prius belongs to the motor-main form. Power batteries are mainly used in the field of hybrid cars. Internationally, power batteries of different capacities from 3.6Ah to 8Ah have been developed and mass-produced, with a power-to-weight ratio greater than 4500 W/kg. China has developed a 6Ah ternary high-power battery for hybrid cars, with an energy density of 108 Wh/kg, a power-to-weight ratio greater than 4000 W/kg, and a cycle life of more than 8000 times (2C charge and discharge); a 6Ah lithium iron phosphate high-power battery with an energy density of 79 Wh/kg, a power-to-weight ratio greater than 5600 W/kg, and a cycle life of more than 20,000 times (50% discharge depth), has been mass-produced. Super capacitors have seen limited application in the field of hybrid vehicles, mainly divided into double-layer capacitors and lithium-ion hybrid capacitors.

In terms of double-layer capacitors, foreign countries have increased the specific power from several 10 k W/kg to a maximum of 78 k W/kg through the surface treatment technology of the collector, electrode manufacturing technology, positive and negative ratio technology, and electrolyte technology. In terms of lithium-ion capacitors, the energy density reaches 20 Wh/kg, and the voltage range is 2–4 V.

The lithium-ion capacitors developed domestically have an energy ratio of 100 Wh/kg, a power ratio of 15 k W/kg, and a lifespan of 100,000 times. The developed 12000F double-layer capacitor has an energy ratio of 9.67 Wh/kg and a power ratio of 8 k W/kg.

In October 2020, the "Energy Saving and New Energy Vehicle Technology Roadmap 2.0" was released, which proposed that by 2025, hybrid passenger cars will account for 50% of traditional energy passenger cars, and 75% and 100% respectively in 2030 and 2035. HEV, due to no mileage anxiety, low fuel consumption, and low cost, is currently the preferred choice of energy-saving vehicles for consumers. Affected by multiple factors such as the dual-integral policy and the National Six emission standards, it is expected that several hybrid vehicle models will be launched in 2021–2022, and the annual sales of hybrid (HEV) vehicles in China will be about 2 million by 2025.

According to the different voltage platforms, the HEV system equipped with lithium-ion batteries can be divided into 48 V systems and high-voltage systems. The former mainly adopts air cooling or natural cooling solutions, the system power is about 10–13 kW, the vehicle fuel saving rate can reach 10–15%, and the corresponding battery capacity is mostly 8–10 Ah, and the power density is 5000–6000 W/kg; the latter system voltage is 200–350 V, the power is 1–3 kWh (mostly 1.5–2 kWh), the battery capacity is 5–Ah, the discharge depth is 30–50%DOD, the 2 s peak charge and discharge power can reach 80–100 kW, and the 10 s peak charge and discharge rate can reach 60–80 kW, and air cooling or liquid cooling is used for heat dissipation.

The current high-power battery system uses 5-series secondary spherical ternary materials as the positive electrode, high interlayer spacing graphite as the negative electrode, and single-sided coated polyethylene separator, and matches high conductivity and low film impedance electrolyte. The product capacity for commercial vehicles is 25 Ah, and the product capacity for passenger cars is 6 Ah. The HEV battery systems using the above two types of batteries have completed verification and entered the mass production stage on passenger cars, heavy trucks, and fuel-cell commercial vehicles. The road test results of heavy trucks show that the comprehensive fuel saving rate of the 25 Ah–14.9 kWh battery system can reach 15–20%, and the life can reach 800,000–1 million kilometers; the comprehensive fuel saving rate of the passenger car with the 6 Ah–2.1 kWh battery system can reach 38%.

1.2.3 Plug-in Hybrid Electric Vehicles

Plug-in hybrid electric vehicles (PHEV) refer to vehicles that are designed to charge the power battery from an external power grid, and can also be driven by an engine when the battery power is low. This type of vehicle has the same components as traditional hybrid vehicles, including engines, motors, batteries, power coupling devices, and gearboxes, etc., and can be divided into series, parallel, and mixed configurations, as shown in Fig. 1.5.

The series structure (shown in Fig. 1.5a) is also known as the APU (Assistant Power Unit) structure, which has a relatively simple structure. The whole condition adopts pure electric drive, and the engine is only used to drive the generator to generate electricity. This structure completely decouples the engine working mode from the vehicle condition, which can control the engine to work at the optimal economic condition point, but because the energy from the engine to the drive motor needs to go through multiple conversions, the system efficiency is low, and the vehicle economy is slightly better than traditional vehicles. This configuration is suitable for congested road conditions.

The parallel system (shown in Fig. 1.5b) requires only one motor compared to the other two configurations, but it requires a simple power coupling device. In this configuration, both the engine and the motor can drive the vehicle. The motor

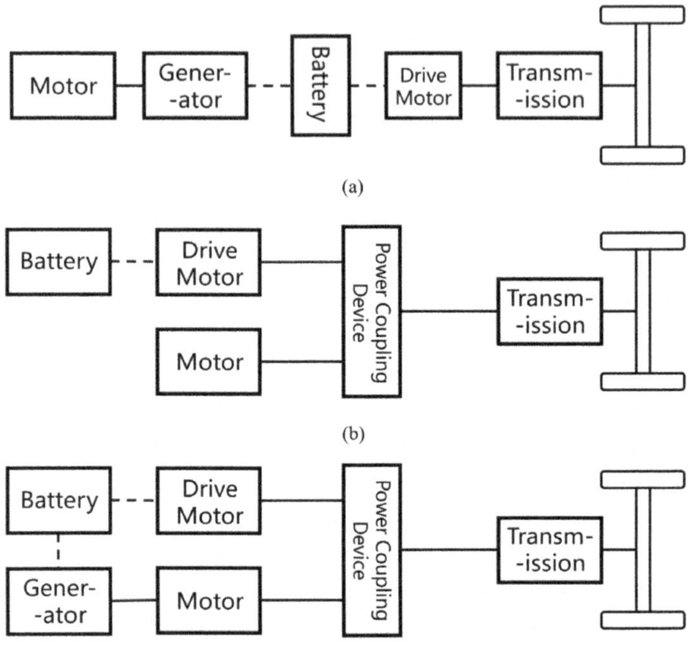

Fig. 1.5 Hybrid power system configuration **a** Series structure, **b** Parallel structure, **c** Mixed structure

mainly drives at low speeds, and the engine mainly drives at high speeds. This system avoids multiple energy conversions and can optimize the engine's working condition point. The economy is significantly better than the series structure under high-speed conditions.

The mixed structure (shown in Fig. 1.5c) has a complex structure and has the advantages of both series and parallel modes. Compared with other configurations, this configuration has good power, economy, and comfort, and can adapt to various conditions.

PHEV is essentially a hybrid vehicle. Because it has a large battery capacity, it can store more electrical energy, and it can charge the battery from an external power grid, so it has special features. The working mode of PHEV can be divided into 4 working modes, as shown in Fig. 1.6, including pure electric working mode (EV), power consumption mode (CD), power balance mode (CS), and charging mode.

EV mode. Because the battery SOC (State of Charge, (Battery Charge Status) is high, the battery has enough energy, it can fully provide all the energy required for the vehicle to move, and quickly reduce the battery level, which is conducive to recovering as much braking energy as possible. In this mode, the engine is shut down, the motor drives the vehicle alone, the emissions are zero, the fuel consumption is zero, and it is an energy-saving and environmentally friendly working mode.

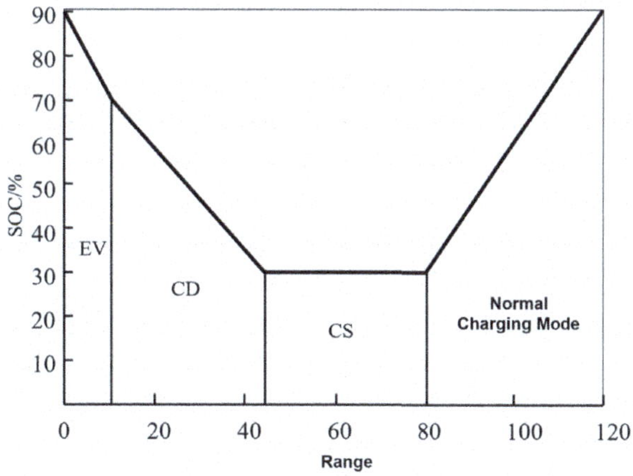

Fig. 1.6 PHEV working mode

CD mode. When the battery level is at a higher level, the CD mode is generally used to drive the vehicle. This mode mainly uses the motor to drive the vehicle, and the engine plays an auxiliary role. The battery SOC shows a downward trend, but the rate of decline is smaller than that of the EV mode. Since the SOC can have a large range of changes, all low-efficiency operating points of the engine can be driven by the motor, and the engine idle time can be completely eliminated, while avoiding frequent engine start and stop, so the fuel consumption at this stage is less.

CS mode. This mode is the same as the traditional hybrid mode, the battery SOC changes within a smaller range, and the engine provides the main driving energy, in order to avoid large fluctuations in the battery level, the engine needs to often participate in the driving process, and in order to avoid frequent start and stop processes, there will inevitably be idle fuel consumption, so the fuel consumption of this mode is generally higher than that of the CD mode.

Regular charging mode. In this mode, the external power grid is used to charge the battery, so that the battery SOC is restored to the highest level. This mode generally appears at night or other times when the vehicle does not need to be used. When the power grid charges the battery, the battery level quickly rises to the highest level designed by the battery.

Energy management strategy is a problem of general concern, it directly relates to the economy, power and emission performance of the hybrid system. Compared with the traditional power system where the accelerator pedal is directly associated with the engine throttle, the accelerator pedal of the hybrid system is decoupled from the engine, the accelerator pedal no longer directly controls the engine, but after obtaining the accelerator pedal opening, the vehicle controller converts it into a demand torque, and according to the current state of the drive source, reasonably allocates torque to the engine, thus achieving engine control. This control method improves the flexibility of engine control and plays an important role in optimising engine operating conditions. How to reasonably distribute the driver's demand torque to the two power sources, to meet the vehicle's power demand and reduce energy loss, reduce pollutant emissions, this is the problem of research on the energy management strategy of the hybrid system.

Energy power balanced type power battery is mainly used in the field of plug-in hybrid cars (Plug-in Hybrid Electric Vehicle, PHEV) and extended-range hybrid cars. Internationally, the single cell-specific energy of the energy power balanced type power battery reaches up to 195 Wh/kg, the specific power reaches 1600 W/kg, and the life reaches 3000 times. At present, the ternary soft pack battery produced in China has a specific energy of 200 Wh/kg, a particular power of 1500 W/kg, and has achieved mass production matching; a high specific power fast charging ternary battery with a specific energy of 130 Wh/kg and 6C charging not less than 80% of the rated capacity has been developed.

1.2.4 Fuel Cell Vehicles

Fuel cell vehicles are also electric vehicles, but the "battery" is a hydrogen–oxygen mixed fuel cell. Compared with ordinary chemical batteries, fuel cells can be refuelled, usually with hydrogen. Some fuel cells can use methane and petrol as fuel, but they are usually limited to use in power plants and forklifts and other industrial fields.

Fuel cell vehicles (FCV) are cars that use the electricity generated by the on-board fuel cell device as power. The fuel used by the on-board fuel cell device is high-purity hydrogen or high-hydrogen reforming gas obtained from hydrogen-containing fuel reforming. Compared with the usual electric vehicles, the difference in power is that the electricity used by FCV comes from the onboard fuel cell device, and the electricity used by electric vehicles comes from the battery charged by the power grid. Therefore, the key to FCV is the fuel cell. The fuel cell is a highly efficient power generation device that directly converts the chemical energy of the fuel into electrical energy through an electrochemical reaction without burning the fuel. The basic principle of power generation is: that the anode (fuel pole) of the battery inputs hydrogen (fuel), and hydrogen molecules (H_2) are dissociated into hydrogen ions (H^+) and electrons (e^-) under the action of the anode catalyst, H^+ moves through the electrolyte layer of the fuel cell towards the cathode (oxidation pole) direction, e^- cannot pass through the electrolyte layer and flows to the cathode through an external circuit; oxygen (O_2) is input at the battery cathode, oxygen is dissociated into oxygen atoms (O) under the action of the cathode catalyst, and combines with the e^- flowing to the cathode through the external circuit and the fuel passing through the electrolyte H^+ to form stable structure water (H_2O), completing an electrochemical reaction that releases heat. This electrochemical reaction is completely different from the violent combustion reaction of hydrogen in oxygen. As long as hydrogen is continuously input into the anode and oxygen is continuously input into the cathode, the electrochemical reaction will continue uninterrupted, and e^- will constantly flow through the external circuit to form a current, thereby continuously providing power to the car. This is completely different from the traditional principle of mechanical power generation by cutting magnetic lines with conductors. This electrochemical reaction is a static power generation method that obtains power without physical movement. Therefore, fuel cells have the advantages of high efficiency, low noise, and no pollutant emissions, which ensures that FCVs become truly efficient and clean cars. To meet the requirements of car use, automotive fuel cells must also have high specific energy, low operating temperature, fast start-up, no leakage, etc. Among the many types of fuel cells, proton exchange membrane fuel cells (PEMFC) fully possess these characteristics, so the fuel cells used by FCVs are all PEMFC.

The working principle of a fuel cell vehicle is that hydrogen, as a fuel, undergoes a redox chemical reaction with oxygen in the atmosphere in the fuel cell carried by the vehicle, generating electrical energy to drive the motor. The motor drives the mechanical transmission structure in the car, which in turn drives the mechanical structure of the front axle (or rear axle) of the car, thereby driving the electric car forward.

The core component is the fuel cell. The reaction result of the fuel cell will produce very little carbon dioxide and nitrogen oxides, and the main by-product is water, so it is called a green new environmentally friendly car. A fuel cell vehicle is a type of electric vehicle, and its core component is a fuel cell. Through the chemical action of hydrogen and oxygen, instead of combustion, it directly becomes electric energy power. The hydrogen fuel of fuel cell vehicles can be obtained through several channels. Some vehicles carry pure hydrogen fuel directly, while others may be equipped with fuel reformers that can convert hydrocarbon fuels into hydrogen-rich gas. Individual fuel cells must be combined into a fuel cell stack to obtain the necessary power to meet the requirements of vehicle use.

1.3 Research Status at Home and Abroad

1.3.1 Failure Modes and Fault Mechanisms of Power Batteries

The frequent accidents of electric vehicle power battery systems catching fire or even exploding under vehicle-mounted conditions in recent years have shown that the losses caused by battery failure are huge and heavy. Therefore, the safety and reliability of power battery systems are essential. The research on the failure mode and fault mechanism analysis (Failure Mode and Fault Mechanism Analysis, FMFMA) of the power battery system can provide a theoretical basis for the failure prevention research of the power battery system, and ensure the reliability and safety of the power battery system through various improvements, preventions, and protections. In addition, using fault tree (Fault Tree Analysis, FTA) analysis to determine common faults and causes of power battery systems, and using FMFMA analysis to make preliminary judgments on the possible consequences and countermeasures of various potential faults, can provide data accumulation and support for the verification of various battery faults and the improvement of battery BMS fault diagnosis strategies.

From the perspective of failure modes, the failure modes of power battery systems can be divided into safety failure modes and non-safety failure modes. Safety failures mainly include positive and negative short circuits, battery cell leakage, negative lithium deposition, and cell swelling. These failure modes are all serious problems that may cause casualties, and the longer the cell is used, the greater the risk of failure of the power battery system. Non-safety failures of the power battery system mainly affect the performance of the battery, mainly manifested as poor capacity consistency, excessive self-discharge, reduced low-temperature discharge capacity, and capacity decay. Among them, capacity decay is an inevitable problem of the battery. At present, the first thing that major battery manufacturers need to solve should be the safety failure problem of the power battery system and the battery consistency problem. On this basis, consider improving the durability of the battery and extending the cycle life of the battery.

From the perspective of battery structure hierarchy, power battery systems are usually composed of battery cells, BMS, and Pack systems. Therefore, the failure modes of power battery systems can generally be divided into three different levels of failure modes, namely cell failure, BMS failure, and Pack system integration failure. The failure of a single battery cell is not only related to the battery itself, but is usually also related to BMS failure. BMS failure can sometimes cause serious safety accidents, such as BMS voltage detection failure leading to battery overcharging or over-discharging, BMS current detection failure leading to excessive current, BMS temperature detection failure causing excessive temperature, large SOC estimation deviation causing excessive charging and discharging depth of the battery, resulting in poor battery consistency or reduced cycle life, in addition to insulation monitoring failure, electromagnetic compatibility communication failure and other issues. Pack system integration failure modes include busbar failure, main circuit connector failure, high voltage contactor sticking, fuse overcurrent protection failure and other modes. In response to these various failure modes, researchers and battery manufacturers need to continuously improve battery manufacturing processes to enhance the safety of battery cells, BMS manufacturers need to fully understand the performance of batteries, design safer and more reliable BMS systems based on safety design principles, and correct and reasonable use of the battery system is the last and most significant barrier to ensure battery safety.

At present, there are very few studies on the comprehensive failure modes of power batteries at home and abroad, most of which are based on a small number of experimental studies under specific failure modes. Zhang Huajie and others from Sun Yat-sen University first analysed the dynamic behaviour of 18,650 lithium-ion batteries based on stress wave theory, conducted experimental studies on the fracture dynamic behaviour of lithium-ion batteries, and obtained battery failure criteria related to fracture speed, proving that the battery's fault displacement and short-circuit location depend on the range of fracture speed. M. Held and others from the Swiss Federal Laboratories for Materials Science and Technology analysed battery faults based on FMFMA and FTA methods and designed battery system-level experiments, focusing on the behaviour of internal short circuits in batteries and their impact on battery systems and vehicles. Foad H. Gandoman and others from the Université Libre de Bruxelles introduced various battery faults used to evaluate the reliability and safety of lithium-ion power batteries for electric vehicles, studied the degradation laws of lithium-ion batteries under five major fault modes and capacity/power attenuation, and proposed a reliability evaluation model to address existing challenges. Wang Hongwei and others from the China Institute of Mechanical and Electrical Products studied various common failure modes and causes of commercial lithium manganese oxide ($LiMn_2O_4$) power batteries, and the results showed that temperature is a critical stress causing capacity attenuation, thermal runaway, gas leakage or liquid leakage of lithium manganese oxide power batteries, and the corrosion behaviour of the current collector is mainly affected by the electrolyte salt. Dandan Lyu and others from the University of California reviewed the failure mechanisms of lithium-ion batteries and their possible solutions, and proposed future research directions and challenges in experimental directions and computational models for lithium-ion battery fault analysis. Zou Yufeng and others from Tianjin Lishen Battery proposed a theoretical analysis of a failure mode of lithium-ion power battery modules, and conducted a large current discharge test on three batteries in different SOC states in series according to actual usage conditions. Through the simulation and analysis of the failed battery's state, suggestions were made for battery sorting and grouping and the battery management system to extend battery life. Binghe Liu and others from the Beijing University of Aeronautics and Astronautics described the complete mechanical-electrochemical-thermal coupling behaviour of lithium-ion batteries by reviewing experimental, theoretical and modelling studies at each battery evolution stage under mechanical abuse loads, and outlined the latest modelling framework for multi-physical field characteristics of power battery systems.

From the perspective of the failure modes of power battery systems, the main inducing factors include overheating, overcharging, over-discharging, short-circuiting and mechanical destruction, among which thermal runaway is the ultimate manifestation of battery safety problems caused by power battery system failure. Therefore, starting from the analysis and research of the thermal runaway phenomenon of power battery systems, we can delve deeper into the failure modes and fault mechanisms of power battery systems. Thermal runaway of power battery systems is usually caused by mechanical (such as puncture, extrusion, etc.), electrical (such as overcharging, short circuit, etc.) and thermal (such as overheating, fire, etc.) abuse situations or abnormal battery ageing, and the internal factors are mainly caused by intense internal short circuits. In addition, the packaging form (steel shell, aluminium plastic film, etc.), geometric shape (cylindrical, square, etc.), group connection method (series, parallel), fixed connection structure, power battery system parameters (capacity, voltage) and usage conditions, environment and other factors will also have a certain impact on the thermal runaway of power battery systems.

In the field of thermal runaway mechanism and characteristic research, scholars at home and abroad have carried out extensive research on the chemical and physical changes of the internal materials of power batteries during thermal runaway and the fire and explosion process triggered by thermal runaway, as well as the changes in external parameter characteristics. Mendoza-Hernandez and others from Nagaoka University of Technology in Japan studied the thermal runaway behaviour of 18,650 batteries with cathode materials of $LiCoO_2$ and $LiMn_2O_4$ under different charging states. The results showed that the thermal stability of $LiMn_2O_4$ is superior to that of $LiCoO_2$. Using an accelerating calorimeter to provide an adiabatic environment, Feng Xuning and others from Tsinghua University studied the failure mechanism of NCM ternary material batteries under excessive temperature before thermal runaway and the temperature characteristics when thermal runaway occurs. Cui Zhixian and others from the University of Science and Technology of China targeted the internal short-circuit behaviour that leads to thermal runaway of lithium-ion batteries, and used finite element numerical simulation methods to simulate the behaviour of mechanical impact, collector edge burrs and lithium dendrite growth that may cause internal short circuits in the battery by piercing the separator. They compared and analysed the thermal response of the battery under different lithium dendrite radii, quantities and centre distances. Fu Yangyang and others from the University of Science and Technology of China conducted observational research on the thermal runaway process of 18,650 batteries, analysing their mass loss rate, ignition and explosion time, heat release rate, surface temperature and gas composition. They found that different SOC and heat fluxes can affect the ignition and explosion time of the battery. In terms of thermal runaway experimental and model research, Feng Xuning and others from Tsinghua University conducted puncture experiments on lithium-ion power battery modules, testing the trigger time, order, temperature field distribution, and heat propagation pathways of thermal runaway, and found the characteristics of sudden voltage drop and sudden temperature rise during the thermal runaway process. Donal Finegan from the National Renewable Energy Laboratory in the United States used computer tomography scanning and X-ray photography technology to observe key degradation patterns such as battery material delamination, electrode layer collapse and structural degradation caused by gas during the battery thermal runaway process. Gi-Heon Kim from the National Renewable Energy Laboratory in the United States established a three-dimensional model of lithium-ion power battery thermal runaway, and simulated the heat conduction under the condition of battery thermal runaway, as shown in Fig. 1.7a. Luo Xingna and Zhang Qingsong from the Civil Aviation University of China proposed a domino effect model of lithium-ion power battery thermal runaway transmission, and proved the feasibility of delaying thermal runaway and practical engineering application by adding flame retardant partitions between batteries through experiments and simulations, as shown in Fig. 1.7b. A Zhoujian from Beijing Jiaotong University developed a thermal runaway analysis model to predict the thermal runaway of square and soft pack batteries caused by external short circuits or ultra-high discharge rates Yikai Jia from the University of North Carolina in the United States studied the thermal runaway propagation behaviour between two batteries and built a battery

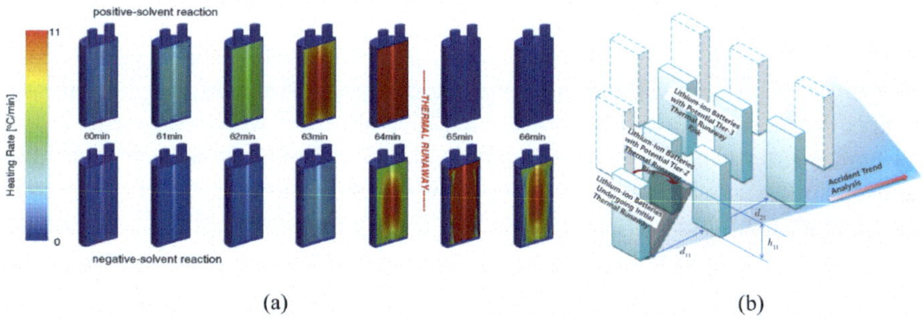

Fig. 1.7 Three-dimensional model of power battery system thermal runaway and domino effect model

thermal runaway model triggered by mechanical abuse load through experiments by combining a 3D thermal runaway model with an electro-thermal conduction model They then observed different thermal runaway propagation patterns according to different battery spacings, and concluded from the simulation results that higher overall temperature and local overheating are the two main ways of thermal runaway propagation, and SOC is also a key factor determining the probability and speed of propagation. The research also provided a simplified thermal runaway propagation mathematical model, laying the foundation for the development of an efficient battery module safety design calculation framework.

Thermal runaway and thermal diffusion prevention and control technologies can be divided into intrinsic, active and passive safety technology Intrinsic safety technology starts from the battery material level, aiming to fundamentally solve the problem of thermal runaway. Active safety technology starts from the system control perspective, using the battery management system to strictly control the use boundary of the battery and give early warning of possible battery swelling. Passive safety technology is aimed at managing and controlling the heat release of the thermal runaway battery based on the battery system design after the battery undergoes thermal runaway, to achieve the purpose of the battery system not catching fire.

Recently, significant breakthroughs have been made in the intrinsic safety technology of batteries. Firstly, there has been new progress in the field of high-concentration electrolytes. The violent reduction reaction of high-concentration electrolytes at high temperatures, which generates a large amount of heat, is one of the key mechanisms triggering thermal runaway in batteries. High-concentration electrolytes have always been considered one of the most promising safety electrolytes for lithium-ion batteries due to their low flammability, non-flammability, and good electrochemical properties. In previous studies, the main factors affecting the safety of battery electrolytes were focused on the flammability of the electrolytes. The latest research suggests that flammability only represents the ability of the electrolyte to be oxidised. For the high safety of battery electrolytes, the interaction between the active materials inside the battery (fully charged positive electrode, negative electrode, electrolyte) is more worthy of attention. Studies have shown that high-concentration electrolytes can regulate the sequence of key chemical reactions inside the battery, cutting off the electrolyte oxidation reaction path triggered by the oxygen release caused by the phase change of the positive electrode. However, the lithium salt LiFSI in the electrolyte and the lithiated negative electrode with strong reducibility will undergo a violent exothermic reaction (LiC_6 + LiFSI), which advances the reaction sequence and triggers thermal runaway in the battery. Furthermore, even if a flame retardant is used as the solvent for the non-flammable high-concentration battery electrolyte, the battery will still undergo thermal runaway. Through heat calculation, the main exothermic reaction of thermal runaway originates from the material crosstalk between the positive and negative electrodes. Therefore, for high-concentration electrolyte systems, the triggering of thermal runaway and the primary exothermic reaction are unrelated to the flammability of the electrolyte. In the safety assessment of batteries, the interaction between the positive and negative electrodes and the electrolyte under the charged state should be systematically considered. These findings provide valuable insights into the thermal runaway mechanism of high-concentration electrolytes in organic systems and aqueous systems (Fig. 1.8).

In addition, the latest research shows that batteries with a TiO_2 coated NCM523 positive electrode have better safety and fast charging performance. The TiO_2 coating isolates the active material from the electrolyte, slowing down parasitic oxidation and the dissolution of TM (NiF^{4-}, CoF^{4-} and MnF^{4-}), delaying phase changes during

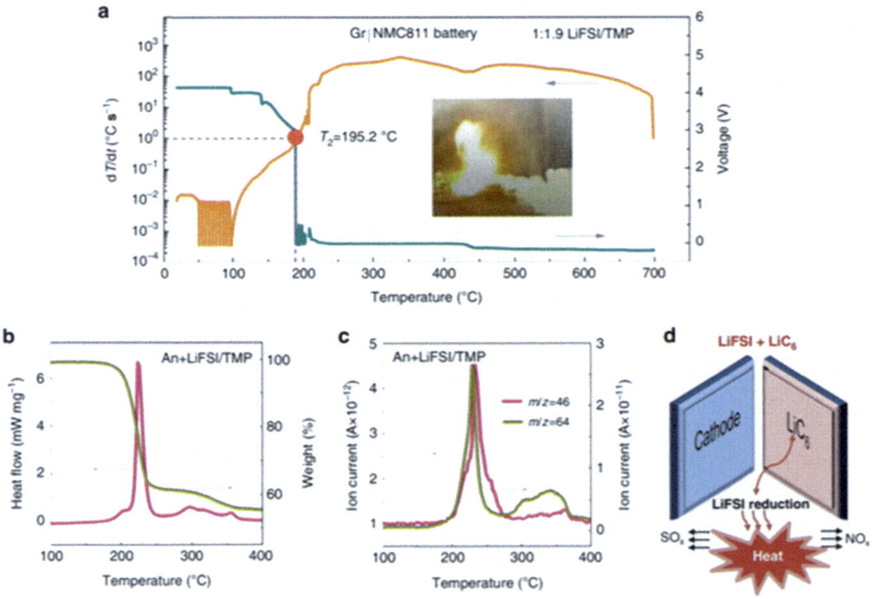

Fig. 1.8 Thermal runaway characteristics of non-flammable high-concentration electrolyte LiFSI/TMP (1:1.9 by molar)

the heating process, and greatly improving the safety of the battery. Electrochemical impedance spectroscopy analysis shows that the TiO_2 coated NCM523 positive electrode gives the battery a more stable charge transfer impedance. Transmission electron microscopy shows that the TiO_2 coating reduces the accumulation of the positive electrode electrolyte interface layer on the particle surface. Time-of-flight secondary ion mass spectrometry analysis shows that the TiO_2 coating significantly improves the interface stability of the positive electrode particles and protects the particles from severe electrolyte corrosion. Accelerated thermal analysis shows that the thermal runaway trigger temperature of the battery with TiO_2 coated NCM523 as the positive electrode material is 257 °C, which is higher than that of the uncoated NCM523 positive electrode. In-situ X-ray diffraction during the heating process shows that this enhanced safety is due to the coated positive electrode material inhibiting phase changes (Fig. 1.9).

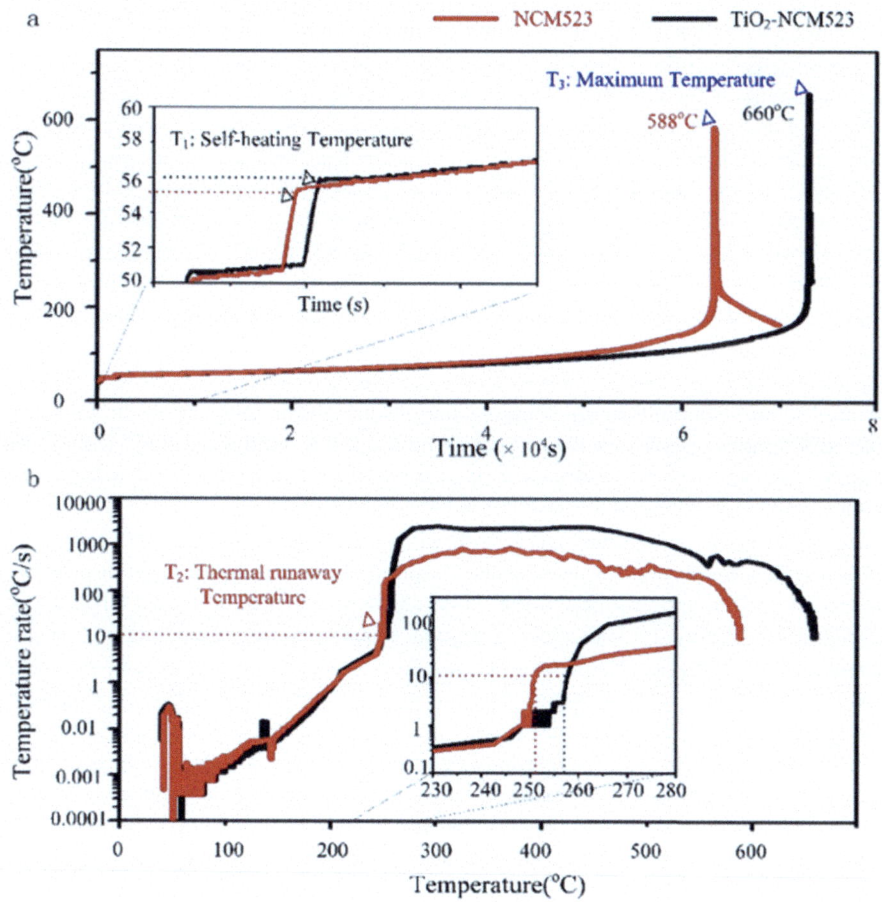

Fig. 1.9 Thermal runaway characteristic curves of NCM523 and TiO_2-NCM523 measured by ARC

The safety mechanism research of high nickel NCM positive electrode materials has also been further deepened. In the NCM positive electrode, Ni is usually considered the cause of poor stability, but the role of Mn and Co in the NCM positive electrode structure has always been controversial. Recently, research has shown that Co dominates the chemical and structural stability of the deeply delithiated NCM positive electrode. When heated, the deeply delithiated manganese-rich sample shows lower stability compared to the cobalt-rich sample. By adjusting the content of minor TM cations (Co, Mn), the oxygen release behaviour during the degradation process of the NCM positive electrode can be regulated, thereby improving the safety of the battery. In the high nickel NCM positive electrode, the reduction of the main Ni^{4+} is the most intense and releases most of the oxygen. Since the reduction of Co^{4+} occurs after Ni^{4+} Previously, by occupying the tetrahedral sites, the migration process of Ni was prolonged, thus inhibiting the release of oxygen. Although Mn itself is stable, it did not significantly stabilise or slow down the reduction of Ni^{4+}. This result overturns long-standing research cognition and is of great significance for the design of high energy density and safe high-nickel cobalt-free NCM cathodes. After 2017, due to the bottleneck in the innovation of ternary battery materials, the increase in the specific energy of the battery slowed down, and the increase in the system's specific energy depended on the structural design of the battery system. Therefore, innovations in battery system structures emerged like spring bamboo shoots after rain in 2018, such as the launch of Volkswagen's VDA and MEB modules. From 2019, Chinese companies have leveraged their advantages in cell manufacturing, accumulated over time, and continuously launched Cell to Pack (CTP), blade batteries, Jellyroll to Module (J2M) and other technologies on the cell structure, surpassing Volkswagen's VDA and MEB cell size standards, leading the innovation in battery system structure technology, as shown in Fig. 1.10. Both CTP and blade batteries can significantly improve the volume/mass grouping efficiency of the battery system, especially the blade battery can increase the grouping efficiency by up to 50%, and both grouping methods can simultaneously reduce the manufacturing cost of the battery system by 20–30%. However, the emergence of new structures brings new challenges to the means of thermal runaway suppression.

In the battery system, when a cell undergoes thermal runaway, the thermal propagation failure mode in the system does not simply spread in sequence, but exists in three modes: "sequential propagation", "synchronous propagation" and "disorderly propagation". Figure 1.11 shows the sequence statistics of a CTP battery system thermal propagation test. Sequential propagation, as the name suggests, is when the module follows a sequence during thermal propagation; synchronous propagation often occurs in thermal propagation between models in the system, manifesting as multiple cells undergoing thermal runaway at the same time in a very short time; and disorderly propagation mostly occurs in the late stage of system thermal propagation, its main feature is that the order of thermal propagation of cells in the module is irregular. Among the three propagation modes, synchronous propagation is the most harmful, as the thermal runaway energy of multiple batteries is released at the same time in a very short time, which accelerates the thermal propagation of the entire

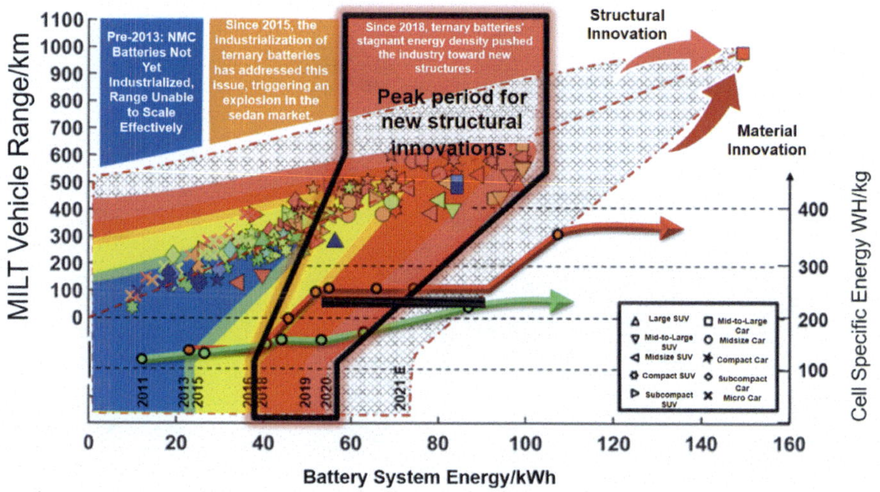

(a)

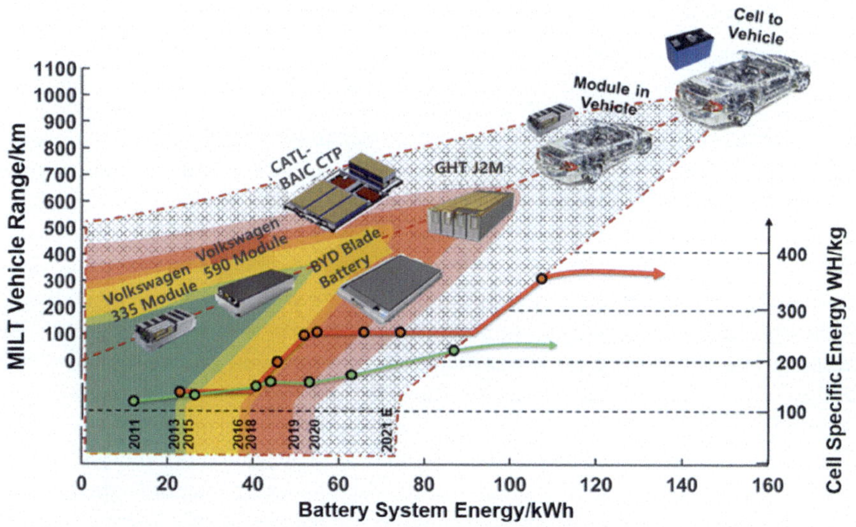

(b)

Fig. 1.10 **a** Development trend of power battery system specific energy enhancement technology. **b** Summary of innovative technologies in battery system structure

battery system. Hence, it needs to be taken seriously in system thermal propagation suppression.

In terms of thermal runaway suppression measures, research shows that pure insulation materials alone can no longer solve the thermal runaway propagation problem of high specific energy ternary battery systems, and it is necessary to carry out liquid

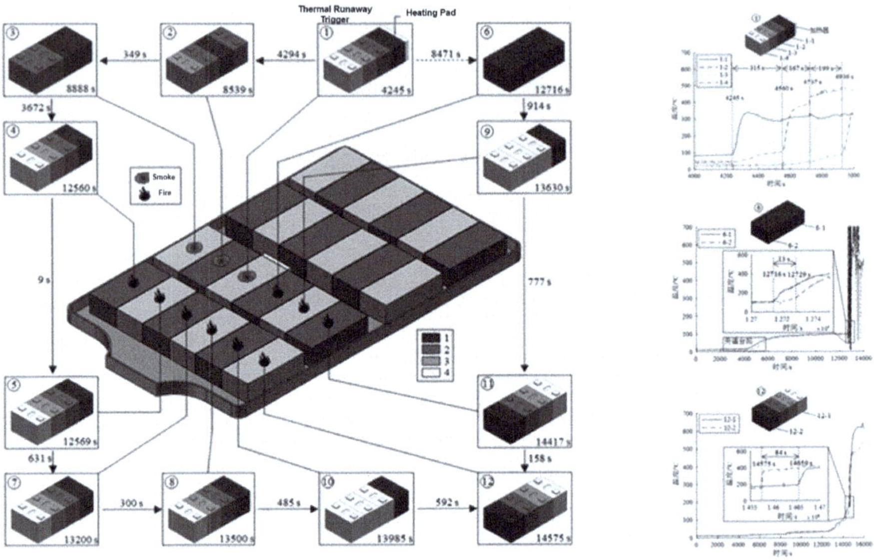

Fig. 1.11 Three thermal propagation failure modes of the battery system

cooling and insulation cooperation to possibly achieve thermal runaway suppression effect. As shown in Fig. 1.12, with the decrease in heat dissipation, relying solely on insulation materials for thermal runaway suppression, the thickness of the material tends to infinity, making it difficult to meet the design requirements of the module. At this time, new composite insulation materials need to be proposed, such as developing thermal runaway suppression materials with "insulation + cooling" multiple suppression effects by combining PCM materials, as shown in Fig. 1.13. In actual experimental verification, it was found that the composite suppression material combined with PCM material can well achieve the effect of "insulation + cooling", and has significant advantages compared with traditional aerogel insulation materials.

Therefore, to prevent and control thermal runaway propagation in battery systems, it is still necessary to trigger from the intrinsic safety of battery materials, the passive safety of battery systems, and the active safety of battery management, in order to design a safe and reliable power battery system. In summary, scholars at home and abroad have summarised the failure modes and fault mechanisms of power battery systems based on the structure of power batteries. At present, the research on the failure modes and characteristics of power battery systems is mostly based on a small number of experimental studies under specific failure modes, especially the research on the ultimate manifestation of battery failure—thermal runaway is becoming increasingly rich; the research on fault mechanisms is mainly based on years of battery reliability tests, structural disassembly and failure analysis research; a large amount of research work on the thermal runaway characteristics of power

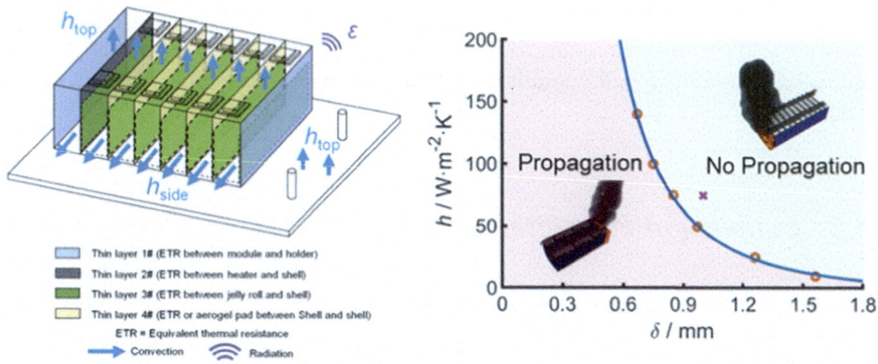

Fig. 1.12 Boundary of "insulation + liquid cooling" thermal runaway suppression in the battery system

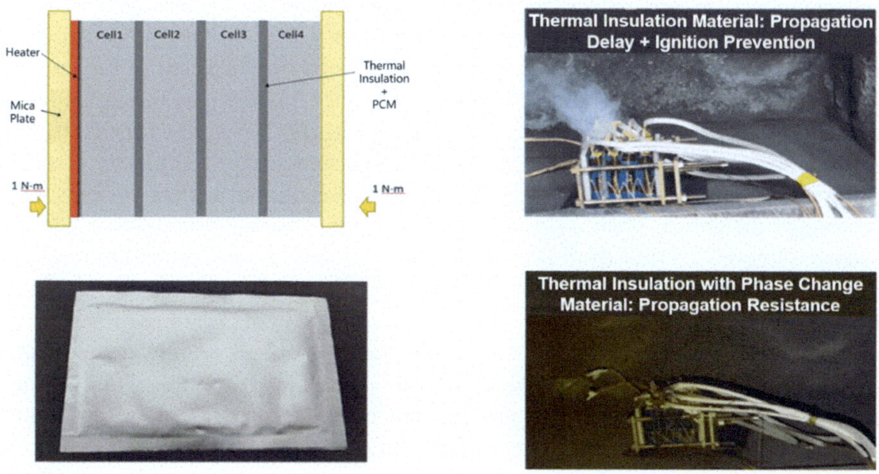

Fig. 1.13 Material with "insulation + cooling" multiple thermal runaway suppression effects

battery systems has been carried out in the four aspects of thermal runaway mechanism analysis, thermal runaway characteristic simulation and emulation, experimental testing and mathematical modelling, which has played an important theoretical and experimental support role for further exploring the fault mechanism of batteries and verifying the impact of battery failure. However, although the battery is small, the aspects it involves are extremely complex. In the analysis of the failure modes of power battery systems, the materials, processes, working environment, usage and host equipment of the battery should be fully examined according to the actual situation, and the external failure modes and consequences of the power battery system should be analysed and corresponding measures proposed. When designing a power battery system, various failure modes should be considered to improve the safety and reliability of the power battery system. Therefore, this topic intends to carry out a comprehensive analysis of the safety characteristics of power battery systems based on the actual operation monitoring data of the real vehicle operation big data platform.

1.3.2 Coupling of Power Battery Safety and Consistency

The inconsistency of individual cells during battery grouping and actual work is one of the root causes of safety issues in battery systems. Therefore, in recent years, scholars at home and abroad have carried out extensive research on the coupling mechanism of power battery safety and consistency. Wang Zhenpo and others from Beijing Institute of Technology, based on the application data of lithium-ion power batteries, have established a consistency distribution probability model and a mathematical model of power battery group life under the influence of consistency. Experimental data have verified that the worse the consistency, the faster the capacity decay of the battery group. Yuang-Shung Lee from Fu Jen Catholic University in Taiwan and Thanh Hai Phung from Grenoble University in France have proved through a large number of experimental studies that consistency directly affects the service life of power battery systems. Controlling the voltage consistency of power battery systems within a certain range can effectively extend the service life of battery systems, and the voltage difference is an essential factor affecting the safety of power battery systems. Gregory J. Offer from Imperial College London analysed the difference in voltage consistency of power batteries during normal operation and fault occurrence, and proposed that monitoring the voltage difference and temperature difference of power battery systems can ensure that power battery systems are always in a safe and controllable state. Zheng Yuejiu from Tsinghua University, in response to the problem that the life of the battery group is much smaller than the life of the single cell, studied the evolution mechanism and influencing factors of the battery group life, and proposed the consistency identification and corresponding balancing algorithm and fault diagnosis method for the constant current charging condition and dynamic operating condition of electric vehicles. Shang Yunlong from Shandong University, in response to the absolute inconsistency and complex non-linear characteristics of lithium-ion power batteries, proposed a real-time multi-fault diagnosis method for early battery faults based on improved sample entropy. By detecting the modified sample entropy of the battery voltage sequence in the moving window, this diagnosis method can diagnose and predict different early battery faults, including short circuit and open circuit faults, and can accurately predict the time of fault occurrence. Liu Peng from Beijing Institute of Technology, based on theoretical analysis and experimental data, established a dynamic stress degradation model of power battery single cells according to the theoretical method of accelerated life test. On this basis, based on the analysis of inconsistency, the charging and discharging process of series, parallel and hybrid power battery groups was analysed, and the corresponding battery degradation model was established, revealing the mechanism that the life of parallel battery groups is superior to that of series battery groups.

In summary, scholars at home and abroad have carried out extensive research on the coupling relationship between power battery safety and consistency, but most of these studies are still based on engineering practice experience and a small amount of experimental data, so the amount of data used for research is far from sufficient for practical engineering applications; in addition, due to the artificial controllability of battery voltage, current, temperature and other parameters under experimental conditions, the reproduction effect of research results for complex and variable real vehicle driving conditions is not good.

1.3.3 Prediction of Power Battery System Safety Status

The accurate estimation or prediction of the SOS of the power battery system is of great importance to the safety and reliability of electric vehicles. Eliud Cabrera Castillo from TUM-CREATE in Singapore and Andreas Jossen from the Technical University of Munich, based on the concept of safety being inversely proportional to abuse, defined and described the associated parameters or attributes of the SOS of the power battery system, including battery voltage, temperature, current, internal resistance, SOC, SOH, mechanical deformation and some other derived parameters. In this paper, the SOS parameters such as voltage, temperature, SOC, and SOH, which are easier for the research of real vehicle power battery systems, are taken as the research objects, and the research status and progress of each will be summarized separately.

(1) Battery Voltage Prediction

Due to the difficulty in accurately predicting the voltage of power battery systems through current physical experiments and mathematical models, there has been little research on battery voltage prediction over the past few decades. Since the battery life ends when the battery voltage drops below a certain critical value, modelling the voltage-current relationship is crucial for extending battery life Therefore, as early as 2005, Daler Rakhmatov and others from the University of Victoria in Canada proposed a predictive model that can accurately predict battery voltage using a given discharge current. This model can accurately predict the battery voltage as the given current changes over time. Felipe Tobar and others from the University of Chile proposed a novel core adaptive filter suitable for multiple inputs for battery voltage prediction, and used standard linear filters and adaptive linear filters as benchmarks for short-term and long-term predictions of actual battery voltage data for comparison verification. Li Kaiyuan and others from Nanyang Technological University in Singapore combined the battery model with the analysis model to accurately predict battery terminal voltage characteristics, State of Energy (SOE), and remaining operating time. Laboratory-scale experimental test results show that the battery model used to estimate the battery SOE and terminal voltage has extremely high accuracy and reliability under dynamic load and battery ageing conditions. Farid Karbalaei and others from Shahid Beheshti University in Iran proposed a new method for predicting voltage trajectories based on voltage jump points and stable states, and confirmed in a case study of a bus power system that this method can significantly improve calculation speed while maintaining system accuracy. Wang Zhenpo and others from the Beijing Institute of Technology proposed a real-time diagnosis method for battery voltage faults based on modified Shannon entropy. The set anomaly coefficient can carry out real-time evaluation and prediction of voltage abnormal states, but it cannot predict the actual value of the battery voltage. In recent years, with the gradual development of machine learning, research on battery status based on machine learning has also been gradually carried out. However, most of the methods proposed in these literatures are based on experimental data, and it is difficult to take into account the complex and variable weather, road conditions, and driver driving behaviour during the actual vehicle operation process, so there are still many difficulties in applying these methods to actual operating electric vehicle power battery systems. For this reason, part of this project uses Long Short-term Memory (LSTM) to carry out multi-step joint prediction research on battery voltage, temperature, and SOC during actual vehicle driving, and uses a large amount of vehicle historical operation data as the training data set. The test results show that the proposed LSTM model can accurately predict the battery voltage, temperature, and SOC for a future period of time.

(2) Battery Temperature Prediction

In recent years, hot research on preventing thermal runaway of power battery systems has further promoted related research on Battery Thermal Management System (BTMS) and temperature prediction. Under normal circumstances, the safety and durability of the battery system have a strong dependence on the actual environmental temperature of the vehicle. For example, excessively high battery temperature and large battery temperature differences will significantly affect the driving safety of electric vehicles. Therefore, a stable and efficient BTMS is essential for a high-security power battery system. In order to accurately predict the trend of battery temperature changes to alleviate potential thermal safety risks, this research project proposes a real-time warning method for battery temperature faults based on the modified Shannon entropy algorithm. Abnormal temperatures and potential temperature faults in the power battery system can be accurately diagnosed through real-time temperature anomaly evaluation, thereby effectively preventing thermal runaway of the power battery system and ensuring the personal safety of vehicles and passengers. Sun Jinlei and others from the Harbin Institute of Technology proposed a method for online estimation of battery internal temperature based on a simplified thermal model of the Kalman filter. As an improvement, the effects of entropy change and over potential on battery heating were quantitatively analysed, and the model parameters were identified through current pulse testing. Charging/discharging experiments were carried out at different current rates on the same battery, and the verification results showed that the maximum estimation error of the battery temperature was about 1 K. However, the above method evaluates the known current battery temperature or future temperature change trend based on historical battery temperature, but it is impossible to obtain the unknown battery temperature value at a future point or period. Feng from Nanyang Technological University in Singapore Xue and others proposed a system-level simulation model for commercial lithium-ion batteries that takes into account temperature prediction. Based on the typical battery equivalent circuit, four equations were derived to characterise the battery discharge behaviour. As the thermodynamic description considers changes in internal resistance and entropy heat variables, this model can predict battery temperature, thus avoiding safety hazards caused by unwanted temperature increases. Researchers such as Chen Zeyu from Northeast University studied the temperature rise characteristics of lithium batteries caused by external short circuits and proposed an online prediction method for maximum temperature rise. By conducting battery external short circuit abuse tests under changing environmental temperatures, it was found that the heat generation of lithium batteries mainly manifests in two modes: Joule heat mode and reaction heat/Joule heat mixed mode, with gas leakage during battery thermal runaway being an external manifestation of the latter. A two-step prediction method for maximum temperature rise based on support vector machines was then proposed, and it was verified through experimental data that the proposed method can predict the maximum temperature rise 22.3 s in advance, with an average prediction error of 3.05% for eight test units. Great efforts have been made to accurately predict the temperature of battery systems, but most of these studies are still conducted in laboratory environments, so their effectiveness in actual vehicle applications remains to be verified, and few people have attempted to make real-time temperature predictions and related thermal fault warnings based on actual vehicle operation data.

(3) Estimation/Prediction of Battery SOC

SOC is one of the most critical state parameters of a power battery system because it is directly related to the remaining capacity and energy of the battery. Accurate battery SOC estimation or prediction is a tedious task because the battery system is in a complex and unpredictable vehicle operating environment during actual operation, and the uncertain nonlinear relationship between SOC and other battery parameters (such as voltage, current, battery internal resistance, etc.) has not yet been definitively answered. So far, apart from using some estimation algorithms, there is no effective method to obtain the real battery SOC through measurement. A considerable number of studies on SOC estimation have been published, among which the open circuit voltage method (OCV) based on characterisation parameters and the Coulomb counting method based on ampere-hour integration are the two initial estimation techniques widely used. However, due to their inherent limitations in SOC estimation, more effective data-driven and model-based methods have been gradually developed. For example, M. A. Hannan from the National Energy University of Malaysia and others detailed the advantages and disadvantages of various SOC estimation methods and concluded that although the next generation of BMS needs to develop SOC estimation methods based on electrochemical models of battery mechanisms, SOC estimation based on equivalent circuit models and other battery models is still the current and next research hotspot. Professor Gregory L. Plett from the University of Colorado, USA, and others proposed a SOC estimation method based on the Kalman filter (KF) for the state lag model and self-correcting model he proposed, and continued to extend this method to obtain the Extended Kalman filter (EKF). Later, He Zhiwei from Hangzhou University of Electronic Science and Technology and others also tried to use the Sigma-KF algorithm for battery SOC estimation, laying a basic theoretical framework for subsequent power battery SOC estimation. According to KF theory, KF and EKF are generally only applicable to linear or weakly nonlinear models, but the power battery system is a strongly time-varying and strongly nonlinear system. In order to achieve accurate battery SOC estimation based on KF, Tian Yong from Tsinghua University and He Zhigang from the University of Maryland, USA, used the Unscented Kalman Filter (UKF) algorithm to carry out power battery SOC estimation research. Since UKF can propagate the mean and covariance of states through the unscented transformation without having to solve the Jacobian matrix of the model, it is more accurate than the linearisation of standard KF and EKF models. After this, some studies further improved UKF and proposed algorithms such as square root UKF and spherical UKF, but all UKF-based battery SOC estimators are only approximate nonlinear estimators, and the estimation results for the strong nonlinear characteristics of discharge in the low temperature and low SOC range are still not ideal. In addition to KF, other filtering algorithms such as Particle Filter (PF) and Bayes Filter (BF) have also been used in research on lithium battery SOC estimation. To improve the estimation accuracy of PF, David E. Acuña from the University of Chile and others proposed a method based on the PF algorithm that can correctly characterize the uncertainty of the evolution of nonlinear power systems over time, and tested and verified the prediction effect of this algorithm on lithium-ion battery SOC using experimental data. Ye Min from Chang'an University and others proposed an online SOC estimation method for lithium-ion batteries based on dual-scale and adaptive PF, and the estimation accuracy of SOC is generally better than the results without considering noise adaptation. Dong Guangzhong from the University of Science and Technology of China and others revealed fundamental flaws in existing SOC estimation schemes based on BF in considering state constraints. In order to improve the accuracy and robustness of the estimation, they designed a constrained Bayesian dual filter framework for SOC estimation and verified through extensive comparative experiments that the proposed method is superior to existing methods in terms of SOC estimation accuracy and robustness. Hou Jing from Northwestern Polytechnical University and others proposed a method based on variational BF and Huber M estimation. The adaptive filter of M-estimation and HME simultaneously estimates the measurement noise covariance and battery SOC. Variational Bayesian approximation can improve adaptability, and HME can enhance robustness against abnormal values of current and voltage measurements.

In recent years, research on battery SOC estimation based on various observers has been flourishing, such as Luenberger Observer (LO), Sliding Mode Observer (SMO), H-infinity Observer (HIO), and Proportional Integral Observer (PIO). For example, Hu Xiaosong from Chongqing University, Yun Haitao from Qingdao University of Technology, and He Ting from Tsinghua University have carried out a large amount of battery SOC estimation research using LO and adaptive LO algorithms, achieving high estimation accuracy. Li Jiahao from the Solar and Hydrogen Energy Research Centre in Baden-Württemberg, Germany, compared the SOC estimation accuracy of LO, EKF and other methods, finding that the estimation accuracy of LO is somewhat lower than EKF. Unlike LO, SMO uses nonlinear high-gain feedback to reduce estimation errors. Il-Song Kim from the National University of Transportation in Korea proposed a new technique for lithium battery SOC estimation and health status assessment based on standard SMO and Dual-Sliding-Mode Observer (DSMO). Zhong Fuli from the University of Electronic Science and Technology and Chen Xiaopeng from Swinburne University of Technology in Australia introduced adaptive theory and proposed a joint estimation method for lithium-ion battery SOC based on Adaptive Sliding Mode Observer (ASMO) and fractional order equivalent circuit model. By adaptively adjusting the switching gain to compensate for modeling errors, the estimation accuracy is maximized. In response to the problem that a single Equivalent Circuit Model (ECM) cannot provide accurate SOC during the entire discharge period, Chen Ning from Central South University proposed a lithium-ion battery SOC estimation algorithm that can integrate multiple models to improve the practical performance of SOC estimation results. Yang Xiaolong from Hunan University and Xu Jun from Xi'an Jiaotong University used an improved PIO for precise estimation of battery SOC, with the SOC estimation error of the proposed method being less than 1%.

According to filter theory, various filter-based estimation algorithms not only reconstruct the estimated SOC but also filter noise. However, various observer-based methods do not model noise, so the most commonly used SOC estimation method in current battery systems is still filter-based. However, these methods are usually computationally intensive and often require additional battery parameters or need to adapt to different battery working environments through multiple model fusion, which hinders the application of existing SOC estimation methods in actual operating vehicles. In recent years, with the gradual increase of various studies oriented towards actual operation big data, machine learning techniques for SOC estimation have also made great progress. For example, considering the strong coupling, multivariate and nonlinear characteristics of batteries, Zheng Lichuan used the Support Vector Machines (SVM) method to predict battery SOC, and the simulation results on MATLAB showed that the estimation error can be controlled within 5%. Akram Eddahech from the Renault Technology Centre in France applied Artificial Neural Networks (ANN) to develop an adaptive SOC predictor that considers battery operating conditions, which can make very accurate SOC estimates, so that vehicle controllers can confidently use the entire working range of the battery pack without worrying about overcharging or undercharging. However, the aforementioned forward-propagation neural networks (FNN) based on a unidirectional multi-layer structure cannot learn and remember long-term historical information, let alone perform forward multi-step SOC prediction, so they can only estimate the current SOC. For this reason, RNNs, such as LSTM-RNN neural networks with memory functions, may be more ideal candidates for modelling multi-step SOC prediction. Ephrem Chemali from McMaster University in Canada introduced LSTM-RNN for accurate SOC estimation of lithium-ion batteries, without the need for any battery model or inference system such as a filter, and established an LSTM-RNN model that can correctly estimate SOC under different environmental temperature conditions, achieving an average absolute error of 0.573% under fixed environmental temperature and an average absolute error of 1.606% on a dataset where the environmental temperature rose from 10 °C to 25 °C. Song Xiangbao from Google Technology (Shenzhen) Co., Ltd. proposed a CNN-LSTM neural network that combines Convolutional Neural Network (CNN) and LSTM for SOC estimation. The proposed CNN-LSTM network has the advantages of both CNN and LSTM networks, and can extract spatial and temporal features from input parameter data (measurable data such as current, voltage and temperature) to infer the SOC of the battery. Experimental results show that both networks can capture the nonlinear relationship between SOC and measurable variables well, and have better tracking performance than LSTM and CNN networks. In the case of an unknown initial SOC, the proposed network quickly converges to the real SOC, then presents smooth and accurate results, with the maximum root mean square error being less than 2%. However, these methods are all based on preset pulse charge and discharge conditions under experimental conditions, and the predictive ability of the model under complex and variable actual vehicle operating environments still needs to be verified.

(4) Estimation/Prediction of Battery SOH

The working performance of the power battery system will continuously decline during long-term use due to battery aging, and various known/unknown factors such as environmental temperature, road conditions, weather, driving behaviour and other adverse changes in state can potentially reduce the available capacity of the battery, accelerate battery life aging, and even cause safety hazards such as thermal runaway. Therefore, it is vital and necessary to accurately estimate or predict the State of Health (SOH) of the power battery system. Although the detailed mechanism of battery aging is still under continuous exploration, current research generally believes that the cause of battery aging is due to the decomposition of the Solid Electrolyte Interphase (SEI) film, the loss of active materials, and the formation of lithium dendrites. Usually, the SOH of a lithium-ion battery is determined by the ratio of the current actual value after use to the initial value before use of a directly measured or indirectly calculated battery performance parameter (such as the battery's capacity, OCV, internal resistance (IR), etc.). Daniel-Ioan Stroe from Aalborg University in

Denmark and Zhang Lingling from Harbin Institute of Technology, through a large number of laboratory calendar aging tests, found that the decline in battery capacity and the increase in internal resistance/impedance are state changes that usually accompany the battery aging process, and proposed and verified a semi-empirical battery life model based on the laboratory results, which can accurately predict the decrease in capacity and increase in the internal resistance of lithium-ion batteries over a wide temperature range. The internal resistance of the power battery is closely related to other battery state parameters such as SOC, temperature, and current. Although it can be calculated or identified in real time through algorithms, it is still unrealistic to carry out related SOH real-time evaluations on actual vehicle power battery systems that often accelerate or brake, because accurate measurement of the battery impedance spectrum (Electrochemical Impedance Spectroscopy, EIS) requires the help of an electrochemical workstation and real-time measurement is difficult. Therefore, the commonly used battery SOH characterization parameter indicators are battery capacity, remaining available capacity or remaining useful life (RUL). However, to obtain an accurate battery capacity, it is necessary to perform constant current and constant voltage (CCCV) tests on the battery under a stable environment, which is also very difficult for a vehicle with a very complex operating environment, so the problem of power battery SOH estimation has generally been converted into the estimation of the battery capacity or RUL. In recent years, there has been a large amount of research on battery SOH estimation or prediction methods at home and abroad. For example, Xiong Rui from the Beijing Institute of Technology divided the battery SOH estimation methods into two major branches based on experiments and based on empirical models, among which the most popular methods in recent years are indirect analysis methods, electrochemical models, empirical models and data-driven methods. Zheng Linfeng from the University of Technology Sydney in Australia developed a capacity increment analysis (ICA) and differential voltage analysis (DVA) method for on-board battery capacity estimation, which solved the sensitive problem of battery resistance and polarization changes during the battery aging process, and can achieve a relative error of 2% for battery capacity estimation. In addition, Matthieu Dubarry from the University of Hawaii, Caihao Weng from the University of Michigan Ann Arbor, Liu Guangming from Tsinghua University, and Meiner Lewerenz from RWTH Aachen University in Germany have respectively adopted the ICA method and DVA method to carry out battery SOH estimation research. However, although the ICA and DVA methods can reflect the aging mechanism of power batteries to a certain extent and have good accuracy, the experimental requirements are generally small-rate constant current, the experiment takes a long time and the acquisition accuracy requirements for voltage sensors are high. Liu Kailong from the University of Warwick in the UK and Zheng Yong from the University of Science and Technology Beijing studied the impact of battery charging and discharging on battery capacity loss, and based on the proposed battery modeling and multi-objective constrained nonlinear optimization technology, they derived a suitable charging mode for lithium-ion batteries, and found that the dissolution and destruction of the solid electrolyte interface is the main reason for the capacity loss of lithium-ion batteries under over-discharge/low-voltage conditions. The existing battery electrochemical model research requires a deep understanding of the electrochemical structure of the battery, and even when combined with indirect analysis methods, it still needs to use specific experimental equipment to obtain the internal structure or parameter changes of the battery during the experiment, the detection time is long, the efficiency is low, so it is difficult to meet the requirements of real-time fast SOH estimation. Using experimental data to summarize or calculate the relationship between various battery state parameters (such as temperature, cycle times, charge/discharge current, etc.) and battery SOH, and then obtaining the empirical model of battery SOH estimation through data fitting. Compared with other battery models, empirical models have a simpler model structure and more flexible practical applications, but they still need to obtain data through experimental means, and the estimation results usually depend too much on the setting of working conditions. Han Xuebing from Tsinghua University analyzed the aging mechanisms of different types of lithium-ion batteries and proposed a battery SOH estimation method based on online models and regular calibration. Sebastian Paul from the German BMW-Peugeot Citroen Electric Company tested 96 LiFePO4 batteries. A power battery system composed of single cells is studied for capacity degradation during cycle life testing. The results show that the difference in aging of single cells within the battery system is due to uneven temperature and SOC within the battery system, as well as other different battery loads. Battery life is affected by temperature, usage time, and SOC, and these three parameters follow the famous Arrhenius rule. In recent years, based on the assumption that battery aging is a stable random process, there have been many data-driven studies on battery reliability and RUL estimation. Data-driven modeling methods focus more on the relationship between input stimuli and target responses, rather than the decay mechanism of the battery system, so it is possible to establish the relationship between battery SOH and input parameters without a deep understanding of the internal mechanism. Data-driven SOH estimation methods are usually based on collected data, and then use various data analysis methods to study hidden effective information, such as SVM, Related Vector Machine (RVM), grey system, PF and BF algorithms, etc. However, there are still some shortcomings in the above methods: firstly, most of the algorithms or models used for battery SOH or RUL estimation still require a lot of computation, and the accuracy of the estimation results is usually poor due to the limitations of experimental data. Secondly, all the studies that use electrochemical models or mathematical models to estimate the SOH or RUL of lithium-ion batteries are carried out under constant or preset current/temperature experimental conditions, which are difficult to reproduce for complex electric vehicle operating conditions, so their effectiveness and practicality on actual running vehicles are yet to be verified.

1.3.4 Power Battery Fault Diagnosis and Risk Warning

"Fault" is usually considered as an unforeseen or undesired change in the behavior of the target system, which degrades the performance of the system or affects its normal operation, usually the former effect is called a latent fault, and the latter fault is a system failure. If the cause and location of the degradation can be diagnosed in the early "latent fault" stage of the fault, also known as fault warning, it can to some extent avoid system failure; system failure is usually due to the inability to warn of early faults, so the system evolves to a more severe degree and may lead to dangerous situations. For a system, faults can be divided into actuator, component sensor, and controller faults. The highly electronic controllers currently in use have a small probability of failure, while components and sensors have a higher probability of failure. For power battery systems, the main components are the power battery and the connecting board/plate between the batteries. Because component faults (especially battery body faults) cause more direct harm than sensor faults, it should be first determined whether the battery has failed when diagnosing sensor faults.

The fault diagnosis of power battery systems is mainly to ensure the reliability and safety of power batteries in complex vehicle operating environments. In recent years, scholars at home and abroad have carried out in-depth research on power battery fault diagnosis and fault warning methods, which can be generally divided into three directions based on electrochemical mechanisms, model-based and data-driven, and explained. For example, Wu Chao from the Harbin Institute of Technology reviewed the research and development of lithium-ion battery aging mechanisms in recent years, summarised the causes and manifestations of lithium-ion battery faults, and the latest research results of related fault diagnosis technologies and their advantages and disadvantages. They proposed the relationship between battery performance and aging factors in battery aging tests, and found that the cross-dependence between aging factors can be quantified. Chen When from Wayne State University in the United States used the classic LO observer to design a method that can diagnose and isolate individual faults in series battery packs with uncertainties. This method uses the fault detection residuals generated by the classic LO observer to achieve fault isolation, and establishes a batch of learning observation factors to estimate isolated faults, and then proves the effectiveness of this design through experiments and simulation studies of the battery pack model. Liu Zhentong from the Beijing Institute of Technology proposed a model-based sensor fault diagnosis method for lithium-ion power battery cells and battery packs. Based on the difference in cell voltage under current or different voltage sensor faults, by real-time and long-term offline monitoring of different cells, based on the equivalent circuit model and adaptive KF algorithm, a battery model was established, as shown in Fig. 1.14a, and finally achieved online detection and separation of current sensor and multiple voltage sensor faults in series power battery packs. Satadru Dey from the University of California, Berkeley, based on a two-state thermal model, proposed a thermal fault diagnosis algorithm for lithium-ion batteries, using the surface temperature measured by a nonlinear observer and the reconstructed core temperature to generate feedback residual signals to detect faults. In order to avoid invalid diagnosis caused by model uncertainty, an adaptive threshold combined with an observer was used to design the diagnosis scheme, as shown in Fig. 1.14b. A thermal radiation model was used and a Lyapunov analysis observer (Lyapunov analysis) was built on this basis observer, LAR) model, and compares the residuals of the observer with an adaptive threshold to enhance the robustness of the fault detection system. Yang Sheng from Shanghai University and others proposed a state monitoring and fault diagnosis system for lithium battery packs, which communicates with a PC via a CAN bus, providing a wealth of important information for the battery management system. Experimental results show that this system can effectively judge the type and level of faults, and can take relevant protective measures in a timely manner, thereby improving the reliability and safety of the power battery pack. Zhao Yang from the Beijing Institute of Technology and others proposed a new power battery system fault diagnosis method based on big data statistical methods. According to machine learning algorithms and 3σ multi-level screening strategies, abnormal changes in the terminal voltage of the battery pack can be detected and calculated in a probabilistic form, and then the neural network algorithm is used to combine the fault diagnosis results with the big data statistical rules to establish a complete battery system fault diagnosis model. Yao Lei from Zhengzhou University of Light Industry and others used the voltage fluctuations of the power battery system under different vibration frequencies as input, and accurately detected the time and location of the battery false connection using the overall Shannon entropy (OSE) algorithm, achieving real-time prediction of lithium-ion power battery connection failure. Subsequently, a lithium-ion battery fault detection method based on a wavelet neural network was proposed, which decomposes and reconstructs the discrete wavelet transform (DWT) to eliminate voltage signal noise, and uses parameters such as voltage, voltage difference, covariance matrix, and variance matrix as input for the general regression neural network, significantly improving the efficiency and accuracy of battery fault degree classification.

In order to prevent serious faults from occurring and optimise the battery maintenance cycle, in addition to being able to accurately diagnose faults in the power

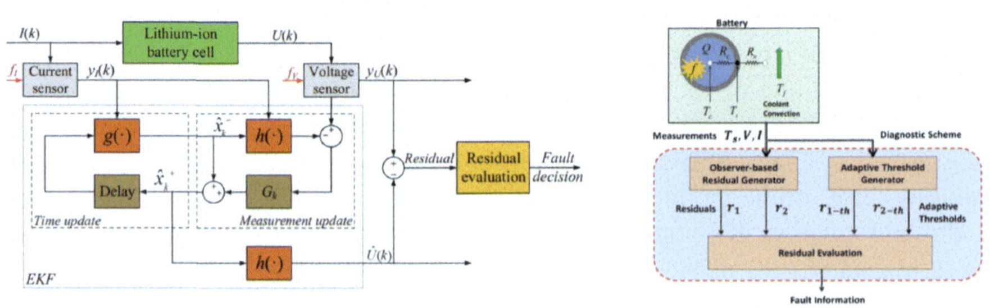

(a) (b)

Fig. 1.14 Model-based power battery sensor fault diagnosis scheme and thermal fault diagnosis mechanism battery system, breakthroughs must be made in risk prediction for the power battery system. Accurate advance prediction and warning of potential battery safety risks are extremely important for ensuring vehicle driving safety and the personal safety of drivers and passengers. Seyed Mohammad Rezvanizaniani and Zhang Jingliang from the University of Cincinnati in the United States reviewed the research progress in health monitoring and risk prediction technology for power batteries, focusing on the unmet needs of battery manufacturers, car designers, and electric vehicle drivers in this field, and summarised the current international SOC estimation, current/voltage estimation, capacity estimation, and RUL prediction technologies, algorithms, and models for monitoring and warning battery health performance. Lu Mingzhe from Beijing University of Technology estimated the battery SOH using the SVM method in data-driven methods based on laboratory data and battery online operation data, obtained the trend of battery internal resistance and other parameters based on knowledge-based methods, and then predicted the possible safety risks of the battery based on empirical methods and gave the real-time health status of the battery. Hu Xiaosong from Chongqing University and others carried out research on machine learning-type power battery system safety risk prediction. Using the sample entropy of short voltage sequences as an effective feature of battery capacity loss, the basic correspondence between capacity loss and sample entropy is obtained using the advanced Sparse Bayesian Prediction Model (SBPM). By combining guided sampling concepts and SBPM, it is possible to predict and warn of severe battery safety states. Shi Yushuai from the Guangdong University of Technology and others analyzed the common fault risks of power battery SOC jump, single-cell overvoltage, single-cell under voltage, single-cell over temperature, excessive pressure difference, excessive temperature difference, etc. from the two levels of short-term safety warning and long-term health warning in detail, and gave the fault trees of single cell overvoltage, single cell over temperature, and SOC jump as shown in Fig. 1.15, and analyzed the root causes and warning methods of these safety risks.

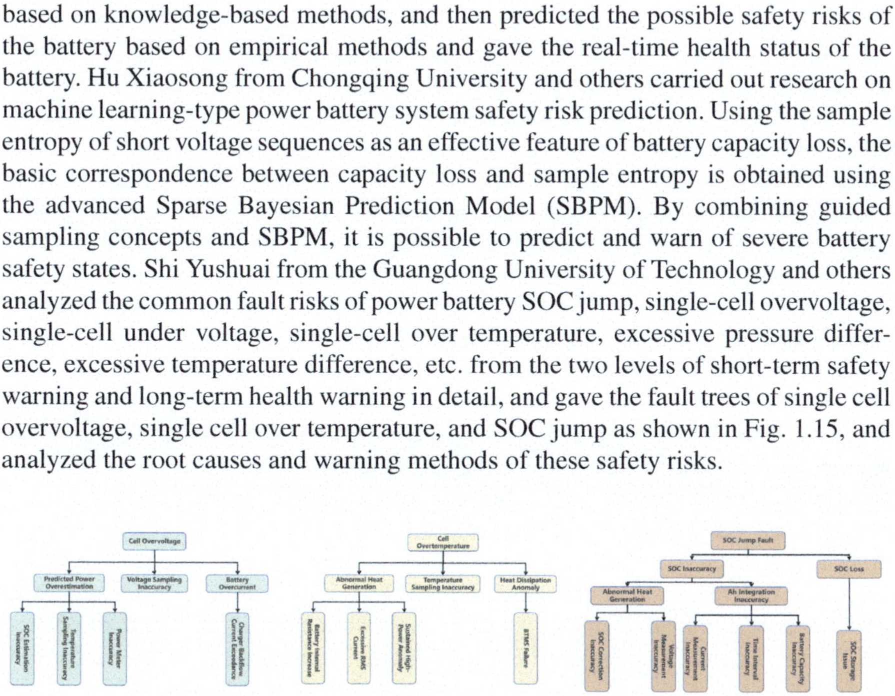

(a) (b) (c)

Fig. 1.15 Fault trees of several typical battery faults: **a** single cell overvoltage; **b** single cell over temperature; **c** SOC jump

1.3.5 Power Battery System Safety Control Management

With the deepening of research on major special topics in electric vehicles around the world, scholars and corporate institutions at home and abroad have carried out research work and achieved certain results in areas such as safety accident early warning mechanisms, safety control strategies, and high safety structural design. In terms of safety control management of power battery systems, some of the research results of this study are based on the big data of electric vehicle monitoring on the big data platform of real vehicle operation. Using the improved Shannon entropy and *Z*-score method, real-time monitoring and early warning of voltage and temperature faults of power batteries have been achieved. At the same time, a multi-level evaluation method of abnormal coefficients has been used to formulate corresponding power battery system safety management strategies. Zhou Yang, in response to the demand for reliability and rapid fault identification and diagnosis in the large-scale application of electric vehicles in Beijing, analyzed the types, characterization forms, and fault mechanisms of electric vehicle electrical system faults, focusing on the diagnosis methods of power battery system faults. He established a power battery fault early warning model using fuzzy theory, providing theoretical support for the design of safety control strategies for power battery systems. Sun Fengchun from Beijing Institute of Technology, Hu Yao from Wuhan University of Technology, Wu Jianrong from Jilin University and others have carried out a large amount of research on the design of remote monitoring systems for pure electric vehicles and fault diagnosis methods. At present, a three-level early warning mechanism for power battery systems has been formulated based on the big data platform for new energy vehicles, and corresponding trigger thresholds have been set, which have been successfully applied to the national electric vehicle operation monitoring system. Yuan Hong and others from Shandong Electric Power Research Institute have collected real-time parameters of electric vehicle power batteries through onboard terminals, used cloud computing for data analysis, and achieved early warning of early faults in electric vehicle power battery systems. In terms of high-safety power battery system structural design, Tesla has achieved high-safety battery box design by using methods such as power battery module partitioning, adding isolation layers to the power battery box partition and vehicle passenger compartment, and guiding flame direction. Wang Zhenpo from Beijing Institute of Technology and others have achieved the design of safe battery boxes for commercial electric vehicles by quickly separating abnormal situations inside and outside the box, and spraying high-temperature fireproof paint on the cover plate. Zhang Haiyan from Liaocheng Normal University and others have designed an overvoltage automatic protection system for power battery systems using high-voltage insulation and normal voltage conductivity of voltage-sensitive membranes, and have achieved good application effects. In recent years, there have been some progress and breakthroughs in the research of battery safety management control related technologies, such as James M. Zumstein from Lawrence Livermore National Security Company and Tetsuo Sato from Renesas Electronics Corporation have designed a battery safety management method and device, using internal sensors can effectively monitor and detect various faults in the battery, and transmit signals from the sensor to the control and data acquisition module through wireless transmission. Hu Xiaosong from Chongqing University and Zou Changfu from Chalmers University of Technology in Sweden have investigated the main roles, types and management needs of battery technology, and proposed a model-based control method to manage the charging operation of battery systems. Based on a fully coupled electrothermal model, the fast charging strategy is formulated for the first time as a linear time-varying model predictive control problem, which can prevent overcharging and overheating of the battery according to specified constraints; at the same time, using noise measurements of current, voltage and surface temperature, the unmeasurable internal safety state of the battery including the charging state and core temperature is estimated through a nonlinear observer.

1.4 Current Research Problems and Deficiencies

Through the above research and analysis results, it can be seen that domestic and foreign research on the SOS prediction, fault diagnosis, risk prediction and safety control management of power battery systems has been carried out in depth and has achieved fruitful theoretical and practical results. However, research on the safety control management of power battery systems based on actual vehicle operation big data has just begun and is not yet perfect, which can be summarized as several problems and deficiencies:

(1) Coupling Relationship Between Safety and Consistency of Power Battery Systems

Scholars at home and abroad have carried out extensive research on the coupling relationship and evolution law between the safety and consistency of power battery systems, but most of these studies are still based on engineering practice experience and a small amount of experiments in voltage, current, temperature and other directly measurable electrical signals. It is difficult to take into account complex influencing factors such as environment, road conditions, charging behavior, driving behavior, etc., in the actual vehicle driving process. Therefore, the existing research methods do not reproduce the complex and variable actual vehicle driving conditions well, and the research on the coupling mechanism between the safety and consistency of the power battery system is not yet perfect.

(2) SOS Prediction of Power Battery Systems

Accurate prediction of the SOS of power battery systems is of great importance to the safety, reliability, and durability of electric vehicles. Existing research based on physical experiments and mathematical models can to some extent make reasonable speculation about the future trend of SOS parameters, but it is difficult to accurately predict their numerical changes over a period of time. At present, there is little research on online SOS prediction for actual vehicle power battery systems. Existing power battery system SOS prediction models require high accuracy of model parameters and often involve complex calculations. Various state observers and filters do not model environmental noise well and struggle to effectively replicate the actual operating environment of power battery systems.

(3) Estimation of SOH of Power Battery Systems

To achieve accurate estimation of battery SOH, existing experimental research requires a stable experimental environment, preset charging and discharging conditions, small sampling intervals, and long testing times to complete a full set of battery capacity or internal resistance tests. However, for actual vehicle power battery systems, the complex operating environment and dynamic changes in charging and discharging depth are not conducive to extracting accurate conventional SOH features such as battery internal resistance, charging capacity, and discharging capacity under a unified standard. Existing equivalent circuit models and other battery modeling methods suffer from reduced modeling and battery impedance characteristic identification effects in complex actual operating environments. Therefore, existing SOH estimation methods are mostly limited to laboratory research and are still difficult to apply to actual operating vehicles.

(4) Fault Diagnosis of Power Battery Systems

In recent years, scholars at home and abroad have conducted in-depth research on power battery system fault diagnosis methods based on battery electrochemical mechanisms, physical experiments, mathematical models, and data-driven methods. However, these studies are mostly based on a small number of experiments, and the algorithm structures are complex, the calculations are large, and the real-time diagnosis effect of battery faults is poor. In addition, because existing fault diagnosis model research struggles to consider the complex working environment of actual operating vehicles, the model's diagnostic accuracy for actual vehicles is low. How to achieve robustness analysis of power battery system fault diagnosis for unknown initial values such as system noise, environmental noise, and system uncertainty, and to complete highly accurate real-time online fault diagnosis on the basis of small calculations, remains a focus and difficulty of this research.

(5) Risk Prediction of Power Battery Systems

To prevent serious faults from occurring and optimize the maintenance cycle of power battery systems, accurate advance prediction of potential safety risks of power battery systems is extremely important for ensuring their safety and reliability under complex conditions. However, the characteristics of various safety risks of power battery systems, such as "mechanical–electrical-thermal", are complex and variable, with multiple factors interdependent, making early abnormal signal extraction difficult. Existing power battery system risk prediction methods have poor online recognition and localization capabilities for early abnormal signals, and the robustness of prediction effects under actual vehicle operating conditions is not good. Multi-level risk prediction strategies for actual vehicle power battery systems need to be improved.

(6) Safety Management Strategies of Power Battery Systems

Scholars and corporate institutions at home and abroad have conducted extensive research on safety control strategies for power battery systems, mainly focusing on online diagnosis of battery faults, safety accident pre-warning mechanisms, and high-safety structure design. However, the prediction accuracy of various battery safety risk prediction/warning models integrated in existing BMS systems is low during the actual operation of power battery systems, and their universality is poor. The real-time monitoring of the SOS state of power battery systems and the reverse tracking of accident causes are not good. Safety control strategies for power battery systems with high safety, high reliability, and high stability throughout the life cycle of actual vehicle power battery systems need to be developed.

1.5 Main Research Content of This Book

This project aims to improve the safety control management capabilities of power battery systems. Considering the dynamic complexity of the actual vehicle operating environment, based on large data from actual vehicle operation, it addresses the problems and deficiencies in current research. It conducts systematic research from the aspects of the safety and consistency coupling mechanism of power battery systems, safety status prediction, fault diagnosis, risk prediction, and safety control strategies.

Chapter 2
Research on the Coupling Mechanism of Safety and Consistency in Power Battery System

2.1 Overview

The frequent occurrence of fires and explosions in electric vehicle power battery systems in recent years has proven that the safety and reliability of power battery systems are extremely important. This chapter analyses the structure and working principle of the power battery system, summarises the safety hazards of the power battery system, and then analyses the coupling relationship between the safety and consistency of the power battery system through actual vehicle operation data and practical experience. The failure modes and mechanisms of the power battery system are explained from three aspects: cell failure, BMS failure, and Pack system integration failure. On this basis, research on the fault mechanism and diagnosis technology of the power battery system is carried out, and a preliminary judgment is made on the severity and treatment measures of various battery failure modes and faults, providing a theoretical basis for the design of high-reliability power battery system fault diagnosis schemes and the improvement of BMS system fault diagnosis strategies in the future. The content of this chapter is shown in Fig. 2.1.

2.2 Structure and Working Principle of the Power Battery System

The mainstream power battery technologies on the market currently include lead-acid batteries, nickel-hydrogen batteries, fuel cells, and lithium battery technologies. Among them, lithium-ion power batteries have developed rapidly based on disposable lithium battery technology. They have many advantages such as high specific energy, high specific power, low pollution, and low discharge rate. They are currently the most favored vehicle-mounted power battery system by research institutions, battery manufacturers, and car manufacturers, and they are also the main direction of research

J. Hong, *Big Data Analysis and Application of Power Battery Operation Safety*, Key Technologies on New Energy Vehicles, https://doi.org/10.1007/978-981-96-4399-8_2

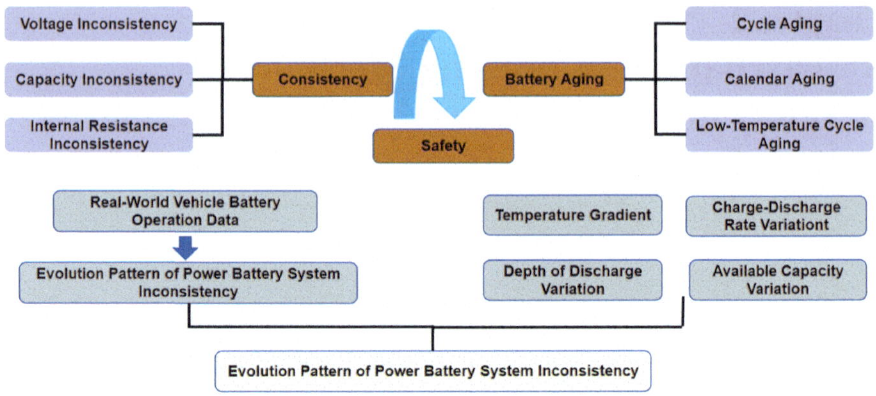

Fig. 2.1 Content of this chapter

and development of power battery systems for pure electric vehicles. Below, we will elaborate on the working principle and safety issues of lithium-ion power batteries as an example.

The structure and working principle of the lithium-ion battery are shown in Fig. 2.2. It mainly consists of a positive electrode, a negative electrode, an electrolyte, a separator, packaging components, and external connections. The positive and negative electrodes are evenly coated on the copper foil and aluminum foil current collectors. Because the positive electrode potential is high, the positive electrode material is often a lithium intercalation transition metal oxide or a polyanion compound, such as lithium cobaltate ($LiCoO_2$) shown in Fig. 2.2, as well as lithium manganate ($LiMn_2O_4$), ternary ($Li(NiCoMn)O_2$), lithium iron phosphate ($LiFePO_4$), etc.; the negative electrode is usually graphite and non-graphite carbon materials; the electrolyte is mainly a non-aqueous solution composed of an organic mixed solvent and a lithium salt, the solvent is mostly an organic solvent such as a carbonate, and the lithium salt is mostly a monovalent polyanion lithium salt, such as lithium hexafluorophosphate ($LiPF_6$), etc.; the battery separator is mostly a polyethylene ($(C_2H_4)n$) or polypropylene ($(C_3H_6)n$) microporous membrane, which mainly serves to isolate the positive and negative electrode materials, the separator only allows the ions in the electrolyte to pass through, but will prevent the electrons from passing through and causing a short circuit in the battery.

During the charging and discharging process of the battery, lithium ions migrate back and forth between the batteries. During the charging process, the lithium ions inside the battery are removed from the positive electrode, pass through the electrolyte and the separator, and are embedded in the negative electrode; the external electrons migrate to the negative electrode through the external circuit. During the discharge process, the lithium ions inside the battery are removed from the negative electrode, pass through the separator and are embedded in the positive electrode; the external electrons migrate to the positive electrode through the external circuit.

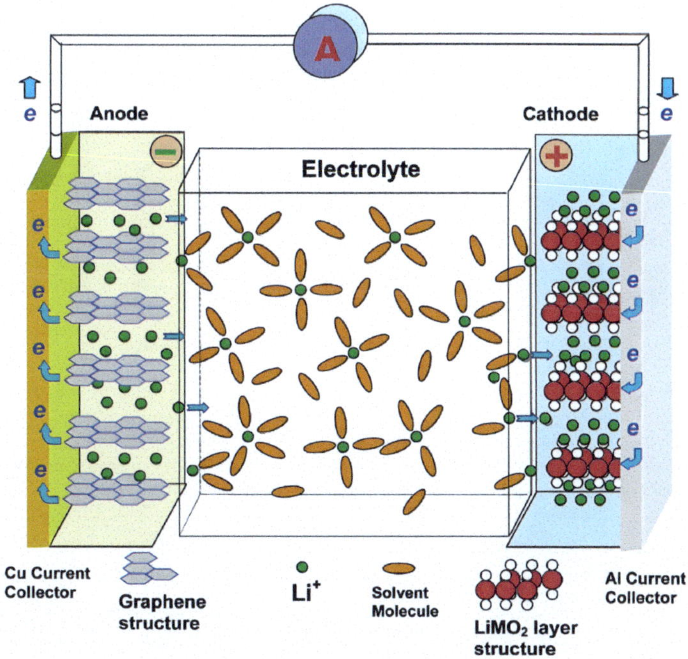

Fig. 2.2 Working principle diagram of lithium-ion battery

2.2.1 Background of the Development of Power Batteries

Under the background of major countries and regions such as China, the United States, Japan, and the European Union vigorously promoting the development of new energy vehicles, the global lithium-ion power battery market has maintained a high growth trend in recent years. SNE Research predicts that by 2025, the matching volume of power batteries in the field of new energy vehicles will reach more than 1.1 TWh, and by 2030 it will reach more than 2.9 TWh; the shipment volume of power batteries is expected to reach 1.4 TWh by 2025 and 3.5 TWh by 2030.

In 2021, the cumulative installed capacity of global power batteries was 296.8 GWh (data source: SNE Research, a South Korean market research institution). In 2021, the domestic sales of new energy vehicles were about 3.521 million units, a year-on-year increase of 157.5%, and the market share increased to 13.4%, indicating that the new energy vehicle market has shifted from policy-driven to market-driven. China's cumulative power battery output was 219.7 GWh, a year-on-year increase of 163.4%; China's cumulative power battery sales reached 186.0 GWh, a year-on-year increase of 182.3%; China's cumulative power battery installed capacity was 154.5 GWh, a year-on-year increase of 142.8% (as shown in Fig. 2.3). China's power battery installed capacity accounts for about 52% of the world.

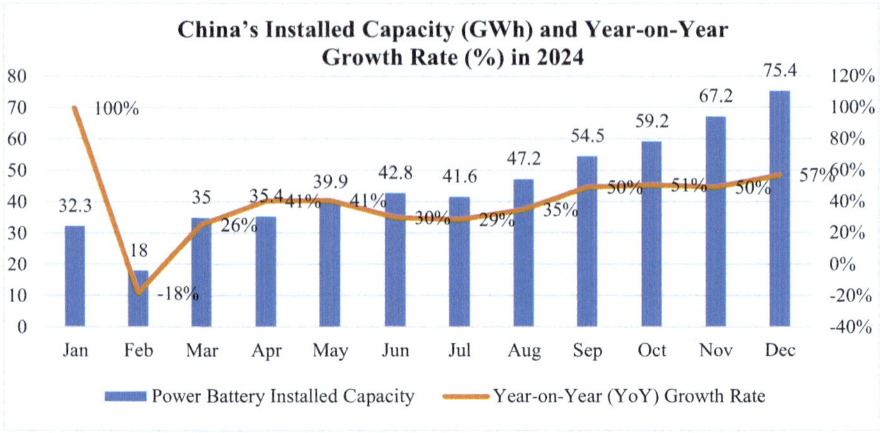

Fig. 2.3 2024 domestic power battery installed capacity (*Data source* China Automotive Power Battery Industry Innovation Alliance)

The installation capacity and market share of domestic battery companies in 2021 are shown in Fig. 2.4. In 2021, a total of 58 power battery companies in China's new energy vehicle market achieved vehicle matching, a decrease of 13 from the same period last year. The top 3, top 5, and top 10 power battery companies had power battery installation capacities of 114.6 GWh, 128.9 GWh, and 142.5 GWh respectively, accounting for 74.2%, 83.4%, and 92.3% of the total installation capacity respectively. CATL ranked first with a domestic market share of 52%; BYD ranked second with a market share of 16%; and the third, fourth, and fifth places were occupied by CENAT, Guoxuan High-tech, and LG Chem respectively.

2.2.2 *Principles and Classification of Power Batteries*

As the basic unit of energy storage and conversion devices for new energy vehicles, the electrical performance, safety and reliability, and service life of power batteries are crucial to vehicle design, development, operation, and maintenance, directly affecting the market application of new energy vehicles and the acceptance of ordinary consumers. The level of technological development is a key support for the global automotive industry's transition to electrification.

Looking at the cost structure of new energy vehicles, the power battery system costs about 30–40% of the total vehicle cost. The power battery system has become the core component of new energy vehicles, and power battery technology has become the core technology reflecting the R&D strength and main advantages of various companies. To enhance the promotion and application of new energy vehicles, power batteries should have the following key characteristics: high energy density,

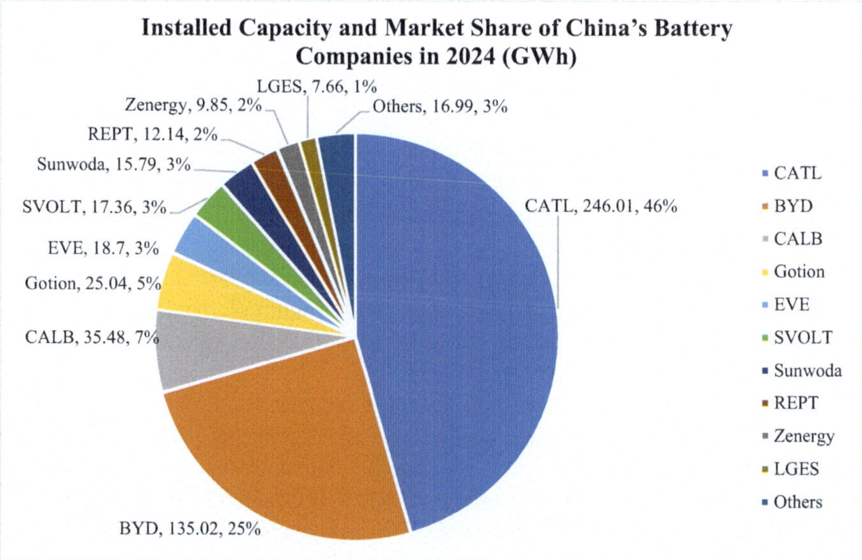

Fig. 2.4 Installation capacity and market share of domestic battery companies in 2021. (*Data source* China Automotive Power Battery Industry Innovation Alliance)

high power density, long cycle life, wide working temperature range, low cost, and high safety.

Overall, the current power battery technology route is diverse and shows a diversified trend. In addition to the mature mass-produced lithium iron phosphate batteries, ternary material batteries, cobalt-free batteries, and quaternary material batteries, there are also sodium-ion batteries, solid-state batteries, lithium metal batteries and other battery technologies that have formed samples and are about to be mass-produced. There are also new system battery technologies such as lithium-sulfur batteries and lithium-air batteries that are still in the laboratory stage.

Lithium iron phosphate battery: Born in 1996, the United States and Japan took the lead in developing lithium iron phosphate batteries. Currently, its specific energy is about 180–190 Wh/g, and leading companies can reach 200–210 Wh/kg. Lithium iron phosphate batteries are presently widely used in mid-to-low-end new energy vehicles and a small number of high-end models.

Ternary material battery: Ternary materials can be divided into NCM (lithium nickel cobalt manganese oxide) and NCA (lithium nickel cobalt aluminum oxide). The NCM ternary material battery was invented in 2008 and has now become one of the main types of new energy vehicle power batteries. The NCA ternary material battery was invented in 1999, with fewer applying companies, mainly Panasonic. Currently, both types of batteries can reach a specific energy of 300 Wh/kg, and show two technological trends, one is the high-nickel/ultra-high-nickel ternary battery developing towards high specific energy, the currently mass-produced 811 (nickel cobalt manganese ratio) ternary battery, 90 series and higher 95 series ultra-high nickel products are gradually industrializing; the other is the medium-nickel ternary battery that balances high safety and high specific energy, such as the 523 ternary battery.

Cobalt-free battery: The narrow sense of cobalt-free battery refers to the battery of layered nickel manganese binary positive electrode material. It was invented in 2019, and Honeycomb Energy took the lead in achieving mass production of this battery globally. Currently, its specific energy reaches 240 Wh/kg, which is comparable to the medium-nickel ternary battery.

Quaternary material battery: refers to the power battery with NCMA as the positive electrode material, invented in 2019. It adds an aluminum element on the basis of NCM material, which can reduce the cost of the positive electrode material, and the current specific energy can reach 300 Wh/kg.

Sodium-ion battery: The research on sodium-ion batteries originated in the 1980s, but the electrical performance was poor. Since 2010, with the progress of battery material technology, the performance of sodium-ion batteries has improved significantly. It has now reached a specific energy of 160 Wh/kg and has good fast charging ability and low-temperature performance. The first-generation sodium-ion battery has been mass-produced.

Solid-state battery: The research on solid-state batteries originated in the 1970s, and now a solid–liquid mixed battery has been developed and industrialized, and the solid–liquid mixed battery that is about to be mass-produced can reach a specific energy of 360 Wh/kg. However, there are still obvious defects in all-solid-state batteries that have not been resolved, and it will take a long time to industrialise.

Lithium metal battery: refers to the battery with lithium metal as the negative electrode material, its research can be traced back to the 1970s. But because lithium metal batteries pose safety risks during charging, they have not been commercialized. Until recently, products of lithium metal batteries have gradually been launched, and their specific energy can now reach 550 Wh/kg.

Lithium-Sulphur Batteries: Lithium-sulphur batteries are a type of lithium battery that uses sulfur as the positive electrode and metallic lithium as the negative electrode. Its reaction mechanism differs from the ion deintercalation mechanism of lithium-ion batteries and instead operates on an electrochemical mechanism. Currently, it is still in the experimental stage, with an energy density that can reach up to 609Wh/kg.

Lithium-Air Batteries: The negative electrode material of lithium-air batteries is lithium, and it uses oxygen as the positive porous carbon electrode, which has the potential for high energy density. Currently, it is still in the experimental stage, with a theoretical energy density limit of up to 11,000 Wh/kg.

2.2.3 Lithium Iron Phosphate Power Batteries

Lithium iron phosphate batteries are a type of lithium-ion battery that uses lithium iron phosphate ($LiFePO_4$) as the positive electrode material and graphite as the negative electrode material. The single-cell nominal voltage is 3.2 V, and the charge cut-off voltage is 3.6~3.65 V. During the charging process, some lithium ions are released from the lithium iron phosphate, transferred to the negative electrode through the electrolyte, and embedded in the negative electrode material; at the same time, electrons are released from the positive electrode, reaching the negative electrode through the external circuit, maintaining the balance of the chemical reaction. During the discharge process, lithium ions are released from the negative electrode, and reach the positive electrode through the electrolyte, and the negative electrode releases electrons, reaching the positive electrode through the external circuit, providing energy to the outside world.

Lithium iron phosphate batteries have the advantages of high working voltage, high energy density, long cycle life, good safety performance, low self-discharge rate, and no memory effect.

Currently, the performance of lithium iron phosphate batteries in China is at the globally advanced level, with energy densities generally reaching 180–190 Wh/kg, and some manufacturers can achieve more than 200 Wh/kg. The gram capacity of lithium iron phosphate positive electrode material can reach 160 mAh/g, and its theoretical specific capacity is 170 mAh/g.

At the same time, lithium iron phosphate has a long cycle life, and the 1C cycle life of lithium iron phosphate batteries generally reaches more than 3000 times, ensuring a service life of 8–10 years. Lithium iron phosphate batteries have good safety performance, and the electrochemical performance of lithium iron phosphate positive electrode material is relatively stable, even under special conditions such as short circuit, overcharge, squeeze, and puncture, it is still relatively safe.

Currently, the leading companies in domestic lithium iron phosphate batteries include CATL, BYD, Guoxuan High-tech, Zhongchuang Xinhang, Lishen, EVE Energy, etc., and the leading companies in lithium iron phosphate materials include Hunan Yueneng, Defang Nano, Hubei Wanrun, Longpan Technology, etc. BYD's developed and mass-produced "blade battery" has further improved the energy density and safety of lithium iron phosphate batteries, promoting the technological progress of lithium iron phosphate batteries.

Foreign power battery companies, including LG Chem, SKI, Samsung SDI, and Panasonic, have mainly focused on ternary batteries as the main technical route and direction in the past. However, as the energy density of lithium iron phosphate continues to increase and its cost-effectiveness is higher, after Tesla, Volkswagen and other leading international car companies have successively adopted lithium iron phosphate batteries, LG Chem and SKI will also enter the field of lithium iron phosphate battery technology, and are currently conducting research and development of lithium iron phosphate batteries, and it is expected to complete trial production in 2022–2023.

In 2021, the cumulative production of power batteries in China was 219.7 GWh, a year-on-year increase of 163.4%. Among them, the cumulative production of lithium iron phosphate batteries was 125.4 GWh, accounting for 57.1% of the total production, a year-on-year increase of 262.9%. It can be seen that due to the lower cost of lithium iron phosphate batteries, the proportion of new energy vehicles equipped with them has gradually increased in recent years, and it will maintain a large market scale in the short term.

2.2.4 *Ternary Lithium Power Batteries*

Ternary material batteries refer to lithium batteries whose positive electrode material uses lithium nickel cobalt manganese oxide ($Li(NiCoMn)O_2$) or lithium nickel cobalt aluminum oxide as the ternary positive electrode material. The ternary composite positive electrode material is made from nickel salt, cobalt salt, and manganese salt as raw materials, and the ratio of nickel, cobalt, and manganese can be adjusted according to actual needs. Batteries with ternary materials as positive electrodes have higher safety than lithium cobalt oxide batteries. Ternary material batteries are relatively balanced in terms of capacity and safety, and their cycle performance is better than that of normal lithium cobalt oxide batteries. Initially, due to technical reasons, its nominal voltage was only 3.5–3.6 V, which was somewhat limited in its range of use. However, with continuous improvements in formulation and structure, the nominal voltage of the battery has reached 3.7 V, and its capacity has reached or exceeded the level of lithium cobalt oxide batteries.

Currently, the performance of ternary material batteries in our country is at a globally advanced level, with energy densities generally reaching 250 Wh/kg, and some manufacturers achieving over 300 Wh/kg. At present, the gram capacity of ternary materials can reach 210 mAh/g, and the theoretical specific capacity of ternary materials with different components varies, roughly around 280 mAh/g.

Currently, the leading companies in the domestic ternary material battery industry include CATL, Zhongchuang Xinhang, Guoxuan High-tech, Lishen, and EVE Energy, while the leading companies in the ternary material industry include Ronbay Technology, Dangsheng Technology, Zhongwei New Materials, Huayou, and Changyuan Lico, etc.

Foreign power battery companies mainly use ternary material batteries, including Japan's Panasonic, South Korea's LG Chem, SK Innovation, and Samsung SDI. However, Panasonic's ternary materials are different from NCM and use NCA materials instead.

At present, the industry mainly improves the specific energy of ternary material batteries by adopting high-nickel/ultra-high-nickel ternary cathode materials and silicon-based anode materials, with the long-term goal of achieving a specific energy of 350 Wh/kg or even 400 Wh/kg.

In 2021, the cumulative production of power batteries in our country was 219.7 GWh, an increase of 163.4% year-on-year. Among them, the cumulative production of ternary batteries was 93.9 GWh, accounting for 42.7% of the global total, an increase of 93.6% year-on-year. For the foreseeable future, high-end models will still mainly use ternary material batteries, and the market size maintains a good outlook.

2.3 Analysis of Big Data Features of Actual Vehicle Power Batteries

2.3.1 Introduction to the Big Data Platform

As one of the hottest research fields in recent years, big data has received widespread attention from the global scientific and academic communities. Big data is not just about large-scale data sets, it usually has the "5 V" characteristics of data scale (Volume), data processing timeliness (Velocity), data diversity (Variety), result accuracy (Veracity), and value depth (Value). Based on massive multi-dimensional heterogeneous data in various fields, the use of advanced big data analysis techniques to mine the intrinsic value of massive data features can greatly promote the development and transformation of various industries. With the gradual improvement of the vehicle information system of new energy vehicles and the increasing degree of informatization, big data technology also shows a good application prospect in the field of new energy vehicles, and the integration of new energy vehicles and big data has become one of the national strategic development directions for the transformation and upgrading of our country's automobile industry.

In the field of new energy vehicles, major companies have established big data platforms, fully utilizing the value of data through the collection, processing, storage, and deep mining of big data from new energy vehicles. To ensure the daily travel safety of residents and promote the development of the new energy vehicle industry, the Ministry of Industry and Information Technology established the NMMP-NEV (National Energy Vehicle Monitoring and Management National Platform) platform in Beijing in 2016, with its monitoring and management mechanism shown in Fig. 2.5. As of July 2022, the number of vehicles connected to the national platform exceeded 9 million, and the establishment of the national platform has played an essential supporting role in strengthening the safety supervision of new energy enterprises and vehicles by government departments. The national platform receives massive real-time operation data of new energy vehicles, forms a multi-source heterogeneous database of new energy vehicles in conjunction with vehicle file data, meteorological data, road construction data, etc., and combines data resources and big data analysis technology to mine the rich value hidden behind the data, to improve the current safety problems of new energy vehicles and promote the development of the new energy vehicle industry.

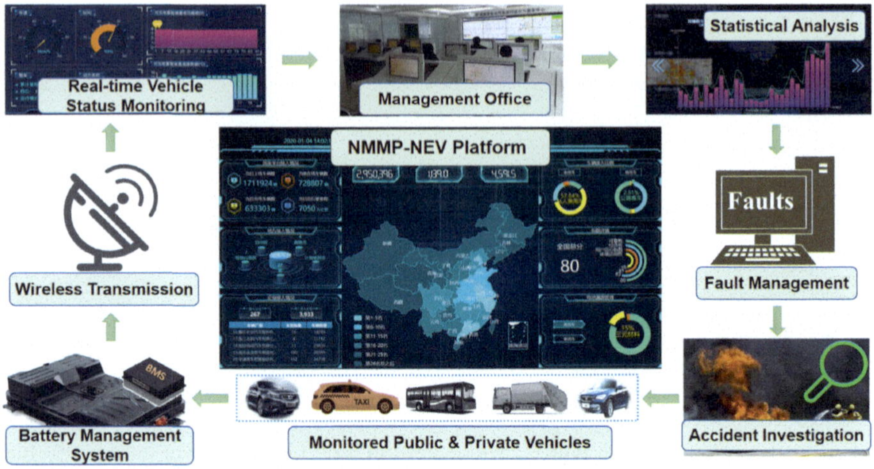

Fig. 2.5 Monitoring and management mechanism of NMMP-NEV

2.3.2 Big Data Feature Analysis of Power Battery Systems

At present, our country has established a three-level new energy vehicle supervision system of the state, local governments, and enterprises, and emphasizes that enterprises, as the first responsible persons for safety, need to take on the main responsibility for safety supervision. In order to achieve effective supervision and safe operation of the power battery system, it is necessary to carry out real-time monitoring and statistical analysis of the operation data of the power battery system, and to carry out real-time evaluation and risk prediction of its safety.

1. **Operational Statistical Features**

   The statistics of the operation big data of the new energy vehicle power battery system mainly include power consumption statistics, driving charging analysis, health status monitoring, fault and accident statistics, etc.

   Currently, new energy vehicles in our country have moved from the nurturing stage to the rapid growth stage, and the market has seen explosive growth. In 2021, the production and sales of new energy vehicles nationwide were 3.545 million and 3.521 million respectively, both increasing by 1.6 times year-on-year. As of the end of 2021, the number of new energy vehicles in the country was 7.84 million, accounting for 2.6% of the total number of vehicles. With the rapid increase in the number of vehicles, new energy vehicle recalls and safety accidents are constantly facing new challenges.

   In 2021, China implemented 59 recalls of new energy vehicles, involving 830,000 vehicles. The number of recalls and the quantity of vehicles recalled increased by 31.1% and 75.9% respectively compared to the previous year. 52.5% of the defects in new energy vehicles were related to power batteries, motors, and electronic control systems. Analysis of new energy vehicle fires in recent years revealed both external factors such as substandard charging facilities, improper user operation, and irregular maintenance, and internal factors such as early product power battery and high-voltage component aging, overall vehicle and battery quality. There are also safety issues with batteries after traffic accident collisions. To reduce accidents involving new energy vehicles, the national platform has fully implemented a 7×24 h duty feedback system, pushing real-time level 3 alarm information to enterprises, regularly feeding back level 1 and 2 alarm information weekly, and regularly reporting to the Ministry of Industry and Information Technology. In addition, the national platform has fully utilised its technical advantages, establishing a power battery system accident warning model and developing corresponding warning software modules, which are currently being trialled by several vehicle manufacturers.

2. **Safety Feature Analysis**

   The safety feature analysis of big data on power batteries mainly includes the following research directions: inconsistency analysis and fault diagnosis, thermal runaway warning, and health status estimation. For example, part of this project describes a voltage fault diagnosis method based on modified Shannon entropy. According to actual vehicle operation data, voltage abnormal cells on electric taxis were successfully diagnosed, and early warning of thermal faults was achieved. In terms of power battery health status estimation, Frisk Erik and others from Linköping University in Sweden used the replacement records of onboard lead-acid batteries to estimate the remaining life of the power battery system using the random survival forest method. Based on the operation data of 33,603 vehicles in five European countries, 291 variables were extracted, including structured data such as temperature and pressure, and unstructured data such as current and voltage probability distribution histograms, which is highly innovative in the application of big data. Current big data analysis techniques are widely used in power battery system fault diagnosis, safety risk prediction, and capacity decay analysis. However, due to the sparse characteristics of real vehicle monitoring data, multi-stress parameter coupling, and strong non-linearity of power battery operation parameters, further in-depth research is needed in safety risk prediction and real vehicle verification based on real vehicle operation big data. In addition, operational big data has a certain application basis in vehicle operation status monitoring, key component detection, new energy vehicle user research, and infrastructure construction layout analysis. For example, the analysis of the impact of driving behavior on energy consumption and power battery life, the statistical analysis of the reliability and fault pattern of new energy vehicles, the research on the location selection method of electric taxi charging stations, and the research on the prediction of new energy vehicle charging demand and grid load. In laboratory-level research on the safety of power battery systems, the battery operation parameters obtained from several groups of cycle charge and discharge are usually used as the safety characterization parameters of the battery, such as the range, mean, and variance distribution of the measured single cell voltage as the measurement indicators of single cell voltage consistency. However, this method is not applicable in the scenario of massive data collected based on the real vehicle operation big data platform, mainly for the following four reasons: (1) Different sampling accuracy: The precision of the power battery system data collected by the big data platform is lower than that obtained in the laboratory environment. At the same time, the power battery system is easily affected by vibration, high temperature, rain, extreme road conditions and other complex environmental factors under actual vehicle driving conditions. The robust noise analysis of the real vehicle power battery system, such as system noise and environmental noise, is still not perfect, so the error of the sampling data of the real vehicle power battery system is still difficult to estimate. (2) Different sampling frequency: Compared with the relatively stable and controllable laboratory environment, the sampling frequency of the big data platform is lower and the sampling interval is longer (for example, the NMMP-NEV platform usually has a sampling interval of 10 s or longer). In addition, the sampling moments are usually discontinuous, and the battery states corresponding to the sampling data points at adjacent moments are also discontinuous, so it is impossible to achieve stable and uninterrupted measurement of the power battery system status as in the laboratory conditions. (3) Different sampling time: Compared with the laboratory environment where accelerated performance tests of power battery systems can be conducted, the time span of data collection on the big data platform is larger. For example, to collect full lifecycle capacity decay data for the same power battery system, it can be completed in as short as one month under high charge and discharge rates through accelerated life testing, while normal monitoring through the big data platform may require more than 5 years. However, the authenticity of the real vehicle operation conditions under the monitoring of the big data platform is also difficult to reproduce through experimental means. The research on the safety mechanism of the power battery system can be completed through accelerated life testing, but its engineering application value still needs to be verified based on real vehicle operation big data. (4) Different sampling samples: The power battery system data collected by the big data platform is more diverse and complex than the data obtained in the laboratory environment. The data includes not only the power battery system of a single vehicle, but also the power battery systems of multiple vehicles of different models, different manufacturers, and different usage scenarios. Therefore, the data collected by the big data platform is more representative and has higher research value. However, due to the high complexity and diversity of the data, the traditional data analysis methods are not applicable, and new data analysis methods need to be developed. Under laboratory conditions, continuous charging and discharging state monitoring can often only be performed on one or more sets of power battery packs. In contrast, the sample size covered by the big data platform is larger, as the monitoring objects are public models and market-sold passenger models. Therefore, there are many sampling samples of the same model and the same type of power battery system, and multiple sets of different operating states of power battery system operating data can be obtained at the same time.

In order to effectively utilize the massive operating data of the real vehicle big data platform, fully explore the potential application value of the data, avoid the decline in the quality of the sampling data due to sampling noise or defects in the preprocessing algorithm, it is necessary to select appropriate statistical models and grouping theories to integrate the collected massive data, extract the effective information therein, and screen the single samples of the power battery system to reduce the impact of system noise and environmental noise during data collection. Then, the consistency of the power battery system can be fitted and described using adaptive parameter identification methods, and the consistency and safety analysis of the power battery system can be carried out.

2.4 Analysis of Inconsistency in Power Battery Systems

2.4.1 Forms of Inconsistency in Power Battery Systems

The inconsistency of the power battery system refers to the existence of certain differences in parameters such as voltage, SOC, RUL, temperature, capacity decay rate, internal resistance and its change rate, self-discharge rate, etc., after the single cells of the same specification and model form a battery pack. The inconsistency of the power battery system mainly manifests in the following two aspects when it is manufactured: (1) Original differences caused by the manufacturing process: errors in the manufacturing process and unevenness of the battery material will cause minor differences in the activation degree, thickness, porosity, etc., of the active material of the battery plate. This inconsistency in the internal structure and material of the battery will cause the capacity, internal resistance and other parameters of the same model and batch of batteries to have certain inconsistencies when they leave the factory; (2) Single-cell degradation differences caused by environmental differences during vehicle use: due to the storage temperature, ventilation conditions, self-discharge level and electrolyte density of each battery cell will have certain differences, this also exacerbates the inconsistency of parameters such as voltage, internal resistance and capacity.

Based on the reasons for the expansion of inconsistency during the use of the power battery system and the way it affects battery performance, the inconsistency of the battery can be divided into voltage, capacity, and internal resistance.

1. **Voltage Inconsistency**

   The main reason for the inconsistency of battery voltage is due to the mutual charging between single cells in parallel battery packs. When the voltage of a cell in a parallel battery pack is low, other batteries will charge this battery. The inconsistency of parallel battery voltage is shown in Fig. 2.6. Assuming that the terminal voltage U_1 of battery 1 is lower than the terminal voltage U_2 of battery 2, a charging circuit is formed between the two batteries, causing battery 2 to charge battery 1 slightly. This parallel connection method will slightly increase

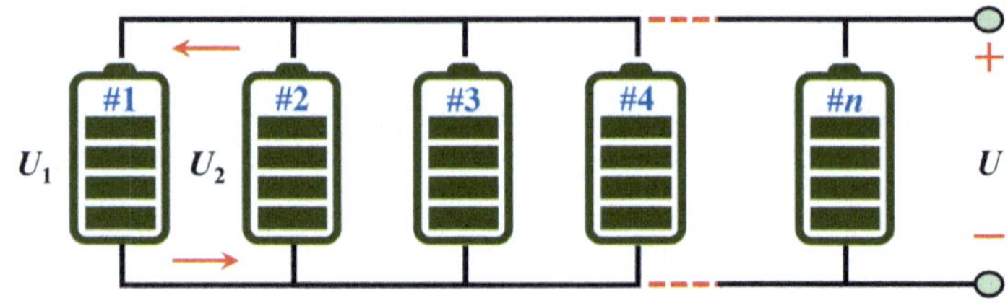

Fig. 2.6 Inconsistency of parallel battery voltage

the capacity of the low-voltage battery while drastically reducing the capacity of the high-voltage battery, and energy loss will occur during this mutual charging process, making the battery unable to achieve the expected output effect. If the low-voltage battery is used together with the normal battery, it will become the load of the battery pack and affect the work of other batteries, thereby accelerating the life decay of the entire battery pack.

In addition, since the battery SOC is linearly related to the OCV within a certain range, the inconsistency of the OCV of the battery under static conditions (generally refers to the battery being stationary for more than one hour) can also reflect the inconsistency of the energy state of the battery to a certain extent.

2. **Capacity Inconsistency**

The inconsistency of the capacity of the power battery system in the actual working process is mainly caused by the inconsistency of the starting capacity and the inconsistency of the discharge current. The consistency of the initial capacity can generally be well guaranteed by the sorting test before leaving the factory. Even in the use of the battery, the inconsistency of the single cell can be made smaller by independent charging and discharging of the inconsistent single cell, so the inconsistency of the initial capacity is not the main contradiction of the power battery system group application.

The inconsistency of capacity will directly cause adverse effects on the health status of the power battery system, causing its available capacity to decrease. According to the barrel effect principle, the highest available capacity of a series battery pack is determined by the battery with the lowest capacity so that capacity inconsistency will reduce the actual available capacity of the entire battery pack. In addition, because of different capacities, different batteries will have different charging and discharging conditions under the same conditions. Overcharging or over-discharging will occur in batteries with smaller capacities and cause irreversible damage to them. Therefore, capacity inconsistency will have a serious impact on the health status of the power battery system during cyclic use.

3. **Inconsistency of Internal Resistance**

Inconsistency in battery internal resistance can cause the heat loss of each cell to be the same during the discharge process, thereby affecting the energy state of the battery cell. Inconsistency in internal resistance can be divided into two situations: series and parallel.

Due to the characteristics of the series connection, the series battery pack makes the discharge current of each cell the same during the discharge process. The inconsistency of internal resistance will make the voltage division of each cell different. The cell with larger internal resistance has higher voltage division, corresponding to larger internal energy consumption and heat generation, and faster temperature rise. However, the internal resistance of the battery will continue to increase with the rise of temperature, and the battery temperature will continue to rise. Once the heat dissipation cannot meet the requirements, it will cause the battery to lose thermal control or even explode. During the charging process, due to the inconsistency of internal resistance, the cell with larger internal resistance will be charged to the highest cut-off voltage (COV) in advance. If charging is stopped at this time to ensure safety, the remaining battery cells will not be fully charged; on the contrary, if charging continues, it will cause overcharging of the battery and pose a safety risk. Therefore, a compromise control method needs to be adopted during the charging process to prevent overcharging.

Due to the characteristics of parallel connection, the parallel battery pack makes the discharge voltage of each cell the same during the discharge process. The inconsistency of internal resistance makes the current of each cell different. The cell with larger internal resistance has a smaller discharge current, and the cell with smaller internal resistance has a larger discharge current, thus making each cell work at different discharge rates. Under different discharge rates, the discharge depth of each cell will also be different, which will have a serious impact on the health of the battery. During the charging process, the current of each parallel branch is different due to the inconsistency of internal resistance, so each cell cannot achieve the same charging effect under the same charging time, and the time for each cell to be charged to COV is also different. To prevent overcharging, a compromise control method also needs to be adopted during the charging process.

2.4.2 Development Law of Inconsistency in Power Battery System

The various inconsistent parameters of the power battery system are not independent of each other, but influence each other and have a cause-and-effect relationship. The most direct manifestation is the inconsistency of battery voltage, which can reflect the inconsistency of other parameters of the power battery system to a certain extent. With the increase of usage time and mileage, the degree of inconsistency in the power battery system will gradually increase, and the most intuitive reaction is the gradual increase in the inconsistency level of single-cell voltage. The following takes a pure electric taxi running in the urban area of Beijing as an example to explore the distribution law of voltage inconsistency in the actual operation process of the vehicle.

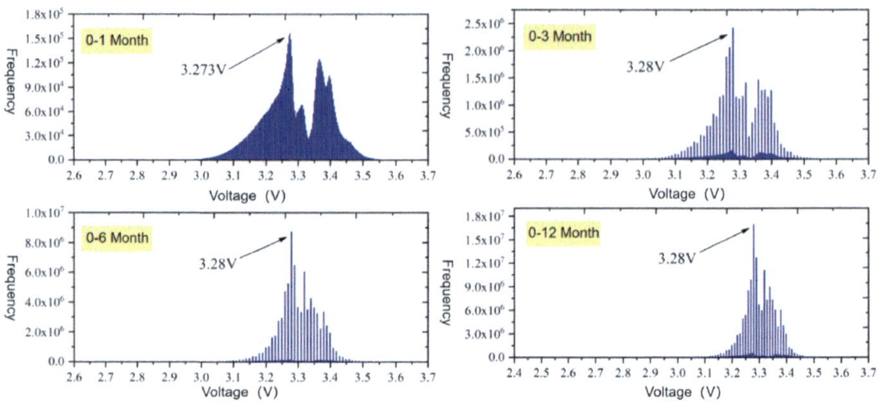

Fig. 2.7 Distribution of single-cell voltage frequency at different times

The distribution of single-cell voltage frequency at different times within a year in the power battery system under study is shown in Fig. 2.7. As can be seen from the figure, the highest frequency of single-cell voltage in January is 3.273 V. With the passage of time, the highest frequency voltage rises to 3.28 V after March, which is likely to be related to the low environmental temperature in Beijing in January; the frequency distribution in the first three months basically shows two peak intervals of 3.25~3.3 V and 3.35~3.4 V, which may be a phenomenon formed by the initial low-temperature working area and the normal temperature working area after a period of driving. With the increase of usage time, the frequency distribution of single-cell voltage gradually concentrates to form a peak after March, and the distribution on both sides of the peak is basically symmetrical and gradually decreases, similar to the normal distribution.

Considering the complexity of vehicle driving conditions, the performance characteristics of the power battery system under driving and charging conditions are completely different. Figure 2.8 shows the actual driving speed and single-cell voltage curve of the power battery system under study on a certain day. As can be seen from the figure, the fluctuation of single-cell voltage is directly affected by vehicle speed and is negatively correlated. The charging and driving data of the vehicle under study are analyzed separately to obtain the probability distribution and mean variance of single-cell voltage under the two states, as shown in Fig. 2.9. Due to the increase of usage time and the complex and changing operating environment, the dispersion degree of single cell voltage under driving conditions is more obvious than that under charging conditions, which is manifested as the probability peak (24.01%) under driving conditions is much larger than the probability peak (19.17%) under charging conditions; the inconsistency of single cell voltage gradually increases, which is manifested as the mean variance reflecting the inconsistency of single cell voltage gradually increases with the passage of time under both charging and driving conditions; in addition, the inconsistency of single cell voltage under driving conditions is much higher than that under charging conditions, for example, the mean variance

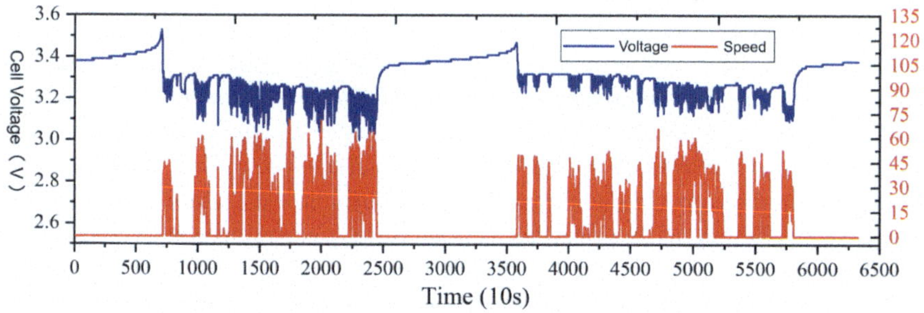

Fig. 2.8 Curve of single cell voltage and vehicle speed

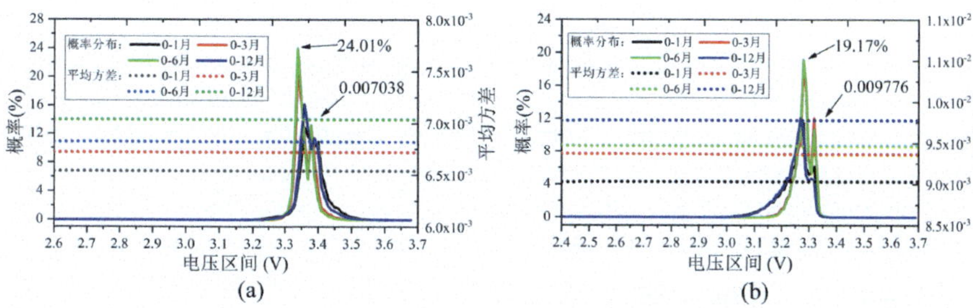

Fig. 2.9 Probability distribution and mean variance of single cell voltage under charging and driving conditions: **a** charging state; **b** driving state

of 0.009776 under driving conditions for the whole 12 months is significantly larger than the mean variance of 0.007038 under charging conditions.

2.5 Coupling Mechanism of Safety and Consistency in Power Battery System

2.5.1 Coupling Relationship Between Power Battery Consistency and Safety

As known from the above, the inconsistency of the power battery system can be expanded from three aspects: voltage, capacity, and internal resistance. The results of the analysis show that battery inconsistency directly affects the life decay and aging speed of the battery. As the battery continues to age, the side reactions inside the battery (such as SEI film thickening, negative lithium precipitation, electrolyte oxidation, etc.) will cause the battery capacity to decay and the internal resistance to increase, thereby leading to a decline in battery safety performance.

Existing research has found that under normal/high-temperature cycle aging conditions, due to the rise in battery internal resistance, Joule heat continues to increase under charging and discharging, the battery's abuse tolerance continues to decline, and the thermal stability of the battery will also change to a certain extent. The change pattern is related to the battery's material system and process level; under normal/high-temperature storage aging conditions, the battery's abuse tolerance will gradually decrease; under low-temperature cycle aging conditions, due to negative lithium precipitation, the thermal stability of the battery will drop sharply. The precipitated lithium metal reacts with the electrolyte at low temperatures, causing the battery's self-heating rate to increase sharply, thereby posing a serious threat to battery safety.

In summary, there is usually no direct correlation between the consistency and safety of the power battery system, but they are indirectly linked through the accelerated decay of battery life. The safety of the power battery system will also deteriorate as the battery life gradually ages. Therefore, to study the coupling relationship between battery consistency and safety, we first need to understand the relationship between battery consistency and battery life. The RUL or available capacity of the power battery system is closely related to the consistency level of the single battery. In the actual vehicle application process, it is mainly reflected in the following aspects.

1. **Temperature Differences**

   Different models of electric vehicles have different requirements for the battery pack layout and installation position of the power battery system. The thermal environment of the single battery will have certain inherent differences. For example, some battery cells or modules may be closer to heat sources such as motors, while other batteries may be in better ventilated positions; in addition, single batteries in the same location area will also have certain temperature differences due to differences in heat source positions and ventilation conditions, which will affect the overall performance and service life of the power battery system.

2. **Charging and Discharging Rate Differences**

   Under normal circumstances, each type of battery has its corresponding optimal charging and discharging rate. If the capacities of the batteries are different, their optimal charging and discharging currents will also be different. In a series battery group, the current is the same. Due to the inconsistency of the battery, some single batteries work at their optimal charging and discharging current, while some batteries may not reach or exceed their optimal charging and discharging current, causing differences in the charging and discharging rate during operation; differences in charging and discharging rates will cause differences in energy consumption and heat generation rates between single cells, accelerating battery life decay and easily triggering thermal runaway of the battery.

3. **Discharge Depth Differences**

   If there is a difference in battery capacity, the discharge depth will also be different after the electric vehicle travels the same distance. At this time, when most single batteries are still in the shallow discharge stage, low-capacity single batteries have already entered the deep discharge stage; when other single batteries start deep discharge, low-capacity single batteries are already in an over-discharge state and have no remaining power to discharge, becoming a load in the circuit, accelerating the energy loss and aging of the battery.

4. **Available Capacity Differences**

   If there is a difference in the available capacity of the battery, during the charging process, low-capacity single batteries will inevitably be charged first. In order to charge other single batteries, these low-capacity batteries will inevitably be overcharged; in the later stage of charging, the charging voltage continues to rise, and when overcharging occurs, it even exceeds the highest COV. Overcharging will seriously affect the charging process and battery life, and is very likely to trigger overcharging safety hazards such as thermal runaway.

2.5.2 Improvement Measures for Power Battery System Inconsistency

Under normal circumstances, the inconsistency of the power battery system has already occurred in its production stage, so the consistency of the power battery system is relative, and inconsistency absolutely exists. In the actual application process, what we need to do is to take effective measures to slow down the trend or speed of this inconsistency expansion. To improve the consistency of the power battery system and slow down the decay speed of the battery pack life, based on a large amount of application experience and experimental research on power battery systems, the following measures can be summarized:

1. **Sorting**

   The purpose of battery sorting is to select individual batteries with similar parameters. In theory, individual batteries from different batches should not be used together, even if they are from the same batch, they need to be selected so that batteries with concentrated parameters are placed in the same battery pack. The methods of battery sorting mainly fall into three categories: static sorting, dynamic sorting, and a combination of both. Static sorting is done by introducing statistical algorithms, setting selection criteria, and selecting based on the open-circuit voltage, internal resistance, capacity and other characteristic parameters of the battery cell; dynamic selection is based on the characteristics exhibited by the battery during the charging and discharging process, or by comparing the relationship between its own charging and discharging curves; the combination of dynamic and static sorting generally refers to using the preliminary grouping of static selection for further dynamic selection, thereby improving the accuracy of battery sorting, but the cost will also rise accordingly.

2. **Balancing**

   To address the inconsistency issue of individual batteries, BMS usually integrates the function of battery balancing. For example, when some individual cells reach the charging COV first and others are significantly lagging behind, the BMS will activate the charging balancing function, usually by connecting a resistor to discharge part of the high voltage cell's energy or transferring energy to the low voltage cell; when the charging cut-off condition is waived, the charging process will restart and more energy will be charged into the power battery system.

3. **Heat Treatment**
   To address the issue of different heat generation caused by the inconsistency of internal resistance in the power battery system, introducing a heat treatment system can effectively regulate the temperature difference of the entire battery system, keeping it within a safe threshold range; the temperature rise of individual batteries with higher heat generation may still be high, but it will not be significantly different from other cells, and the level of degradation between cells will not show a significant difference.
4. **Usage and Maintenance**
   In the actual use process of the power battery system, usage and maintenance is the most direct and effective measures to maintain the consistency of individual batteries, for example: ensuring a good usage environment for the battery, trying to ensure a uniform environmental temperature field, reducing battery vibration, avoiding pollution of the pole column by water and dust, etc.; avoiding overcharging and deep discharging of the battery; regularly performing low current maintenance charging on the power battery system during use, especially for individual batteries with low capacity, to promote the battery's own balance and performance recovery; detecting individual battery parameters, especially the voltage distribution under dynamic and static conditions (during vehicle charging or driving), and adjusting or replacing cells with extremely abnormal parameters in a timely manner.

2.6 Power Battery Consistency and Residual Value Assessment Technology

2.6.1 Current Status of Power Battery Performance Testing for Secondary Use

As batteries vary in brand, model, parameters, and remaining life, how to quickly classify and test the millions of battery packs that retire each year is key to whether they can be successfully used for secondary purposes.

If traditional on-site testing methods are used, from transportation to the testing institution → disassembly of the battery pack → classification testing, the testing cycle is long, and testing 1 million battery packs would waste a lot of time. As the number of retirements continues to increase, this difficulty will further increase. Therefore, large-scale testing needs to be fast and efficient. In addition, traditional testing methods require a large investment in manpower, equipment, and space, with high costs and slow returns. Therefore, large-scale testing needs to be low-cost and require less investment.

Overall, whether the secondary use of power batteries can develop on a large scale depends on whether there is a large-scale, fast, and low-cost testing solution. Therefore, to promote the development of the secondary use market, a rapid testing solution that can solve current problems is needed (Fig. 2.10).

Currently, online big data power battery evaluation services, combined with a large amount of online + a small amount of offline testing methods, achieve batch and efficient testing methods.

This solution is divided into an online evaluation system and an offline testing system. Firstly, the online evaluation system obtains real-time uploaded power battery and vehicle operation data from the platform, and quickly and accurately completes the online evaluation part through the battery multi-dimensional performance evaluation algorithm and big data intelligent analysis technology, and outputs the evaluation results to the offline testing system. 90% of the battery packs can be directly judged as suitable for secondary use or scrapped for recycling through the online evaluation method. This can directly reduce 90% of the time cost and investment compared to traditional testing methods. Therefore, the advantage of the online system is that it can be evaluated on a large scale, quickly, at a low cost, and accurately.

For the remaining approximately 10% of battery packs, offline testing results will be suggested and all will be transferred to the offline testing system. The offline testing system obtains battery data through three types of testing equipment (including large-scale comprehensive testing equipment, small one-stop testing equipment, and mobile portable testing equipment). For the key data obtained, the offline battery health analysis model is used to perform a more accurate health analysis on the incoming batteries, giving the test results of secondary use or scrap recycling. The advantage of offline testing is its accuracy of up to 99%, and it can locate the problem module or single cell.

The new model of online assessment + offline testing can achieve large-scale, fast, low-cost (more, fast, good, save) testing goals while ensuring high testing accuracy.

In the future, for the price assessment of specific batteries, the platform will plan to provide online battery assessment algorithms, which can quickly understand the remaining value of the battery. By introducing this assessment mechanism, combined with industry price standards, historical transaction data, and battery classification application scenarios, a big data assessment algorithm can be established to evaluate the value of retired batteries.

At present, online assessment is a power battery big data health assessment service to provide testing services, and offline is provided by various types of power battery testing laboratories nationwide. The combination of the two has achieved a new model of online assessment + offline testing. The following introduces the testing principles of online and offline respectively.

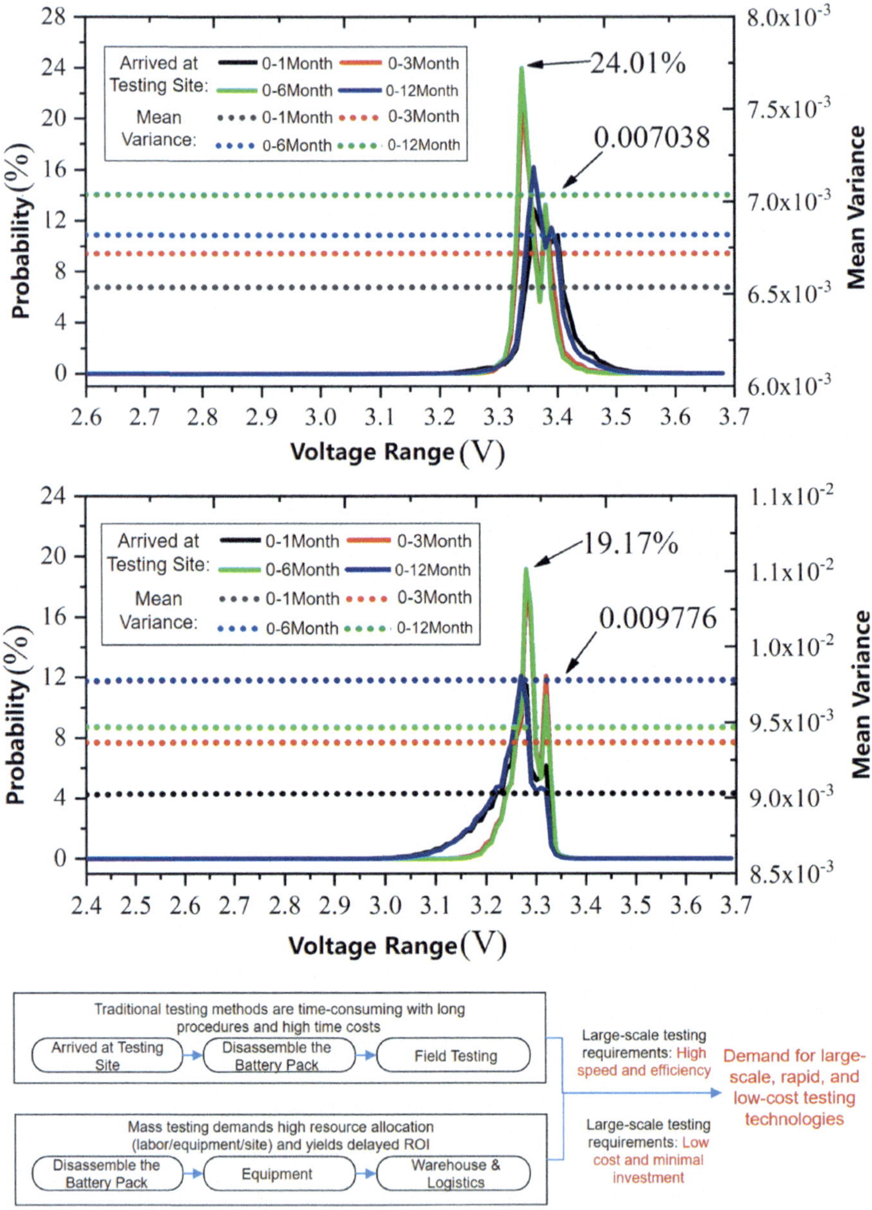

Fig. 2.10 Current status and trends of power battery consistency and residual value

2.6.2 *Principle of Power Battery Big Data Health Assessment*

Based on the development experience of big data, a battery health assessment algorithm has been developed independently. From the perspective of the entire online assessment module, only the battery code needs to be submitted, and the historical big data of battery operation combined with big data intelligent analysis technology, health analysis assessment technology, capacity decay monitoring assessment technology, and battery safety performance analysis assessment technology can quickly evaluate the entire batch of batteries. Finally, an assessment report is given, including detailed assessment results of battery performance, as well as performance grading. For high-quality batteries, it is recommended to use them in stages, and for batteries with severe performance degradation, it is recommended to regenerate and recycle; at the same time, for a batch of batteries with abnormal conditions, offline testing suggestions are given to remind users to further confirm the use value.

The health assessment method is to capture the decay data of the battery under normal conditions, understand the decay data of various indicators from the single cell to the entire pack, and establish a decay characteristic database of the battery under different vehicle types, electrochemical systems, and charging termination conditions.

When evaluating different vehicles, it is necessary to collect real-time operation data of the vehicle, clean the data of the evaluation target vehicle that needs to be monitored through the data cleaning of the preprocessing module, compare the decay curve corresponding to the feature standard database, and pass a set of battery health assessment algorithms and evaluation systems, finally get the health status of the current vehicle battery, and finally achieve the ability to evaluate battery health with big data.

1. **Theoretical Explanation**

   The theoretical basis for health assessment is studied from two dimensions:

   (1) From the single cell level, through laboratory research, the battery decay mechanism is found, and the corresponding decay causes and corresponding decay phenomena are found, so that the corresponding performance feature targets can be set for the single cell, and the current decay situation can be understood according to the change of the feature targets in the monitoring process of the single cell, and the future trend can be predicted.

   (2) From the whole pack level, due to the decay of the single cell, the inconsistency of the whole pack will occur. Therefore, the inconsistency of the whole pack can be discovered by collecting the data of the single cell, and the inconsistency of the whole pack can be improved through thermal management and balancing measures. Thus, the health of the entire battery pack is enhanced and detected.

   (3) Inconsistency of initial conditions of the battery cell: inconsistency of capacity, inconsistency of internal resistance; inconsistency of use environment: inconsistency of temperature, inconsistency of stress, etc. (Fig. 2.11).

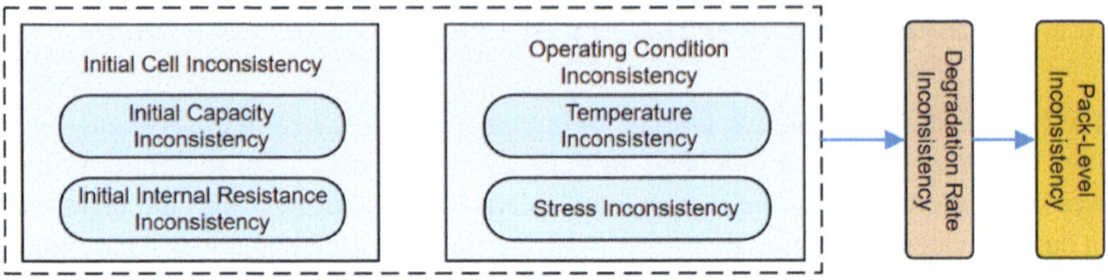

Fig. 2.11 Factors of battery inconsistency

(4) Due to the existence of the barrel effect, the maximum available discharge capacity depends on the cell with the smallest capacity in the whole pack; the maximum chargeable capacity depends on the cell with the smallest chargeable capacity in the whole pack.

$$MaximumAvailableDischargeCapacity C^{B}_{max(dis)} = \min\{C_1^{dis}, \cdots C_i^{dis} \cdots C_n^{dis}\}$$

$$MaximumChargeableCapacity C^{B}_{max(ch)} = \min\{C_1^{ch}, \cdots C_i^{ch}, \cdots C_n^{ch}\}$$

(5) The initial conditions cannot be changed, but the overall inconsistency (characteristics of the whole pack: balancing effect characteristics, voltage consistency, temperature consistency) can be improved through excellent thermal management and balancing measures.

2. **Technical Plan**

The entire plan is aimed at the long cycle and high cost of the current health status detection means of new energy vehicle power batteries, which leads to the inability to accurately and quickly complete the evaluation and detection of the health status of operating vehicle power batteries on a large scale. The technical route is intended to solve the key technology and algorithm model for the rapid and accurate large-scale evaluation of the health of new energy vehicle power batteries, thereby achieving a preliminary assessment of the safety hazards and risks of power battery recycling. The implementation route of the technical plan is divided into three parts, namely the data preprocessing module, the feature standard database module, and the battery health assessment algorithm module. The specific implementation route is first based on the historical operation big data of power storage batteries, through capturing the decay data of the battery under normal conditions, understanding the decay data of various indicators from the single cell to the whole pack, combined with the theoretical basis of power battery decay, using the data preprocessing module to process the historical data, after classification, the establishment of the cell feature standard database and battery feature standard database is completed. Then extract the battery operation data, after the data cleaning of the data preprocessing module, compare the indicator values that need to be monitored by the evaluation target battery with the decay curve corresponding to the feature standard database, and calculate the current health status of the vehicle power battery through a set of battery health assessment algorithms and evaluation systems, and output the evaluation analysis report or serve as an API to the commercial service platform application.

(1) Plan Implementation

Based on the above theoretical research techniques, a feature representation database is constructed:

1. Classify battery packs by vehicle type or electrochemical system;
2. Based on 5000 km as a stage, divide the battery operation data without alarm records within each stage, and establish a classification standard for the cut-off voltage of charging;
3. Construct feature engineering with the data collected from the above vehicles running normally without alarm records;
4. Apply machine learning algorithm technology to establish model training and continuous updates and iterations, thereby forming a series of battery health feature databases;
5. After forming the feature representation database, the vehicle operation data to be evaluated can be extracted, classified according to the vehicle model and electrochemical system, etc., and the preliminary analysis results can be obtained by comparing with the feature representation database by calculating the same feature item values. Then, a comprehensive evaluation is carried out with the risk data such as vehicle fault information to obtain the comprehensive score results of health.

The indicators used to evaluate battery consistency and residual value include the following:

1. Polarisation internal resistance: The voltage difference between the vehicle's full charge and the vehicle's re-powering is used to characterize the polarization internal resistance.
2. Voltage consistency: The relationship between the extreme value of the single-cell voltage at the moment the vehicle is parked and turned off and the SOC.
3. Temperature consistency: The maximum temperature difference during driving; the temperature difference when the vehicle is fully charged.
4. Capacity decay (at the moment of turning off): The relationship between the extreme value of the single-cell voltage at the moment the vehicle is parked and turned off and the SOC;
5. Capacity decay (at the moment of parking): After parking for n hours, the relationship between the extreme value of the single-cell voltage at the moment the vehicle is re-ignited and the SOC.
6. Self-discharge rate (maximum voltage): After the vehicle is parked for n days, the voltage drop value of the maximum single-cell voltage at the moment the vehicle is re-ignited relative to the maximum single-cell voltage at the moment the vehicle is parked and turned off;

7. Self-discharge rate (minimum voltage): After the vehicle is parked for n days, the voltage drop value of the minimum single-cell voltage at the moment the vehicle is re-ignited relative to the minimum single-cell voltage at the moment the vehicle is parked and turned off.

(2) Results display

The online evaluation system can form detailed entries such as battery capacity, internal resistance, voltage and current, consistency, etc., and includes real-time dynamic rolling information of the battery at the beginning of service and at the time of evaluation. Based on the comprehensive score of the detection evaluation items, the final output is the batch online evaluation report of the battery on the right. Based on the online evaluation report, subsequent offline detection is carried out.

Definition of evaluation result scores:

90–100 points: Vehicle level, recommended for vehicle use, good performance;

80–90 points: Recovery level, recommended for secondary use, high value;

70–80 points: Recovery level, recommended for secondary use, average value;

60–70 points: Recovery level, recommended for offline detection confirmation, estimated secondary use value is low;

Below 60 points: Recovery level, no secondary use value, recommended for regeneration and recovery.

2.7 Summary

This chapter starts with the structure and working principle of the power battery system, clarifies the safety hazards of the power battery system under the current technical background, based on real vehicle operation big data and a large amount of experimental data, deeply analyses the inconsistency performance, development law and fault mode of the power battery system, and carries out a variety of inconsistency fault online diagnosis case analyses, clarifies the deterioration trend of the safety of the power battery system with the non-linear expansion of inconsistency, reveals the coupling mechanism between safety and consistency, and provides data and theoretical support for the safety evaluation of the power battery system and the development of high-accuracy, high-reliability fault diagnosis technology.

Chapter 3
Estimation and Prediction of Power Battery SOH

3.1 Overview

The real-time requirement for battery state estimation in the actual operation of electric vehicles is high. How to quickly and accurately estimate the health state (State of Health, SOH) of the power battery system and promote the corresponding estimation methods in engineering applications is a major problem that needs to be solved in the health state estimation of the power battery system in actual vehicle operation. To achieve an accurate estimation of the health state of the power battery system for actual vehicle applications, two main problems need to be solved. The first is to find an evaluation algorithm or health state feature value that can accurately evaluate the battery health state in actual vehicle operation. The second is to explore a health state estimation algorithm that can accurately apply to actual vehicle operation. These two problems are interrelated and promote each other, jointly determining the final estimation accuracy of the battery health state (Fig. 3.1).

3.2 Study on Power Battery Degradation Mechanism and External Characteristics

3.2.1 Mechanism of Power Battery Degradation

Although a lot of research results have been achieved in the estimation of the state of power batteries, and they have gradually been applied in on-board battery management systems, how to achieve high-precision, high-stability, and high-reliability state estimation of power battery systems during actual vehicle operation is still a hot and difficult issue worldwide. Most of the power battery system management systems on the market are mainly based on their physical models to estimate the state under

J. Hong, *Big Data Analysis and Application of Power Battery Operation Safety*, Key Technologies on New Energy Vehicles, https://doi.org/10.1007/978-981-96-4399-8_3

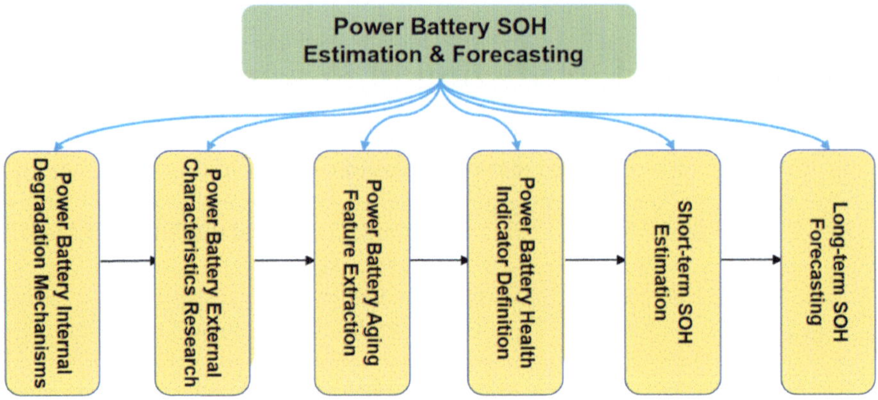

Fig. 3.1 Overview of this chapter

characteristic conditions and some battery states, such as estimating the battery SOC by collecting battery voltage and current. However, these methods make it difficult to consider the complex actual operating environment of power batteries, so it is also difficult to accurately assess the overall health status of the power battery system.

The SOH of the power battery system is inseparable from its degradation mechanism during use. When the lithium battery is charged and discharged, a series of complex physicochemical reactions are always happening inside it. These reactions cause the battery temperature to rise and the irreversible consumption of active substances in the battery, gradually reducing the charging and discharging efficiency of the battery, and eventually leading to rapid decay of the battery capacity until the battery fails. In recent years, many scholars at home and abroad have carried out a lot of research on battery SOH estimation from the perspective of battery degradation mechanism, and found that the main factors affecting battery degradation can be attributed to the formation of SEI film (a passivation film formed on the surface of the battery negative electrode during the migration of lithium ions in the battery charging and discharging process), the deterioration of the internal structure of the battery material, and the impact of environmental temperature. Among them, the deterioration of battery materials will proceed with the continuous cycle of battery reactions, the active materials of the positive and negative electrodes gradually detach from the battery current collector, and the electrolyte itself undergoes oxidation and promotes the decay of battery capacity.

3.2.2 External Characteristic Parameters and Characterisation Methods of Power Batteries

Among all environmental factors, temperature has the greatest impact on the charging and discharging performance of batteries, so its impact on battery capacity decay is also very obvious. The electrode/electrolyte interface is regarded as the heart of the lithium battery, and the electrochemical reaction on the battery electrode/electrolyte interface is closely related to the environmental temperature. Too high or too low a temperature has a very large impact on the actual output of battery capacity. Low-temperature e environment will reduce the activity of the internal electrode material of the battery, increase the viscosity of the electrolyte, and increase the difficulty of ion diffusion, thereby increasing the difficulty of battery charging and discharging. For example, the discharge efficiency of nickel–cadmium, nickel–hydrogen, and lithium batteries will significantly decrease in low-temperature environments (such as below −15 °C), and at −20 °C, the alkali will reach its freezing point, and the charging speed of the battery will also greatly decrease. High-temperature environment will increase the heat generation speed and heat generation of the battery, thereby destroying the chemical balance inside the battery, causing side reactions to accelerate, and eventually leading to a decrease in the effective capacity of the battery. For example, when the temperature rises above 45 °C, the material performance of the lithium battery will degrade during charging, which will greatly shorten its cycle life.

In addition, factors affecting the degradation of the power battery system during actual vehicle operation include:

(1) Discharge depth and rechargeable times. The number of rechargeable times and the depth of discharge are related. The deeper the battery discharge, the fewer the rechargeable times. The number of rechargeable times * discharge depth = total number of charging cycles completed, and the higher the total number of charging cycles completed, the better the battery's SOH, and the longer its life, that is, the number of rechargeable times * discharge depth = actual battery life (ignoring other factors).
(2) Overcharge, over-discharge, and large current charging/discharging. Any form of battery overcharge will severely damage the battery performance, and even cause the battery to thermally run away or even explode. Taking lithium-ion power batteries as an example, deep discharge below 2 V or 2.5 V should be avoided as much as possible, otherwise it will cause rapid irreversible damage to the lithium-ion battery, which may trigger internal metal plating, thereby causing battery short circuits, battery failure or thermal runaway.
(3) The vehicle is parked for a long time and is in a full or no power state. Some experimental data on lithium batteries in notebook computers show that if the battery voltage often exceeds the standard voltage by 0.1 V, the battery life will be halved, and if it is increased by 0.2 V, the life will be reduced to 1/3 of the original. The same is true for lithium-ion power batteries used in vehicles. The fuller the battery is charged, the greater the loss of the battery's SOH. In addition, the long-term low power or no power state of electric vehicles will make the resistance to the movement of electrons inside the power battery larger and larger, causing the battery capacity to decrease gradually. Therefore, when the vehicle needs to be parked for a long time, the lithium power battery is best in a state that is neither full nor undercharged, so as to extend the use life of the power battery as much as possible.

3.3 Definition of SOH

3.3.1 Definition of SOH

Currently, battery life testing schemes are mainly based on the internationally recognized USABC and FreedomCAR test manuals, with the testing process carried out under specific environmental and working conditions. The Hybrid Pulse Power Characteristic (HPPC) test is the most commonly used method for testing the direct current resistance of batteries, which includes Ohmic resistance and polarisation resistance. The State of Health (SOH) of a battery is a physical quantity characterizing the SOH of a power battery. When the battery's SOH cannot meet the usage requirements, we consider the battery life to have ended, i.e., the battery has failed. The standard definition of battery SOH is the ratio of discharge capacity to nominal capacity, which can be defined from the perspective of remaining capacity as:

$$SOH = \frac{C_{Actual} - C_{End}}{C_{Nominal} - C_{End}} \times 100\% \tag{3.1}$$

where C_{Actual} represents the actual capacity of the battery, C_{Nominal} is the nominal capacity, C_{End} represents the end capacity when the battery fails, usually 80% of C_{Nominal}. Therefore, the larger the SOH value, the better the health status of the battery.

3.3.2 SOH Estimation and SOH Prediction

Battery aging is related to many factors, including internal and external factors, such as electrochemical characteristics, manufacturing processes, usage conditions, and external environment. Battery aging can be characterized by two indicators: capacity and internal resistance. The remaining service life refers to the time or number of cycles from the current moment until the battery performance can no longer meet the usage requirements. Both SOH estimation and SOH prediction characterize the degree of battery aging. The commonly used methods for both can be roughly divided into two categories: model-based methods and data-driven methods. The former includes equivalent circuit models and electrochemical models; the latter mainly includes support vector regression, Gaussian process regression, long short-term memory networks, etc. The difference is that SOH estimation uses existing parameters to characterize the current aging status of the battery, while SOH prediction predicts the aging situation for a future period based on existing parameters.

3.4 Estimation of SOH of Power Battery in Actual Vehicle

3.4.1 SOH Feature Extraction Method

Current Capacity (IC) analysis is an indispensable method in battery SOH estimation, with the advantages of being non-destructive and low-cost. It refers to the ratio of capacity change to the corresponding voltage range during the constant current charge and discharge process.

$$\mathrm{IC} = \frac{dQ}{dV} \tag{3.2}$$

During the charging and discharging process, the equilibrium state inside the electrode changes, which is reflected in the charging curve as a gently changing voltage plateau region. The characteristics of the voltage plateau region can reflect the degree of battery ageing. The IC curve can convert the flat curve into an easily recognizable IC peak, and the peak position curve feature has been proven to be related to battery aging, which is different for different types of batteries. Therefore, IC analysis is further used to analyze the determined charging section.

The calculation of the IC curve involves capacity and voltage. The voltage can be directly obtained from the big data platform, and the capacity can be calculated by the Coulomb counting method.

$$Q = \int_0^T i(t)dt \tag{3.3}$$

where, Q represents the change in capacity within the time interval, $i(t)$ represents the charging/discharging current, and T represents the time interval. However, in the actual sampling process, the data is discrete and cannot directly calculate the capacity change. By converting the above formula, the capacity can be calculated.

$$Q = \sum_0^T \Delta Q = \sum_0^T I\Delta t \tag{3.4}$$

where, ΔQ represents the change in capacity within the sampling interval, I represents the charging current within the sampling interval, and Δt represents the sampling interval.

For each charging section, the initial capacity of the first sampling point is considered as zero, and the sampling interval is constant at ten seconds. The total capacity can then be calculated by calculating the capacity change between two sampling intervals. However, due to the presence of data noise, the voltage change between two adjacent sampling points may be displayed as zero or negative, which will reduce the calculation accuracy.

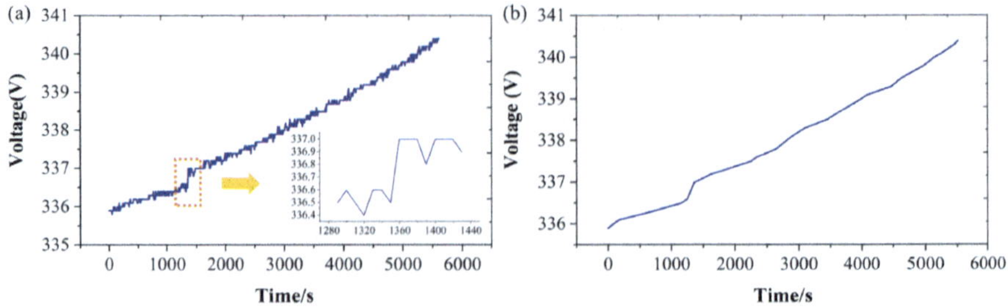

Fig. 3.2 Voltage resampling: **a** original voltage; **b** resampled voltage

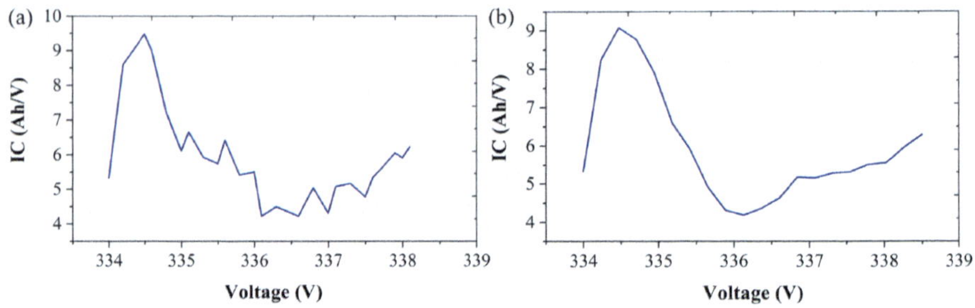

Fig. 3.3 IC curve smoothing: **a** original IC curve; **b** smoothed IC curve

In order to obtain an accurate IC curve, the voltage curve is resampled. The voltage of the first sampling point is set as the initial value and then traversed over time. When the voltage difference between a certain sampling point and the first sampling point is greater than or equal to 0.1 V (i.e., the highest precision of the big data platform), this voltage is set as the second value. According to this method, all data is traversed to obtain the resampled dataset. Figure 3.2 shows the voltage resampling of a charging section. It can be seen that the number of available points after resampling is reduced, causing fluctuations in the IC curve. Therefore, the IC curve is further smoothed by the interpolation method, as shown in Fig. 3.3.

3.4.2 SOH Loss Extraction Technique

To extract aging characteristics, all IC curves are plotted against cumulative mileage in Fig. 3.4. As the mileage increases, the peak position of each IC curve periodically moves within the range of 331~338 V. In addition, the peak significantly drops after 120,000 km, reflecting a decrease in battery capacity. The drop in peak is closely related to the decrease in battery capacity, mainly reflecting the loss of lithium ions, which is the main cause of early capacity decline, corresponding to a capacity range from 100 to 85% or 80%. Therefore, the interval around the peak position is

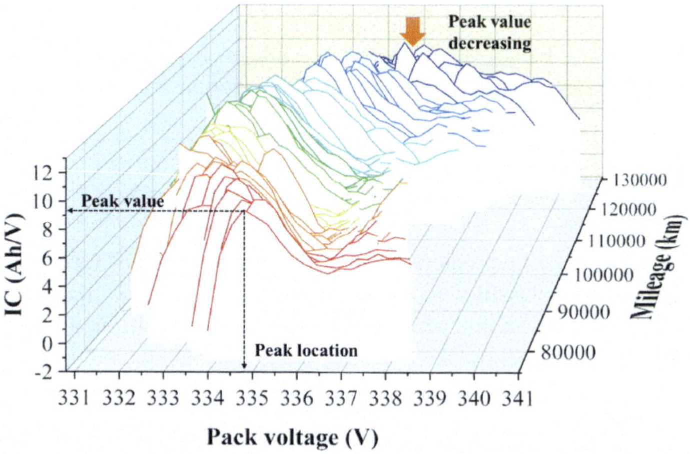

Fig. 3.4 IC curves

determined to calculate the interval capacity, which can fully reflect the health status of the battery.

$$\begin{cases} \Delta V = V_{t_2} - V_{t_1} \\ V_{t_1} = V_p - \Delta V/2 \\ V_{t_2} = V_p + \Delta V/2 \end{cases} \tag{3.5}$$

where ΔV represents the voltage interval around the peak position, V_{t1} and V_{t2} represent the voltage interval boundaries, V_p represents the voltage value at the peak position, t_1 and t_2 are the times at the boundaries. By searching for the maximum value of each IC curve within the range of 331~338 V, the peak is obtained and the peak position is determined. In addition, as mentioned earlier, the measurement of ΔV should cover the entire IC peak, as the peak is closely related to capacity decay, and finally, the voltage interval is determined to be 2 V.

3.4.3 Estimation of SOH Loss

The SOH estimation model based on Catboost, its structure is located in the Gradient Boosting Decision Tree (GBDT) framework, the advantage is that it can effectively handle categorical features. The basic GBDT model can be represented as follows.

$$\Pi(x) = \sum_{m=1}^{M} f(x, t) \tag{3.6}$$

where x represents the set of predictive variables, $f(x, t)$ represents the decision tree, M represents the number of decision trees. After each round of iteration, a new decision tree will be generated, and the prediction function is represented as follows.

$$L(x, t) = \sum_{m=1}^{M} \beta_m f(x, \alpha_m) \tag{3.7}$$

where $L(x, t)$ represents the prediction function of the dependent variable, β_m represents the weight of the classifier, α_m represents the computational factor of the classifier, $f(x, \alpha_m)$ represents an independent decision tree. In addition, the loss function of the gradient is shown below.

$$L_m(x) = L_{m-1}(x) + \beta_m f(x, \alpha_m) = L_{m-1}(x) + \sum_{j=1}^{J} \gamma_{jm} I \tag{3.8}$$

where γ_{jm} represents the input value with j-dimensional space, $L_m(x)$ is the loss function in the gradient direction of $L_{\mathrm{m}-1}(x)$.

The Catboost based on the GBDT framework has improved three key points, including the processing of categorical features, ordered boosting, and the combination of categorical features. Categorical features are often found in strings, which machine learning algorithms cannot use. In Catboost, target statistics are applied to convert categorical features into numerical features.

$$\sigma_{p,k} = \frac{\sum_j^{p-1} \left[x_{\sigma_{j,k}} = x_{\sigma_{i,k}}\right] \cdot Y_{\sigma_j} + \alpha \cdot p_1}{\sum_j^{p-1} \left[x_{\sigma_{i,k}}\right] + \alpha} \tag{3.9}$$

where α is a weight greater than zero. Mileage, temperature, and SOC are different inputs with different units, but the values read by the model are in numerical form. Therefore, feature transformation is not required. Ordered boosting is applied in Catboost to solve the prediction offset problem caused by gradient bias. This process includes three steps, as shown below.

Step one: Randomly sort the training dataset and name the new set as σ;

Step two: Initialise the model for each sample M_i, $M_i = 0$, $i = 0, 1, \ldots$, n;

Step three: For each sample $i = 0,1,\ldots$,n, use the original training dataset (x_j, y_j) to train the next decision tree ΔM: $\sigma(j) \leq i$, and update the result to the new model.

In addition, the Mean Absolute Percentage Error (MAPE) and Root Mean Square Error (RMSE) are used to evaluate the estimation accuracy, as shown in the formulas below.

$$\mathrm{MAPE} = \frac{100\%}{n} \sum_{i=1}^{n} \left| \frac{\hat{y}_i - y_i}{y_i} \right| \tag{3.10}$$

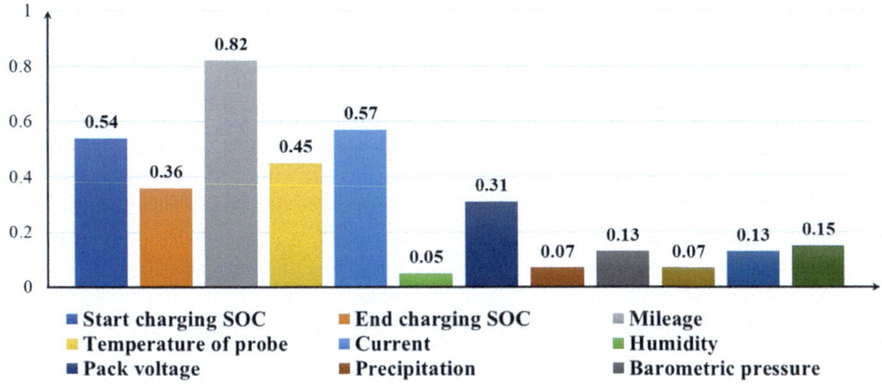

Fig. 3.5 Correlation analysis results

$$\text{RMSE} = \sqrt{\frac{1}{n}\sum_{i=1}^{n}\left(\hat{y}_i - y_i\right)^2} \tag{3.11}$$

where y_i is the actual value, $\hat{y}_i$ is the estimated value.

Through the above analysis, the aging characteristics of the interval capacity within a certain voltage range can be calculated. Then, a correlation analysis is conducted between the interval capacity and various types of vehicle operation data to determine the model input. Usually, highly correlated inputs will improve the accuracy of the data-driven model. The correlation analysis does not involve vehicle speed, motor speed, and brake pedal stroke values, as all these parameters are zero during the charging process.

In addition, the SOC range of the determined charging segment is derived as the start charging SOC and end charging SOC. Figure 3.5 shows the correlation results between the interval capacity and each parameter. Finally, the start charging SOC, end charging SOC, mileage, probe temperature, and current are selected based on the correlation coefficient.

3.4.4 Analysis of SOH Estimation Results

The SOH estimation model based on Catboost discusses two key hyperparameters, including depth and learning rate (LER). The former determines the fitting accuracy, and the latter determines the convergence speed. The structure of the SOH estimation model based on Catboost is located in the oblivion tree, where the depth represents the number of layers of the oblivion tree. LER is a basic parameter in model training, used to determine whether and when the objective function converges to the optimal value. A smaller LER will cause the model to optimise slowly, and a larger LER will cause the model to optimise quickly at the beginning. However, it will produce larger fluctuations when the training is completed, thereby preventing the prediction results from approaching the actual value. According to empirical analysis, the commonly used and recommended depth range is 4~10, and the learning rate is 1.

This study uses the first 70% (66,709.1~118150 km) of the charging segment to train the model, and the last 30% (118,150~131515 km) is used to test the estimation performance. Figure 3.6 shows the interval capacity determined throughout the year, which can obtain a significant decreasing trend, and the decreasing trend is linearly fitted. Figure 3.7 shows the estimation performance under different depths and LERs. When the depth is less than 6, the model accuracy significantly improves with the increase in depth. When the depth is 6, the model achieves the best accuracy, with MAPE being 2.75%, and RMSE being 1.15%. When the depth exceeds 6, unstable changes will occur, indicating overfitting. Then, the depth is set to 6 to analyze different LERs. When LER is less than 0.7, a significant improvement trend can be obtained, and the best estimation accuracy is achieved at 0.7, with MAPE being 2.74%, and RMSE being 1.12%. When LER is 0.9, the accuracy significantly decreases, indicating that the single-step gradient decreases too much, and the model cannot converge to the optimal value. The final results of the SOH estimation model based on Catboost are shown in Fig. 3.8.

In this section, the Catboost SOH estimation model is compared with other commonly used machine learning methods to evaluate the model performance, including Linear Regression (LR), Support Vector Machine (SVM), Random Forest (RF), Adaboost, GBDT, and Light Gradient Boosting Machine (LGBM). The training and testing datasets of these methods are the same as the SOH estimation model based on Catboost, so as to compare under the same conditions. The estimation and accuracy results are shown in Fig. 3.9 and Table 3.1. The accuracy of the Catboost SOH estimation model is much higher than that of other machine learning methods. In

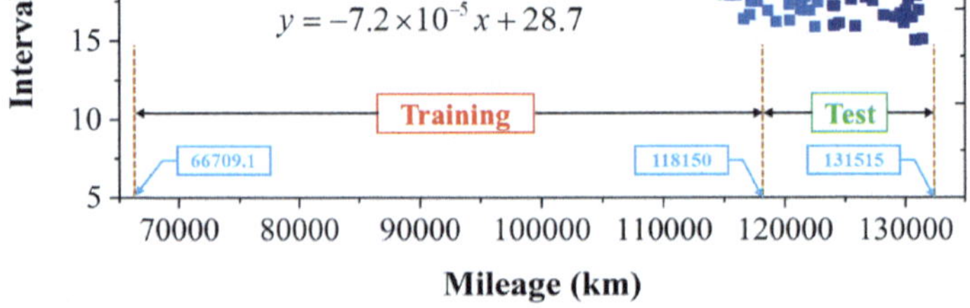

Fig. 3.6 Test data

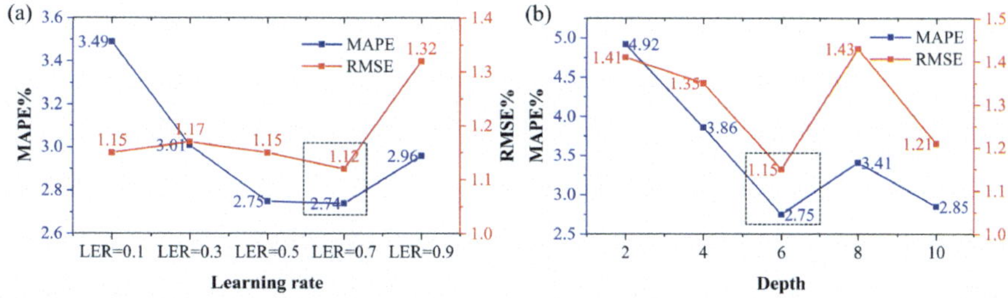

Fig. 3.7 Discussion of different hyperparameters. **a** Different learning rates; **b** Different depths

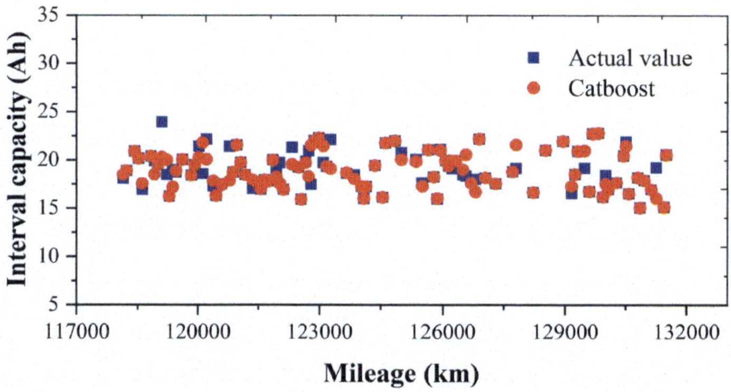

Fig. 3.8 Results of the Catboost SOH estimation model

addition, besides the model proposed in this paper, only RF and LGBM achieved over 95% (with MAPE being 4.29% and 4.18%, respectively).

Considering that the collected data is only the operating data of the studied EV for one year, the interval capacity of the first confirmed charging segment with the least mileage is considered to be 100% of the optimal SOH. Then the interval capacity compared with the initial interval capacity is used to evaluate the SOH.

$$\begin{cases} \text{SOH} = \frac{Q_r}{Q_{ini}} 100\% \\ Q_r = aQ_{ini} \end{cases} \tag{3.12}$$

where Q_r represents the real-time interval capacity, Q_{ini} represents the initial interval capacity in this paper, a is an evaluation value, meaning that when Q_r is reduced to a specific value compared to Q_{ini}, the battery cannot be used, and for new batteries, it is usually set to 0.8.

Due to confidentiality requirements, this study can only use operational data for up to one year. The operational data will not limit the proposed methods and evaluation strategies, and further research plans are shown in Fig. 3.10. In the future, a sufficient database will be built based on a big data platform, involving massive full-life data

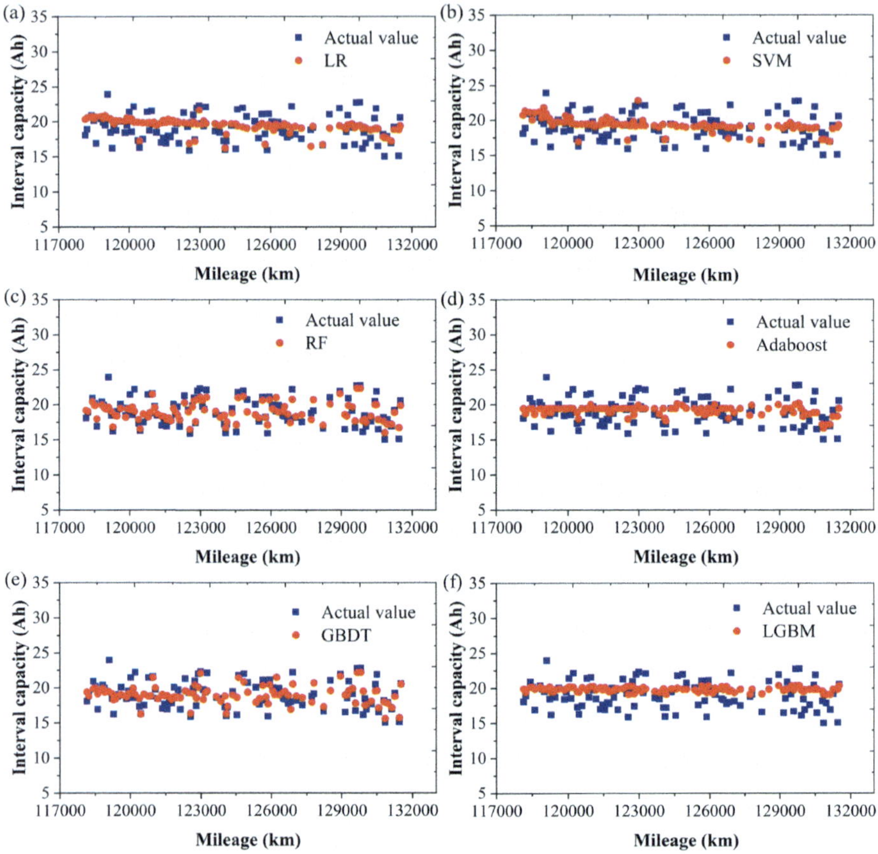

Fig. 3.9 Comparison with other machine learning methods. **a** LR; **b** SVM; **c** RF; **d** Adaboost; **e** GBDT; **f** LGBM

Table 3.1 Accuracy of SOH estimation by different methods

Method	RMSE%	MAPE%
Catboost	1.12	2.74
LR	1.85	8.22
SVM	2.22	10.46
RF	1.26	4.29
Adaboost	2.88	7.53
GBDT	1.32	5.25
LGBM	1.36	4.18

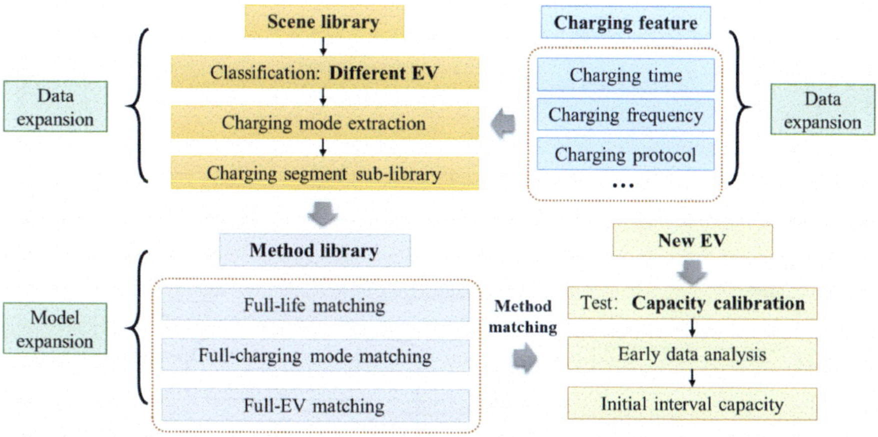

Fig. 3.10 Electric vehicle SOH Prediction research plan

of different electric vehicles. The charging sub-library will be further constructed according to the charging functions of different electric vehicles. Then, different data-driven SOH estimation models will be constructed to form a model library. In the factory test, different interval capacities can be obtained, which serve as the main reference for new electric vehicles. Early charging data (0–5000 km or the first year) will be accumulated for further analysis with the database. Then, the final initial interval capacity is determined by the comprehensive analysis of the early data and the initial test interval capacity.

3.5 Real Vehicle Power Battery SOH Prediction

3.5.1 Ageing Feature Extraction

Existing battery life test schemes are all conducted under specific laboratory conditions, and due to the complexity and robustness of the actual vehicle operating environment, there is currently no method to accurately evaluate the battery SOH or remaining capacity in real-time during vehicle operation. Figure 3.11 shows the capacity cycle test results of a 120Ah lithium-ion battery. When two sets of new/old batteries are charged and discharged at the same time at a 1C rate within one cycle, the old battery charges and discharges significantly faster than the new battery, and discharges completely earlier than the new battery; when three batteries are fully charged at a 1C rate and discharged at 1C, 2/3C and 2.5C respectively, the battery with a higher discharge rate discharges faster. In addition, the battery's capacity decay curve is very close to a steadily declining straight line.

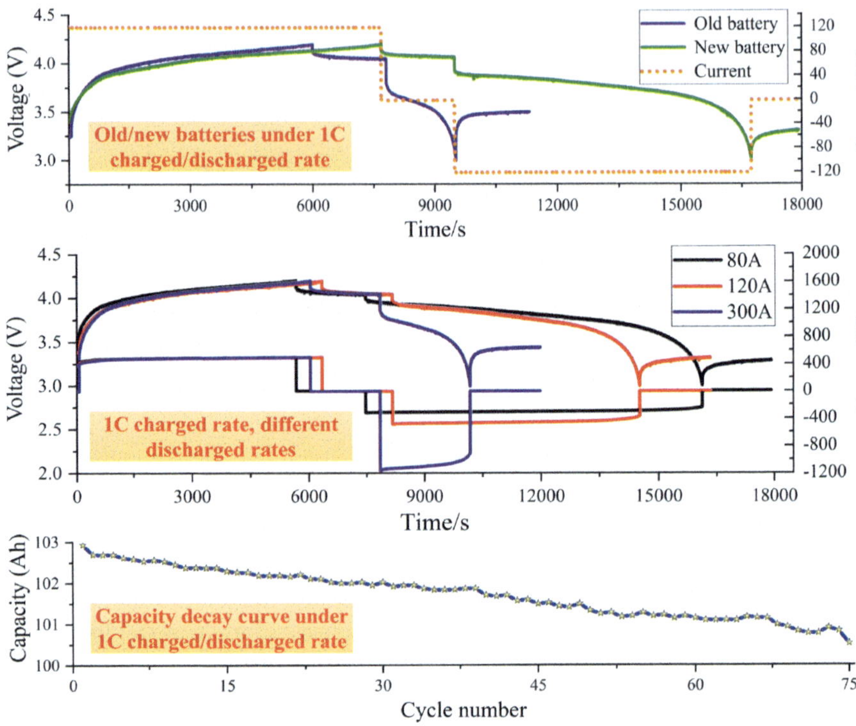

Fig. 3.11 Experimental results of lithium battery capacity decay based on experiments

However, actual operating vehicles are severely affected by complex road/weather conditions and unpredictable driver-driving behavior, and the actual charging and discharging cycles of the power battery system are not always fully charged or discharged at a fixed discharge depth. Figure 3.12 shows the battery SOC, current and total voltage curves of a pure electric taxi throughout the day on March 26, 2016. As can be seen from the figure, the car had three driving stages and two charging stages that day, with both 16A slow charging and 40A fast charging, and different charging and discharging depths.

Completing a complete set of battery capacity tests or HPPC internal resistance tests usually requires a stable experimental environment, set charging and discharging conditions, a small sampling interval (usually less than or equal to 1 s) and a long test time. These inherent factors and various non-negligible factors in actual vehicle operation together determine that existing battery SOH test methods cannot be implemented on an actual driving vehicle: (1) Different actual driving/charging cycles, dynamic environments and variable charging and discharging depths are not conducive to obtaining accurate battery internal resistance, charging capacity or discharging capacity under a unified standard. (2) Existing equivalent circuit models and other battery models can complete battery parameter identification and SOH

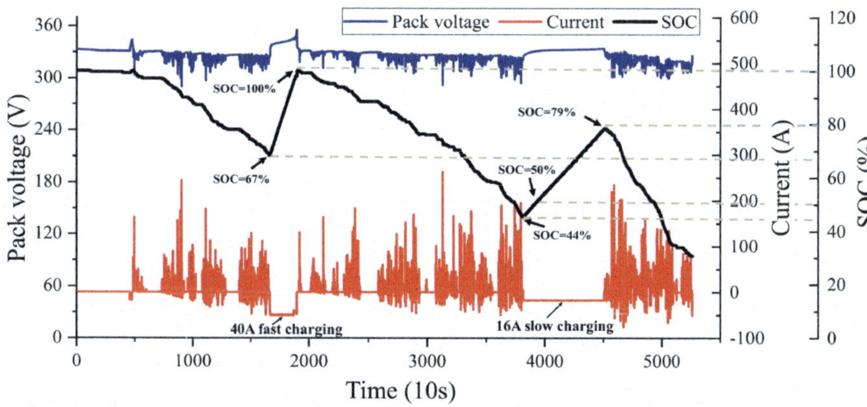

Fig. 3.12 SOC, current and total voltage curves of the power battery system in a day

estimation in an experimental environment, but in a complex actual operating environment, the modeling effect of these models and the identification effect of battery impedance characteristics will be greatly reduced, making them still unable to be applied on actual vehicles. (3) To reduce storage burden, existing electric vehicle operation monitoring platforms usually use a larger sampling interval, for example, the sampling interval in this study is 10 s, and it is planned to extend the sampling interval to 30 s or even one minute in the future. In actual operation, the instantaneous change of battery current is very fast, even within 10 s, the battery polarization resistance will fluctuate greatly, so the actual sampling interval of 10 s or more makes it difficult for us to obtain accurate battery internal resistance values. (4) In the actual vehicle operation process, the battery SOH will be severely affected by weather conditions, road conditions, and driver driving behavior, making it unrealistic for existing experimental-based SOH estimation algorithms to take these factors into account. In summary, it is difficult to obtain a stable operating environment on an actual running vehicle, and it is also difficult to obtain excellent data features as in experimental research, so existing SOH estimation methods are mostly limited to laboratory research and are still difficult to play a role on actual running vehicles.

Although there are already a large number of model-based and data-driven methods, model-based methods are difficult to accurately identify battery parameter characteristics in actual vehicles, and data-driven methods require a large amount of historical data for modelling. There is currently no battery ageing model or key SOH feature parameters that can accurately estimate the power battery SOH in actual vehicle operation. Based on the platform advantage of the national electric vehicle monitoring management platform, we can obtain a large amount of actual vehicle operation data, which provides us with strong data support and the feasibility of future data-driven battery SOH prediction research.

In view of the above problems, considering the complexity of the actual vehicle operation data, this study attempts to extract eight battery SOH feature values under the actual vehicle charging state, and conducts a comparative analysis, in the hope of finding a potential power battery system SOH evaluation method for actual vehicle operation, as follows:

1. Capacity increment analysis (CIA) during each charge

$$SOH_{\text{CIA}} = \frac{\int_{SOC_{sc}}^{SOC_{fc}} Idt}{SOC_{fc} - SOC_{sc}} \times 100\% \left(SOC_{fc} > SOC_{sc}\right) \quad (3.13)$$

where SOC_{SC} is the battery SOC at the start of vehicle charging, SOC_{FC} is the battery SOC when the vehicle stops charging, and I is the real-time charging current. During the constant voltage stage when charging is close to 100% SOC, current decay and SOC correction can cause some calculation errors. To filter out these errors, we use 80% SOC as the maximum charging SOC required for calculating SOH_{CIA} and only take the SOH_{CIA} in the charging SOC range of 50~80% as the final SOH_{CIA} value.
2. Mileage increment analysis (MIA)

$$SOH_{\text{MIA}} = \frac{DistanceTravelled}{SOC_{ed} - SOC_{sd}} \times 100\% (SOC_{ed} > SOC_{sd}) \quad (3.14)$$

where, SOC_{sd} is the battery SOC at the start of the journey, SOC_{ed} is the battery SOC at the end of the journey.
3. Open circuit voltage before charging (OBC)

$$SOH_{\text{OBC}} = OCV_{bc} \quad (3.15)$$

where, OCV_{bc} is the open circuit voltage at the start of charging.
4. Internal resistance before charging (RBC)

$$SOH_{\text{RBC}} = R_{bc} \quad (3.16)$$

$$R_{bc} = \frac{U_{bc} - U_{bc-1}}{I_{bc} - I_{bc-1}} \times 100\% \quad (3.17)$$

where, R_{bc} is the total internal resistance of the battery at the last frame of sampling before charging begins; U_{bc} and I_{bc} are the total voltage and total current at the last frame of sampling before charging begins, U_{bc-1} and I_{bc-1} are the total voltage and total current at the previous frame of sampling.
5. Open circuit voltage after charging (OAC)

$$SOH_{\text{OAC}} = OCV_{ac} \quad (3.18)$$

where, OCV_{ac} is the open circuit voltage at the end of battery charging.
6. Total internal resistance of the battery at the end of charging (Internal resistance after charging, RAC)

$$SOH_{\mathrm{RAC}} = R_{ac} \tag{3.19}$$

$$R_{ac} = \frac{U_{ac} - U_{ac-1}}{I_{ac} - I_{ac-1}} \times 100\% \tag{3.20}$$

where, R_{ac} is the total internal resistance of the battery at the end of charging; U_{ac} and I_{ac} is the total voltage and current at the end of charging, U_{ac-1} and I_{ac-1} are the total voltage and current at the last sampling point before the end of charging.

7. Open circuit voltage when SOC = 50% (OCV when SOC = 50%, OFT)

$$SOH_{OFT} = OCV_{ft} \tag{3.21}$$

where OCV_{ft} is the open circuit voltage when SOC = 50%.

8. Time required to charge into 10%SOC (Time of charging into 10%SOC, TTS)

$$SOH_{TTS} = T_{\Delta 10} - T_{\mathrm{fc}} \tag{3.22}$$

Where T_{fc} is the sampling time at the start of charging, $T_{\Delta 10}$ is the sampling time when 10%SOC is charged.

After a series of complex data preprocessing procedures, we extracted 775 complete driving/charging cycles from the full-year operation data of the vehicle under study (a certain pure electric taxi) in 2016. Figure 3.13 shows some of the main parameter curves in these driving/charging cycles.

Based on the above eight types of battery SOH characteristic values, Fig. 3.14 shows the eight SOH curves of all the driving/charging cycles of the vehicle under study in 2016. In order to facilitate the observation of the trend of data feature changes, we also performed linear fitting on the eight curves, and the trend lines and linear equations after fitting are also shown in the figure.

As shown in Fig. 3.14, the eight SOH curves are not smoothly changing, but fluctuate dramatically, which is closely related to the actual road conditions, weather, and the complexity and variability of driving characteristics during the actual driving process of the vehicle. Apart from the SOH_{CIA} curve and the SOH_{TTS} curve, the other six SOH curves have obvious seasonal characteristics, with clear differences between the SOH in winter and summer, which should be mainly caused by differences in environmental temperature.

In Fig. 3.14, $\mathrm{SOH_{CIA}}$, $\mathrm{SOH_{MIA}}$, $\mathrm{SOH_{OBC}}$, $\mathrm{SOH_{RAC}}$, $\mathrm{SOH_{OFT}}$, and $\mathrm{SOH_{TTS}}$ do not show a clear upward or downward trend, and the slope of their linear trend lines is small, which is inconsistent with the characteristics of a normal pure electric taxi driving for a long time within a year and the SOH decreasing quickly. The linear trend lines of $\mathrm{SOH_{RBC}}$ and $\mathrm{SOH_{OAC}}$ show a clear trend of the total internal resistance of the battery gradually increasing before charging and the open circuit voltage gradually decreasing at the end of charging. Therefore, this paper selects the battery internal resistance, which is a feature that is more obvious and commonly used to reflect the battery SOH, from these two, and uses the battery internal resistance $\mathrm{SOH_{RBC}}$ before

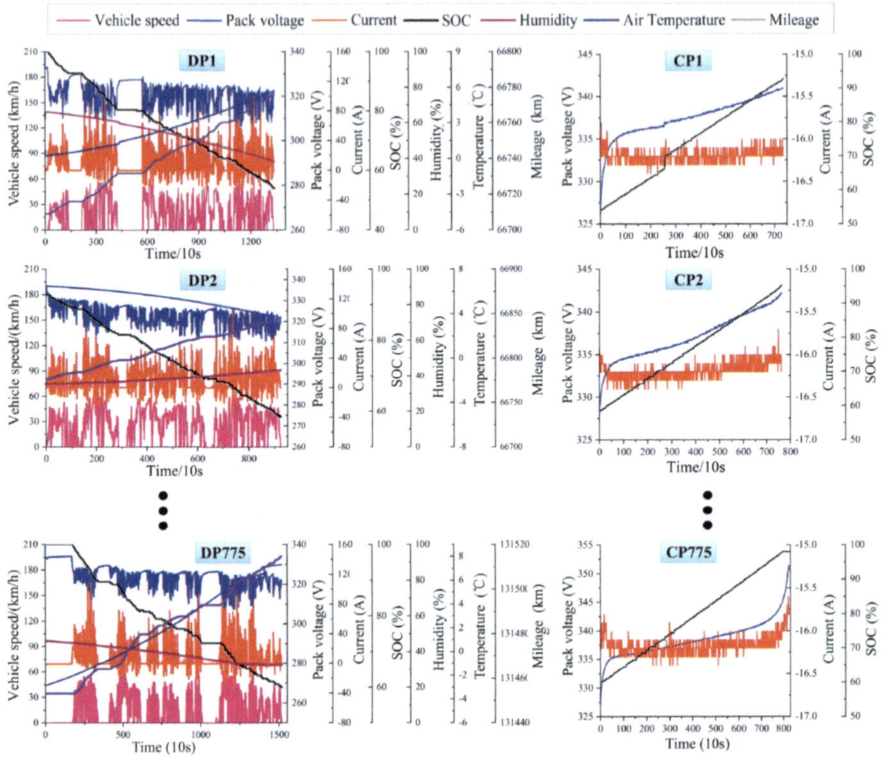

Fig. 3.13 Main parameter curves in the actual vehicle driving/charging cycles

charging as the characteristic value for the study of battery SOH decay characteristics in the following text.

3.5.2 SOH Prediction Model

The National Aeronautics and Space Administration (NASA) in the United States has studied the storage calendar life of lithium batteries for the need for long-life power sources for long-term space missions without relying on acceleration methods. Because the reliability of simulation results is much lower than the cost of space exploration, this test truly placed lithium batteries at a specified temperature for several tens of months and observed the aging process of the batteries by collecting data regularly, as shown in Fig. 3.15. The basic conclusion is that the higher the storage temperature of the lithium-ion battery, the faster the capacity decays, and the temperature has a significant impact on the calendar life of the lithium battery. When the tested battery is stored at room temperature of 20 °C and lower, after 54 months,

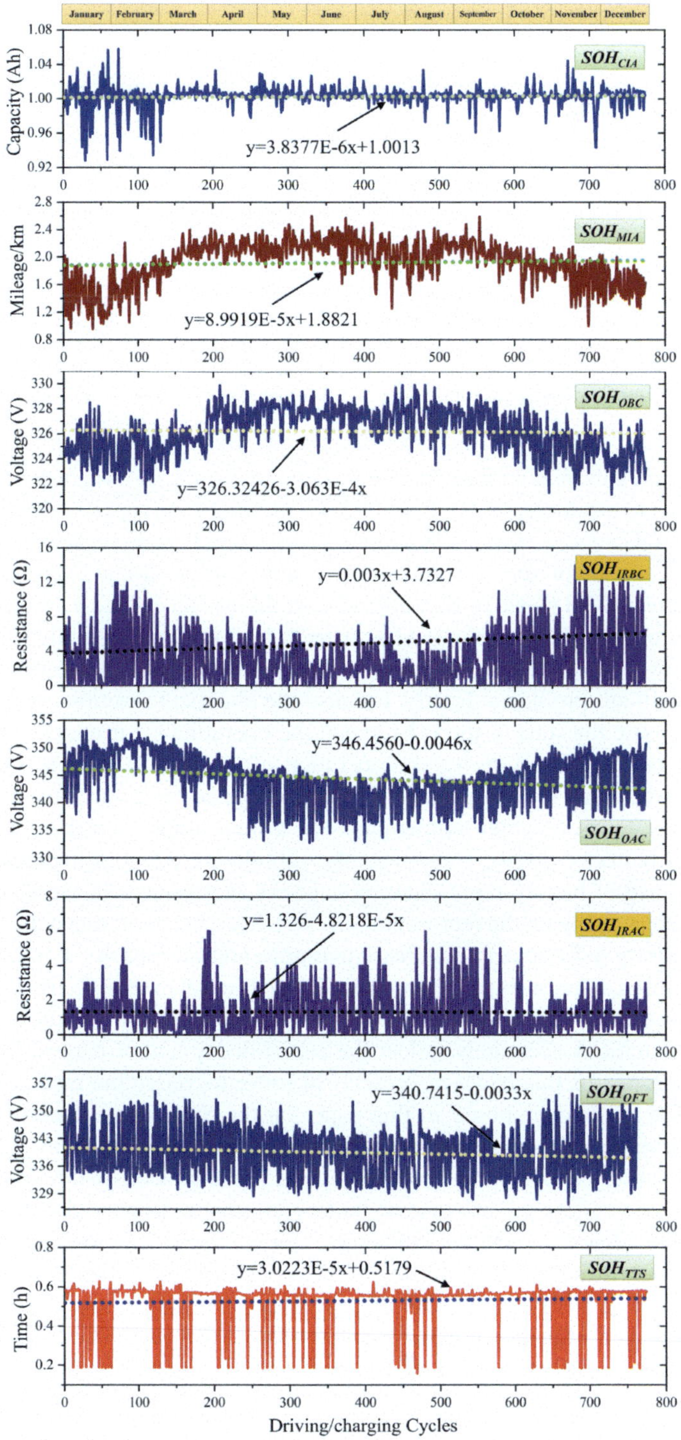
January February March April May June July August September October November December
Capacity (Ah)
SOH_CIA
y=3.8377E-6x+1.0013
Mileage/km
SOH_MIA
y=8.9919E-5x+1.8821
Voltage (V)
SOH_OBC
y=326.32426-3.063E-4x
Resistance (Ω)
SOH_IRBC
y=0.003x+3.7327
Voltage (V)
y=346.4560-0.0046x
SOH_OAC
Resistance (Ω)
y=1.326-4.8218E-5x
SOH_IRAC
Voltage (V)
y=340.7415-0.0033x
SOH_OFT
Time (h)
y=3.0223E-5x+0.5179
SOH_TTS
Driving/charging Cycles

◂**Fig. 3.14** Eight potential SOH curves and trend lines

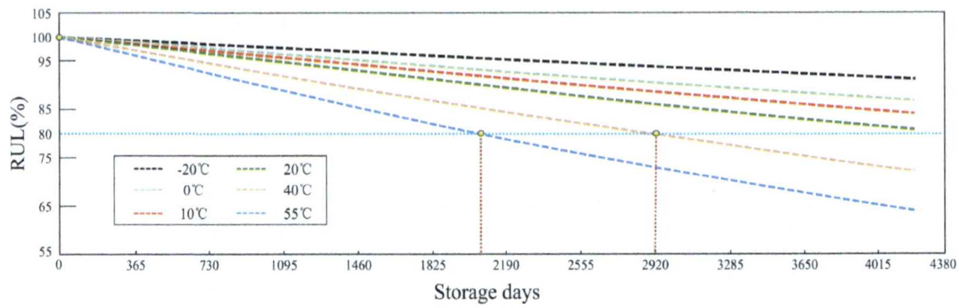

Fig. 3.15 The impact of storage temperature on the aging speed of lithium-ion batteries

the battery can still obtain more than 90% remaining capacity; when stored in high-temperature environments of 40 °C and 55 °C, its remaining capacity decays to 80% of the nominal capacity, which takes about 8 years and 6 years respectively.

A large amount of research has already indicated a strong linear correlation between battery SOH (State of Health) and internal resistance, and a large amount of research on battery SOH estimation has been carried out based on this feature. The estimation of SOH for power battery systems in actual vehicle operation faces the following difficult problems: Firstly, because the chemical reactions of the battery will change with the state of the vehicle and the operating environment, the battery capacity is not conserved during the operation of the vehicle, which is a difficult problem to overcome for our quantitative study of the decay law of battery capacity; Secondly, the environmental temperature has a huge impact on the actual representation of battery capacity, sometimes the stimulating effect of high temperature on the increase of battery representation capacity is greater than the deteriorating effect of battery aging on the representation capacity, which will make the originally decaying battery SOH appear to increase in representation capacity. For the research on the coupling relationship between the decay law of battery SOH and vehicle usage parameters to be carried out in this paper, the curve that cannot be obtained temporarily that can accurately reflect the actual decay law of battery SOH is the biggest challenge of this paper at present. In order to verify the estimation effect of the LSTM method proposed in this paper on the SOH of the battery in actual vehicle operation, an attempt is made to estimate an approximate decay curve of the SOH of the battery under study. From the figure, it can be seen that the fitting result of SOH_{RBC} can reflect the increase of battery internal resistance and SOH decay, but due to data distortion and the influence of the operating environment, the direct use of common fitting methods for the SOH_{RBC} data of the battery under study for the whole year of 2016 cannot obtain its accurate SOH decay law, which is due to the neglect of the influence of many usage parameters during vehicle operation on battery SOH in the fitting process. To make the SOH decay curve estimated in this

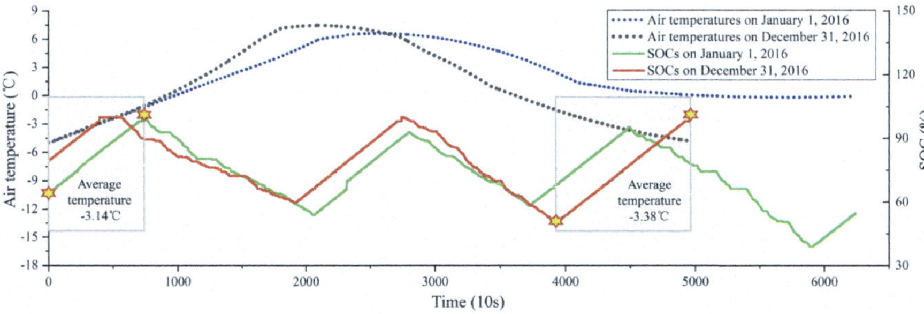

Fig. 3.16 Temperature and SOC curves for January 1 and December 31, 2016

paper as close as possible to the aging characteristics of the battery under study in the real operating environment, the data is processed as follows:

1. **Estimation of SOH loss of actual vehicle power battery system**

In order to avoid the influence of environmental temperature, which has a large impact on the battery, the SOH_{RBC} values (3.6532 Ω) of the first charging stage on January 1, 2016, and the last charging stage on December 31, 2016 (6.5247 Ω) are used as the starting and ending points for calculating the charging SOH of all 775 driving/charging cycles in 2016. The environmental temperatures during these two charges were both around −3 °C (the average temperature during the first charging stage on January 1, 2016, was −3.14 °C, and the average temperature during the last charging stage on December 31, 2016, was −3.38 °C), as shown in Fig. 3.16. According to the idea of the control variable method, it can be approximately considered that the external environment of the vehicle during these two charges is consistent, and the influence of environmental factors such as temperature on the battery SOH representation can be ignored. Therefore, the increase of SOH_{RBC} for the 775 driving/charging cycles in 2016 can be obtained as 2.8715 Ω.

Based on the starting value, ending value, and the increase of SOH_{RBC} for the whole year of 2016, according to the corresponding relationship, we only need to know the starting SOH value of the power battery system of the vehicle under study on January 1, 2016, to be able to calculate the ending SOH value and the SOH loss for this year on December 31, 2016. Due to the current lack of key feature parameters and calculation methods that can accurately reflect the real SOH of the battery in actual vehicle operation, this paper seeks the closest evaluation method based on a large amount of actual usage data of the same model vehicles of the vehicle under study. At present, the range of pure electric vehicles has generally exceeded 400 km, and the life of ternary lithium batteries can reach 10 years and over 240,000 km, such as the publicly announced life cycle of the 2018 BAIC EU5 pure electric vehicle is 10 years and a total driving distance of 150,000 to 200,000 km. The "China Automobile Low Carbon Action Plan" released by the China Automotive Research Institute Data Resource Centre under the guidance of the Energy Conservation Department of the Ministry of Industry and Information Technology in December 2018, in conjunction with industry enterprises, relevant institutions and research institutes in Tianjin, released the energy-saving and environmental protection effects of five pure electric passenger car models throughout their life cycle (Changan Benben EV The 2018 EV260 Deluxe Edition, the 2018 Internet Flagship Edition of SAIC Roewe Ei5, the 2018 EV450 Deluxe Edition of Geely Emgrand New Energy, the single motor version of Tesla Model 3, and the 2017 Glory Edition of Tengshi 400), the total lifecycle mileage of the five cars in this study is calculated as 150,000 km. At the same time, combining a large amount of actual driving history data of the car models studied in this article, the total mileage of the power battery in the entire life cycle of the studied vehicle is determined to be 150,000 km. The car's starting mileage in 2016 was 64,709.1 km, so the initial SOH of the power battery and the 2016 battery SOH loss based on the growth of SOH_{RBC} are as follows:

$$\begin{aligned}\Delta SOH &= 91.37\% * 3.6532 * (\frac{1}{3.6532} - \frac{1}{6.5247}) \\ &\approx 40.21\%\end{aligned} \tag{3.24}$$

$$\begin{aligned}SOH_{起始} &= 100\% - 64709.1/150000*(100\% - 80\%) \\ &\approx 91.37\%\end{aligned} \tag{3-23}$$

The car model under study is an early version of a pure electric mini car with a short driving range. This model was used as an early trial model for pure electric short-distance taxis in Beijing, with daily driving mileage and driving time much longer than private cars, so the life decay rate of the power battery system is faster. In addition, by comparing the historical usage cycle data of other scrapped vehicles of this model, it can be considered that the estimated value of SOH loss is reasonable.

2. **Estimating the impact of environmental temperature on battery SOH decay rate**

According to the results of a study on the impact of lithium battery storage temperature on battery aging completed by NASA, a functional relationship between lithium battery SOH decay and the environmental temperature was fitted. According to the impact of different storage temperatures of lithium-ion batteries on aging speed, an approximate decay equation of battery capacity with storage days under different environmental temperatures was fitted. A large number of statistical results show that the average daily charging times of the car model under study is approximately equal to 2 times. Therefore, according to the idea of controlling variables, under the same other parameter variables, the decay rate of battery capacity with driving/charging cycles is approximately half of its decay rate with storage days. Therefore, the approximate decay equation of battery capacity with driving/charging cycles under different constant environmental temperatures can be obtained as follows:

$$C_{-20} \approx 100\% - 0.001457 * n \tag{3.25}$$

$$C_0 \approx 100\% - 0.001743 * n \tag{3.26}$$

$$C_{10} \approx 100\% - 0.001868 * n \tag{3.27}$$

$$C_{20} \approx 100\% - 0.002715 * n \tag{3.28}$$

$$C_{40} \approx 100\% - 0.003885 * n \tag{3.29}$$

$$C_{55} \approx 100\% - 0.004359 * n \tag{3.30}$$

where, C_{-20}, C_0, C_{10}, C_{20}, C_{40}, C_{55} correspond to the battery capacity at environmental temperatures of -20 °C, 0 °C, 10 °C, 20 °C, 40 °C and 55 °C respectively, n is the number of driving/charging cycles.

Then, the capacity decay rates at different environmental temperatures are linearly and polynomially fitted, as shown in Fig. 3.17. By optimising the fitting parameters, an approximate relationship between the battery capacity decay rate and the environmental temperature was finally obtained. The approximate equation can be expressed as:

$$\begin{aligned} \Delta C_T = &-2.8545E - 10 * T^4 + 4.5626E - 9 * T^3 \\ &+1.1093E - 6 * T^2 + 2.6447E - 5 * T + 0.0017 \end{aligned} \tag{3.31}$$

where, T is the environmental temperature, ΔC_T is the capacity decay rate at the current environmental temperature.

As can be seen from Fig. 3.17, the rate of battery capacity decay is faster between 0 and 45 °C, while the decay rates in low-temperature environments below 0 °C and

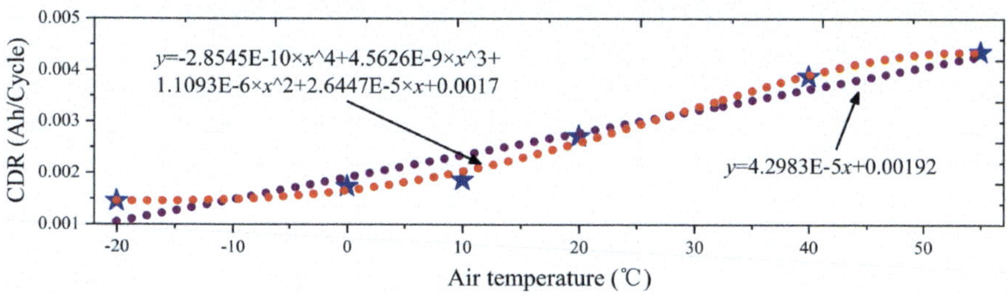

Fig. 3.17 Fitting curve of battery capacity decay rate and environmental temperature

high-temperature environments above 45 °C are relatively low. However, even so, power battery systems for vehicle driving applications cannot work for a long time in these two types of temperature environments with lower capacity decay rates, because the activity of battery materials is lower in low-temperature environments, the available capacity is lower, and the battery performance is worse; in high-temperature environments, the safety of the battery cannot be guaranteed.

3. **Establishing a battery SOH decay model for vehicle operation**

   According to the method of control variables, ignoring the influence of environmental temperature, the relationship between the battery SOH loss amount in each driving stage and the battery operating conditions can be represented by "SOH loss amount = actual battery operating state * actual battery operating duration". Among them, there are many variables in the battery operating state, and the vehicle speed is the most effective feature value reflecting the driver's driving behavior and the battery working state, so this paper selects the average vehicle speed as the feature parameter of the average operating state of the battery in a complete driving stage, so that "actual battery operating state * actual battery operating duration" can also be approximately represented by the driving mileage of the driving stage.

Based on the above analysis, a battery SOH decay model for vehicle operation can be obtained:

$$\Delta SOH_n = \alpha * \Delta M_n * \Delta C_T + \sigma_n \tag{3.32}$$

where, n is the driving/charging cycle number, $\Delta \mathrm{SOH}_n$ is the SOH loss amount of the nth driving/charging cycle, α is the influence factor of driving mileage on SOH decay, ΔM_n is the driving mileage of the nth driving/charging cycle, σ_n represents the influence factor of other vehicle operating parameters of the nth driving/charging cycle, such as maximum vehicle speed, acceleration/brake pedal opening, discharge depth, average SOC and other weather factors, etc.

Since the main factor affecting battery capacity decay in the actual operation of the vehicle is the environmental temperature, in order to simplify the analysis, the influence factor of other vehicle operating parameters $\sigma_n \approx 0$. Therefore, the SOH loss of the nth driving/charging cycle $\Delta \mathrm{SOH}_n$ is only related to the driving distance during the driving stage and the ambient temperature during the charging stage, and they are directly proportional. Although the real value of the factor α affecting SOH decay is still unknown, because it is a constant, the proportionality between the SOH losses of all driving/charging cycles can be obtained only through the driving distance during the driving stage and the ambient temperature during the charging stage. Therefore, the estimated value of the SOH loss of each driving/charging cycle can be obtained as follows:

$$\Delta \mathrm{SOH}_n = \frac{\Delta M_n * \Delta C_{\overline{T}n}}{\sum_{i=1}^{i=775} (\Delta M_i * \Delta C_{\overline{T}i})} * \Delta \mathrm{SOH} \tag{3.33}$$

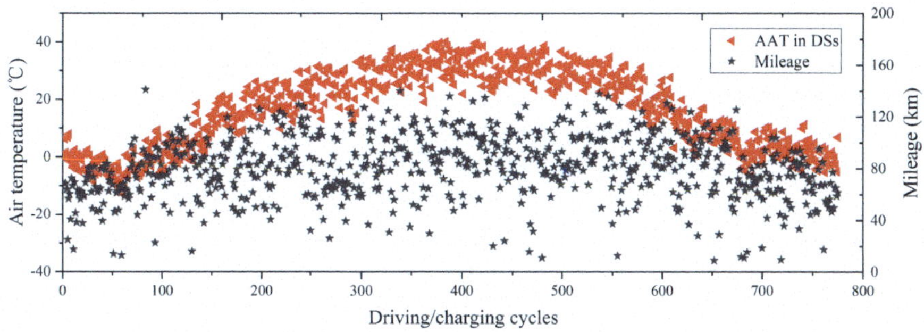

Fig. 3.18 Average temperature and driving mileage during all driving/charging cycles

where, $\overline{T}_n$ is the average temperature of the nth charging stage. According to the above analysis results, the total SOH loss in 2016 is $\Delta\text{SOH} = 40.21\%$. The average temperature during the driving and charging stages of all driving/charging cycles and the driving mileage are shown in Fig. 3.18. The average temperature for the whole of 2016 was between −20~40 °C, with most of the driving mileage concentrated within 40~150 km, and a small amount of short-distance driving below 20 km followed by charging.

Through calculation, the SOH fitting values for all 775 driving/charging cycles in 2016 and the estimated values of the corresponding SOH loss amounts for the subsequent 774 driving/charging cycles were obtained, as shown in Fig. 3.19. From the figure, it can be seen that the points where the SOH loss amount and decay rate are relatively high are mainly concentrated in the high-temperature summer, while the SOH decay rate in winter is relatively slow. This is consistent with the previous analysis of the correlation between the degradation performance of the power battery SOH and the ambient temperature.

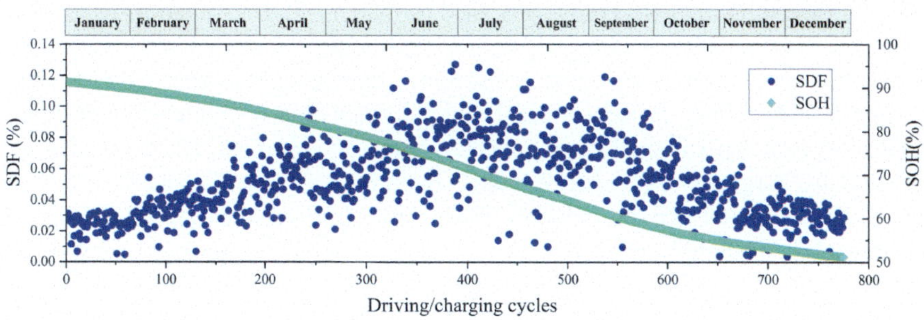

Fig. 3.19 SOH fitting values and loss amounts for driving/charging cycles

3.5.3 Model Validation Using Real Vehicle Data

Currently, there are many methods for evaluating and predicting the SOH of batteries, but these methods cannot truly be applied to vehicles in actual operation, mainly due to the following insurmountable difficulties and challenges.

Firstly, existing studies on the estimation of power battery SOH conducted in laboratory environments have a strong dependence on the stability of the experimental environment, and the application effect of the experimental results on real vehicles has not yet been verified by any feasible technology. More importantly, there is almost no effective SOH estimation method that can accurately and precisely estimate the SOH status of the power battery system on actual driving vehicles in real-time.

Secondly, the capacity estimation for electric vehicles during actual driving is also a major problem that needs to be solved urgently. It is impossible to stop the vehicle at any time during the driving process for capacity testing, which is similar to the laboratory environment. The most important reason is that external environmental factors such as temperature have a significant impact on the SOH of the power battery, but the accurate function relationship between battery capacity and environmental temperature is still unknown, and it is impossible to obtain a stable ecological temperature during the driving process of the vehicle. Therefore, it is currently impossible to obtain the actual value of the SOH of the battery system in real time during the driving process of the vehicle.

Thirdly, external factors such as driving conditions, weather, and driver behavior also have a certain impact on the SOH of the power battery during the actual driving process of the vehicle, and these influencing factors are unlikely to be obtained through simulation in the experimental environment. Therefore, developing a real-time estimation algorithm for the SOH of the power battery system for real vehicle applications that takes into account the dynamic stress effects of various external factors is an important topic in the field of power battery system safety management in the future.

Finally, existing methods commonly used for battery SOH estimation, especially those based on machine learning algorithms, generally require a fixed size for the input, which is determined by the structure and parameters of the network. However, the driving and charging process of a real vehicle cannot be as operable as the charging and discharging experimental steps in the laboratory environment. The actual driving time and discharge depth of a real vehicle is dynamic and variable, and the weather, road conditions, and driving behavior during the driving stage are complex and unpredictable. Therefore, to accurately estimate the SOH loss amount of a specific vehicle's actual driving/charging cycle, the problem of variable length during the driving and charging stages must first be solved. Common solutions include two approaches: one is to operate at the data level, such as data alignment. The simplest is to set an appropriate length, truncate the excess part, and fill in the insufficient part. The second is to handle the network structure. In fact, the ones that really need fixed-length input parameters are fully connected networks. CNN and RNN use a setting of shared parameters between layers, which can handle different lengths of input, but the output length is controlled by the input length. However, we can process the output into a fixed length before sending it to the fully connected network for training. Global average pooling layers and image pyramid methods can achieve this function. The first method is easier to implement, but may lose data features; the second method requires adaptive network structure, but the network structure processing is more complex, and batch training is also more troublesome.

Variable length input is currently widely used in machine learning-based image recognition, emotion recognition, and disease diagnosis in the medical field, but its potential application value in the engineering field has not yet been effectively developed. The frequent occurrence of thermal runaway accidents in electric vehicles in recent years has made the research on power battery fault warnings urgent, and the complex and variable driving conditions are in line with the variable time scale feature of variable input length. Therefore, this paper will focus on exploring the research on SOH estimation and prediction of electric vehicle power batteries using variable input length methods.

Based on the above analysis, this paper has already obtained the SOH loss of each driving/charging cycle of the vehicle under study, and the actual vehicle usage data and operating environment data can also be obtained from the monitoring historical data. Therefore, this paper will continue to use the LSTM neural network algorithm to train and learn the correlation between the vehicle operating status of each cycle and the SOH loss. Since the LSTM neural network algorithm has strong learning and memory capabilities, this paper plans to use the SOH loss of each driving/charging cycle as the output, and train and establish a model with various vehicle usage parameters and operating environment parameters in the corresponding cycle as the input. This will enable precise online estimation of SOH based on the actual vehicle driving environment and driving characteristics.

Furthermore, a major advantage of the LSTM algorithm used in this study is its ability to handle variable-length sequences. When building an LSTM model using Keras, if the LSTM layer is used directly as the first layer of network input, an input size needs to be specified. If variable-length sequences are needed, then a masking layer (Masking layer) or embedding layer (Embedding layer) just needs to be added before the LSTM layer. The method of use is: first convert the sequence into a fixed-length sequence, such as selecting a maximum sequence length, and padding sequences that do not reach this length with 0. Then, specify the filter character in the mask value of the masking layer. When entering the LSTM layer, the 0 s padded in the sequence will all be filtered out. Although the embedding layer also has a filtering function, unlike the masking layer, it cannot specify characters other than 0 for filtering, and because it is an embedded layer, it will map the sequence to a fixed-dimensional space. Therefore, if the demand is simply to allow LSTM in Keras to handle variable-length sequences as in this study, using a masking layer is more suitable than using an embedding layer. Therefore, this paper has designed an SOH estimator for power batteries based on LSTM, the schematic diagram of which is shown in Fig. 3.20.

In order to achieve SOH estimation for the power battery system of actual vehicle operation, the battery charging and discharging cycle corresponds to the driving/

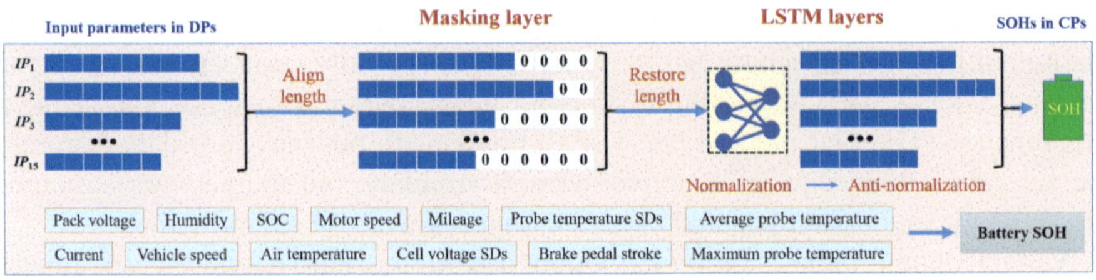

Fig. 3.20 Principle of the power battery SOH estimator based on LSTM

charging cycle of the actual vehicle, similar to existing research on SOH estimation based on experimental batteries, the SOH loss during the stable charging process can be ignored and can be used to calculate or extract the SOH characteristic value of its driving/charging cycle. For machine learning algorithms, each segment of driving data is the input for each training and testing, and the SOH loss in the corresponding driving/charging cycle is the output. However, before establishing the model, two problems must first be faced and solved. First, there are many SOH-related parameters during driving, such as weather factors (temperature, pressure, precipitation, humidity, etc.) and driver-driving behavior factors (speed, acceleration, etc.). Second, the conventional use of neural network algorithms is fixed-length input for continuous training or iterative training, but the driving time of each driving/charging cycle in actual vehicle operation is different, which also results in the input length of each training being different, which the LSTM algorithm cannot recognize. The first problem has been described in the above multi-parameter prediction of the power battery system, and the LSTM algorithm can learn the input parameters shown in Fig. 3.20 well. In response to the second problem, after preprocessing the actual driving/charging data accordingly, the LSTM can be improved by adding a masking layer before the LSTM layer to convert the sequence into a fixed-length sequence, selecting a maximum sequence length (based on a large amount of actual vehicle data analysis, the cumulative driving time of the vehicle between two charges will not exceed 20 h, so the maximum length here is set to 20 h), and padding sequences that do not reach this length with 0. Then specify the filter character 0 in the mask value of the masking layer. When entering the LSTM layer, the 0 s padded in the sequence will all be filtered out.

3.5.4 SOH Prediction Result Analysis

According to the battery multi-parameter prediction research, a set of hyperparameters was initially selected as Batch size = 16, Epoch = 50, the LSTM network consists of 7 layers: 1 Masking layer (padded with 0), 3 LSTM hidden layers (each layer has 100 neurons), a dense hidden layer Dense layer with linear activation, 1 input layer and 1 output layer. Because the training samples for SOH loss in this part are much smaller than the parameter prediction in the previous text, the training effect may be better when the Batch size is smaller. When Epoch = 50, the average relative errors of the SOH loss estimation for Batch size set to 5, 8, and 16 are 0.2686, 0.1503, and 0.1903, respectively, so a Batch size of 8 is more suitable. Then, the model with Batch size = 8, Epoch = 20 was trained, and the average relative error of the SOH loss prediction for the last 50 cycles was 0.3057, that is, the training effect was not as good as when Epoch = 50. The training effects of SOH loss for the four combinations of Batch size and Epoch are shown in Fig. 3.21, and finally, Batch size = 8 and Epoch = 50 were selected, and its SOH training effect is shown in Fig. 3.22, with an average relative error of SOH prediction of 0.0546% (Table 3.2).

This study proposes a new method for estimating the SOH of actual electric vehicle batteries based on variable-length input short-term memory neural networks. It reviews eight SOH evaluation schemes for actual battery systems and discusses their feasibility through comparative analysis, determining the internal resistance before charging as the SOH feature value. On this basis, by fitting the relationship between battery SOH and parameters such as temperature and mileage, it explores

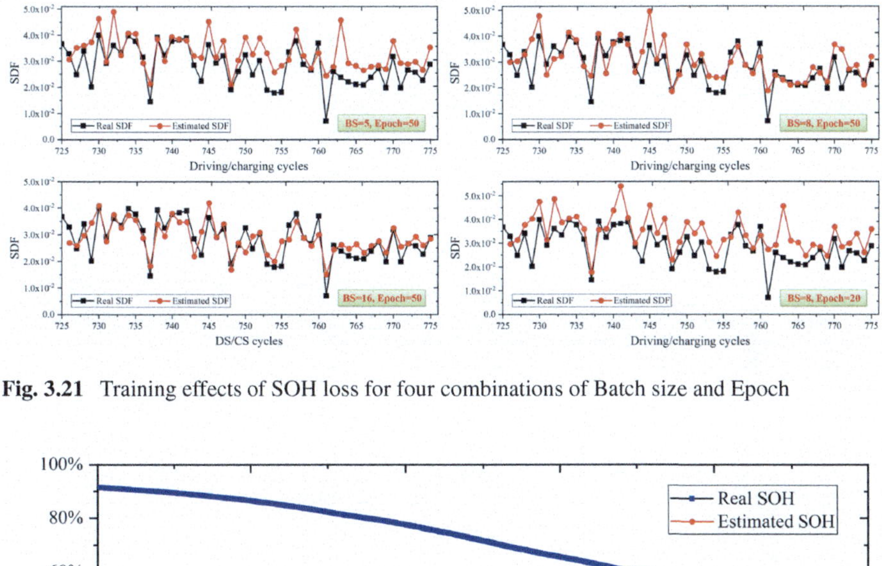

Fig. 3.21 Training effects of SOH loss for four combinations of Batch size and Epoch

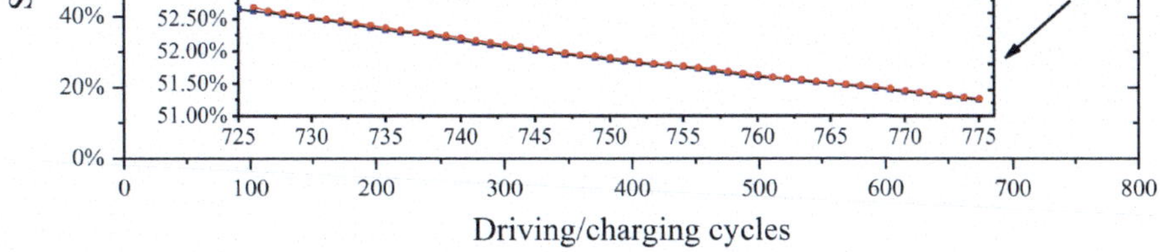

Fig. 3.22 SOH training effect

Table 3.2 Comparison results of various SOH/RUL estimation studies

Method	Battery material	Estimated target	Processing mode	Research environment	Accuracy
Coulomb counting	Li-ion	SOH	Online	Laboratory experiment	MRE ≤ 1/%
Open circuit voltage	Li-ion	SOH	Online	Laboratory experiment	RMSE ≤ 4
Kalman filtering	Li-ion	SOH	Offline	Laboratory Experiment	MAX = 8.3/%
Unscented Kalman filtering	Li-ion	SOH	Offline	Laboratory experiment	MRE < 0.0161/%
Support vector machine	$LiFePO_4$	SOH	Online	Laboratory experiment	MRE < 0.89/%
Iterâtive Kalman filtering	Li-ion	RUL	Online	Laboratory experiment	RMSE ≤ 0.14
Particle filtering	Li-ion	RUL	Online	Laboratory experiment	RMSE ≤ 1.857
Fuzzy logic	Li-ion	SOH	Online	Laboratory experiment	MAX = 5/%
Impedance spectroscopy	Li-ion	SOH + RUL	Offline	Laboratory experiment	MRE = 2.1/%
Composite Gaussian regression	Li-ion	SOH	Offline	Laboratory Experiment	RMSE ≤ 1.5
Wiener process	Li-ion	RUL	Online	Laboratory experiment	MSE < 1
Bayesian	Li-ion	RUL	Online	Laboratory experiment	MRE < 2/%
Capacity increment analysis	Li-ion	RUL	Offline	Electric bus	MRE = 4/%
LSTM + Particle swarm	Li-ion	RUL	Offline	Laboratory experiment	RMSE < 3.62
Extended Kalman filter	VRLA	SOH	Online	Laboratory Experiment	MAX = 2/%
Elman neural network	Li-ion	SOH	Online	Laboratory experiment	MAX = 2.3/%
LSTM in this paper	Li-ion suitable for all battery types	SOH	Online dynamic driving as input	Actual driving vehicle full weather, full state application	MAX ≤ 0.1323% MRE ≤ 0.0546% RMSE ≤ 0.232% MSE ≤ 0.0538

and establishes for the first time an approximate battery SOH degradation model suitable for actual vehicle operation, which can accurately estimate the battery SOH degradation of each driving/charging cycle.

In order to obtain a model for all weather and states, a year's worth of real operation data from an electric taxi was derived. The results show that the offline trained model can accurately estimate the SOH of the battery system integrated on the research vehicle. In addition to directly measurable battery parameters, some parameters related to weather and driving behavior, as well as inconsistencies in battery voltage and probe temperature, are added to the input for collaborative training. Given enough input samples, the trained model can not only perform real-time SOH estimation but also achieve driving/charging cycles, thus obtaining the capacity degradation trajectory of the battery. Admittedly, to establish a comprehensive battery SOH online estimation model, the year-long dataset in this study is far from enough, and the preprocessed data may still have stubborn errors due to the influence of collection noise. Furthermore, facing the gradually accumulated vehicle operation big data, it is necessary to further explore the approximate model of battery SOH degradation and optimize the coefficients of various parameters. The hyperparameters of the established LSTM model also need to be manually/automatically optimized. Therefore, we have further formulated a multi-stage research plan for online SOH estimation of batteries in electric vehicles. Through data expansion and accuracy optimization, a versatile general model can be obtained for full-stage, full-state, and full-weather applications on actual working vehicles. Therefore, this method has a broader practical application prospect in the battery system of actual electric vehicles.

3.6 Summary

By fitting the relationship between battery SOH and parameters such as temperature and mileage, an approximate model of SOH degradation suitable for actual vehicle power batteries was explored and established. This model can accurately estimate the SOH degradation of each battery's driving/charging cycle. In order to obtain a model for all weather and states, a year's worth of real operation data from an electric taxi was derived. The results show that the offline trained model can accurately estimate the SOH of the battery system integrated on the research vehicle. In addition to directly measurable battery parameters, some parameters related to weather and driving behavior, as well as inconsistencies in battery voltage and probe temperature, are added to the input for collaborative training. Given enough input samples, the trained model can not only perform real-time SOH estimation but also achieve driving/charging cycles, thus obtaining the capacity degradation trajectory of the battery.

Indeed, to establish a comprehensive online estimation model for battery SOH, the year-long dataset in this study is far from sufficient, and the preprocessed data may still contain stubborn errors due to the influence of collection noise. Furthermore, facing the gradually accumulated vehicle operation big data, it is necessary to further explore the approximate model of battery SOH degradation, optimize the coefficients of various parameters, and the hyperparameters of the established LSTM model also need to be manual/automatically optimized. Therefore, we have further formulated a multi-stage research plan for online estimation of battery SOH in electric vehicles. Through data expansion and precision optimization, a versatile universal model can be obtained, which can be applied in all stages, all states, and all climates on actual operating vehicles. Therefore, this method has a broader practical application prospect in the battery system of actual electric vehicles.

Chapter 4
Power Battery Safety Status Prediction

4.1 Overview

Currently, the safety risk of electric vehicle power batteries is a significant reason hindering the large-scale promotion and application of electric vehicles. Monitoring key parameters of power batteries, giving early warning of safety risks, and taking corresponding control measures through the Battery Management System (BMS) are key means to improve the safety of power batteries. The prediction of the State of Safety (SOS) is the basis for the BMS to achieve safety functions. The parameters involved in SOS prediction include temperature, voltage, State of Charge (SOC), State of Power (SOP), State of Energy (SOE), etc. The parameters used in SOS prediction can reflect the state of the power battery, thereby characterizing common faults, which is essential for the early warning of power battery faults. This chapter introduces the separate prediction of the above parameters and introduces the joint estimation method of SOS with multiple parameters. This SOS joint estimation method combines various data to analyze the safety of power batteries, and the results are more accurate. The method of predicting the safety status of power batteries is of great significance for battery risk management and safety applications (Fig. 4.1).

4.2 Power Battery SOS Prediction

4.2.1 Temperature Prediction

1. Research Status

   An accurate battery thermal management system is crucial to the power battery system, and the operation effect of the battery thermal management system depends on the design level of the battery system and the working conditions

J. Hong, *Big Data Analysis and Application of Power Battery Operation Safety*, Key Technologies on New Energy Vehicles, https://doi.org/10.1007/978-981-96-4399-8_4

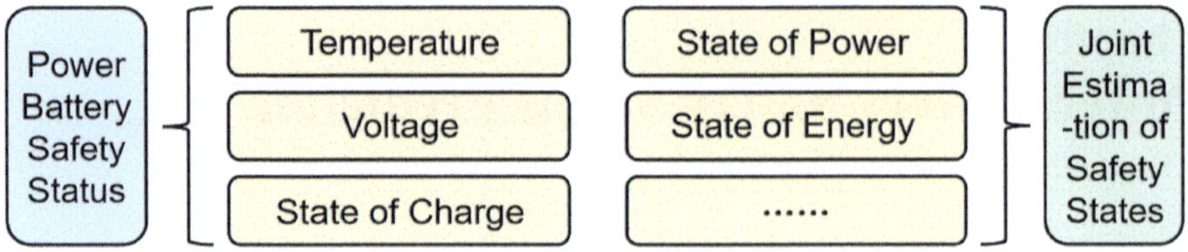

Fig. 4.1 Power battery safety status estimation framework

of the battery. In recent years, battery management research has mainly been used to estimate or predict the battery's SOC, SOP, and SOH.

With the increasing number of electric vehicle safety accidents caused by thermal runaway of power batteries, preventing thermal runaway has attracted the attention of researchers. However, existing research on the thermal safety of battery systems is mainly based on mechanism analysis and models, using experimental data or artificially pre-selected real vehicle data to estimate temperature. Due to the limitation of data volume and the complexity of actual operating conditions, it is difficult to predict the battery temperature of actual operating vehicles in all climates in the future.

Overall, the aforementioned methods all have serious drawbacks. Firstly, most existing battery temperature prediction algorithms or models tend to be computationally intensive and are limited by experimental data, resulting in poor evaluation accuracy. Secondly, these methods mainly predict future temperature trends or values based on known historical data, making it difficult to predict potential temperature anomalies or thermal faults. Lastly, most research is conducted in laboratory environments, with few attempts to perform real-time temperature or thermal fault predictions on actual operating vehicles. Thus, the stability and practicality of battery temperature prediction methods still need to be verified.

With the rapid development of artificial intelligence and computer intelligence technology, machine learning algorithms have gradually taken root in our daily lives. However, few studies have considered the prediction of battery temperature and the pre-judgment of battery thermal faults. Although research on various battery parameter predictions based on machine learning algorithms has gradually developed, the existing literature still has the following unresolved deficiencies.

(1) Most of these works are carried out under constant current discharge/charge conditions in experimental environments, and their application to dynamic operating vehicle battery systems is not ideal.
(2) Due to the limitation of small input samples (Input Samples, ISs), most existing methods perform online training and prediction, which will cause a large computational burden and limit prediction efficiency.
(3) Most existing research is based on recursive neural networks, which have limited long-term memory capabilities. Therefore, they can only make accurate parameter predictions within a short time, resulting in insufficient prediction window length.

(4) There is a lack of sufficient research on battery fault diagnosis (such as thermal faults) based on machine learning algorithms, which is not conducive to the promotion and application of machine learning in the field of electric vehicle battery safety management.
(5) The deep learning models currently used for electric vehicle thermal fault prediction lack seasonal temperature specificity. Therefore, they find it difficult to extract detailed temperature features for each season.

In order to improve the above deficiencies and fill the gaps in existing research, the following introduces a battery system temperature prediction and thermal fault prediction method using a self-attention mechanism (Self-Attention Machine, SAM).

2. Self-Attention Mechanism

RNN is a type of artificial neural network for pattern recognition of continuous data sets. However, the standard RNN has a significant drawback, namely that its gradient may explode or vanish during the backpropagation process. SAM can better capture the long-term dependencies in the sequence. Therefore, SAM has received widespread attention from researchers. The SS-ANN model developed in this work for temperature prediction includes a GRU-GRU-SAM architecture and a network parameter optimizer using the Adam method.

Transformer is a deep learning algorithm based on EncoderDecoder proposed by Vaswani et al. in 2017. Compared with traditional deep learning networks such as BP, RNN, CNN, Transformer is mainly built by SAM, reducing the use of activation functions, having parallel computing capabilities, and better extracting features of long sequence data. SAM is a data feature extraction algorithm with excellent long-range feature extraction capabilities. Self-attention maps the original data through a fully connected layer to obtain query, key, and value matrices. Multiplying the query and key extracts the relationship features between the data, and then multiplying with the value can obtain the attention matrix. SAM has achieved excellent results in long sequence data tasks such as machine translation and semantic annotation, and is one of the leading research directions in the field of natural language processing in recent years.

3. All-Weather Coverage Temperature Prediction Method

The network structure of the Spiral Self-Attentive Neural Network (SS-ANN) is shown in Fig. 4.2a, as can be seen from the figure, this network consists of two GRU layers and a self-attention layer. The processed data from each cycle of the GRU layer is passed to the next layer of the network as a cycle output. After passing through two GRU layers, the time series data is passed to the self-attention layer. The self-attention layer better integrates the output data of the GRU through its ability to process long-range data.

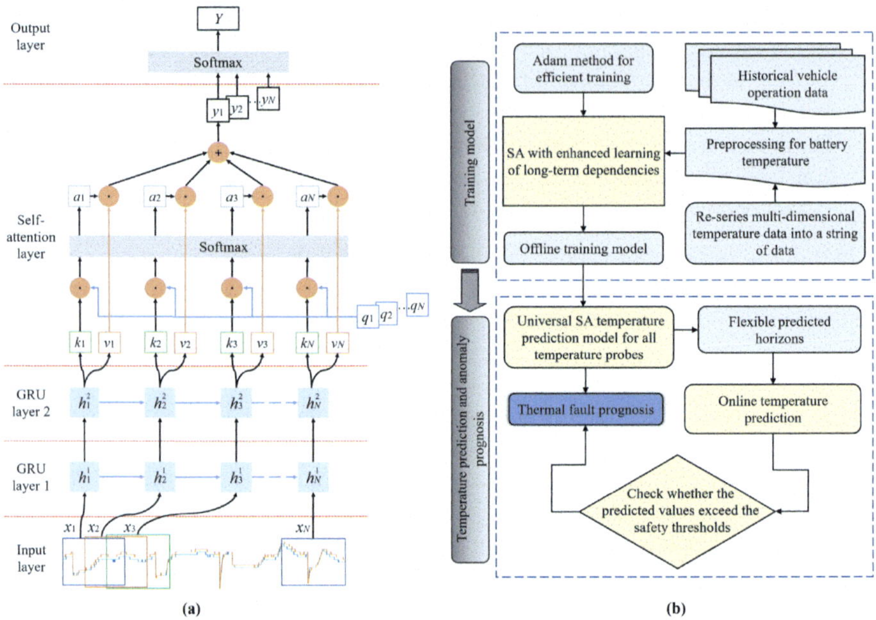

Fig. 4.2 Structure of temperature prediction method: **a** SS-ANN network structure; **b** Process of building SS-ANN model

To establish a temperature prediction model based on SS-ANN, the Keras model was used for training. In configuring the SS-ANN learning process in Keras, the Adam optimizer, which is most widely used in machine learning applications, was used to train the model. Compared with other optimizers, Adam is not only good at handling sparse gradients but also good at handling non-stationary targets, and it has smaller requirements for handling memory, making it more suitable for large datasets than other optimizers.

The abnormal fluctuation of battery temperature is directly related to battery thermal faults, so the prediction of multiple battery thermal faults can be transformed into a temperature prediction problem, by establishing an accurate model, the future battery temperature can be well predicted. The SS-ANN proposed in this section can obtain an accurate model, Fig. 4.2b shows the characteristics and the whole process of building the SS-ANN model. This network is trained based on batch algorithms. The Adam optimization algorithm can be used to replace the traditional stochastic gradient descent optimizer, updating network weights iteratively according to training data. The constructed SS-ANN has enhanced long-term dependency learning ability, can predict battery temperature, and is not affected by the offline training process. After training the temperature of all probes, a universal model can be used for online temperature prediction, and the prediction range can be flexibly adjusted according to the requirements of the expected vehicle. Finally, referring to the safety threshold set based on battery performance, it can be judged whether the thermal fault of the battery system is predicted based on whether the predicted value exceeds the safety threshold.

4. Data Preprocessing

   The vehicle operation data of the actual vehicle comes from the government-level new energy vehicle big data platform. Through the communication of the vehicle to the platform, the motion information and the status of key components of the monitored vehicle can be obtained, with a sampling interval of 10 s. The vehicle under study is an electric taxi, equipped with a 18,650 lithium-ion battery pack, with a total energy of 29.1 kWh and a rated voltage of 326 V. In addition, there are 16 temperature probes (Probe 1~Probe 16) distributed inside/outside the battery pack. The distribution diagram of the temperature probes is shown in Fig. 4.3a.

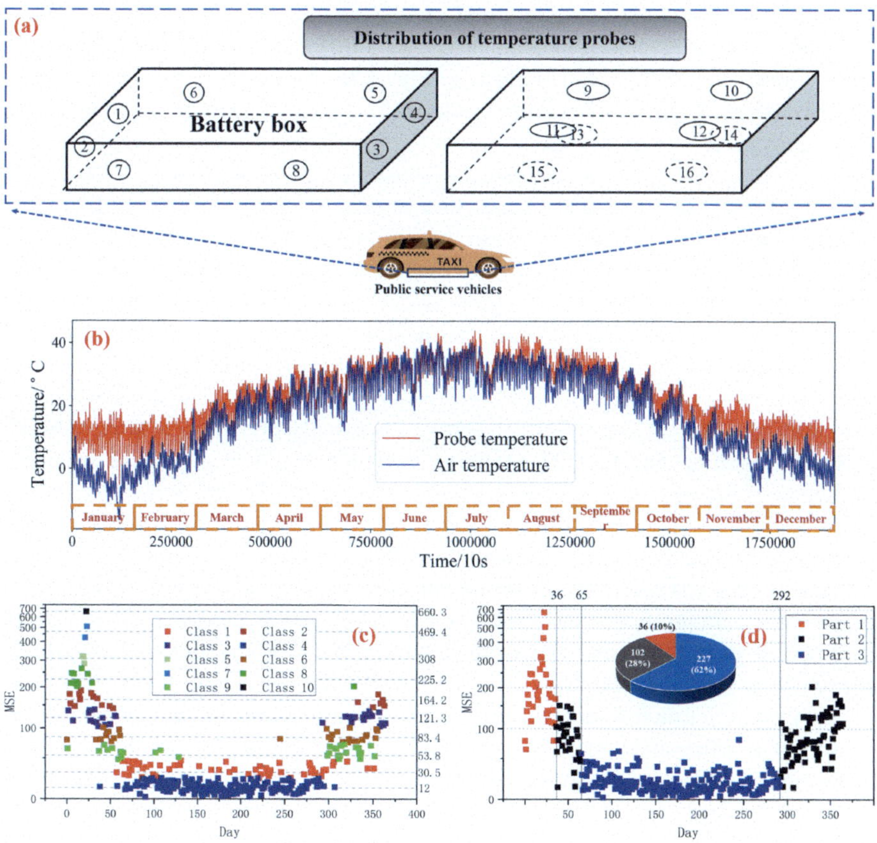

Fig. 4.3 Acquisition and analysis of temperature data: **a** Distribution of temperature probes; **b** 2016 temperature detection and temperature curve; **c** K-Means clustering results of MSE; **d** Data partitioning results

To obtain a full-weather application model, the full-year data of the vehicle under study in 2016 is used as IS. The probe temperature (represented by the temperature of Probe 1) and air temperature curve in 2016 are shown in Fig. 4.3b. As shown in Fig. 4.3b, from April to October, the range of probe temperature and air temperature is the same. However, from November, December, and January to March, the air temperature in Beijing was relatively low, deviating from the appropriate working temperature of the battery system, so there was a significant difference between these two temperature curves. In addition, in order to protect the battery pack and the vehicle body, the operating temperature of the battery is kept above 0 °C all year round.

The data is divided into three parts to facilitate model training according to the degree of difference between probe and air temperature. Each part of the data is used to train the model separately. Specifically, the probe temperature and ambient temperature data were obtained every day in 2016, and the Mean Square Error (MSE) was calculated. The square calculation of MSE can make the data difference clearly displayed, so MSE is chosen to evaluate the data difference. Then, the K-Means is used to cluster the MSEs into ten clusters; Fig. 4.3c shows the K-Means clustering results of MSEs. It can be seen that clusters 5, 7, 8, and 10 form the first layer with MSE values greater than 200. The dates of the first layer are concentrated between 1 and 27. The second layer is composed of clusters 2, 3, 6, and 9, with MSE values between 50 and 200, and the dates are concentrated between 1 and 65 and 292 and 365. The remaining levels 1 and 4 form the last layer, where the probe temperature mainly overlaps with the air temperature, and the dates are between 66 and 291. Considering the above analysis, the training data for each model should not be too small, and the date should be the basis for division. These data are divided into three parts, as shown in Fig. 4.3d, each part of the data includes at least 10% of the total data.

In machine learning algorithms, the dataset is usually divided into three subsets: training, validation, and test. The training dataset is used to adjust the trainable parameters of the model, the validation dataset is used to determine the appropriate hyperparameter settings, it is used to select the final model in the iterative model, and the test data can be used to evaluate the training quality of the model. To ensure the stability and repeatability of the model after training, this study conducted a tenfold cross-validation, and the training dataset, validation dataset, and test dataset accounted for 80%, 10%, and 10% of the entire IS, respectively. In order to evaluate the prediction performance, this section uses the Mean Relative Error (MRE) to evaluate the prediction results of the test data in each cross-validation. The final MRE of complete training and testing is the average MRE of tenfold cross-validation, and the formula of MRE can be expressed as:

$$MRE = \frac{1}{n}\sum_{i=1}^{n}\frac{\left|\widehat{T_i} - T_i\right|}{T_i} \times 100\% \tag{4.1}$$

where, n is the number of test samples, $\widehat{T_i}$ and T_i are the predicted and actual temperature values of the battery system, respectively.

5. Hyperparameter Selection

   When IS is very large, training the model will take a long time, and it is challenging to adjust all the model parameters to the optimal state in a short time. Therefore, it is necessary to determine a set of hyperparameters based on experience. Then, an optimized model can be obtained through parameter optimization. To reduce the consumption of model training in the hyperparameter optimization process, the model's hyperparameters are optimized through grid search. The temperature data for one month in February 2016 is determined as the initial IS, and the data for one day on February 8, 2017, is used as the test data.

Generally speaking, to avoid the computational load of the neural network model being too large, establishing a regular neural network does not require more than three hidden layers. To verify the training effect of SS-ANN with different numbers of hidden layers, the average validation loss is calculated when one and two GRU layers are combined with one SAM layer to form hidden layers, respectively. The results show that three hidden layers can achieve the lowest average validation loss. At the same time, when epoch = 50, the average validation loss can finally converge to a stable minimum value.

In addition, terms like epoch and batch size (BS) are also needed in the machine-learning process. An epoch means that the entire dataset passes through the SS-ANN from front to back, and BS means the number of training instances in a batch. Moreover, the size of different windows will play a key role in training the model. The training window size (TWS) represents the number of each input training sample; the prediction window size (PWS) determines the number of prediction samples for each training window. To learn more historical information and improve the accuracy of the model, the sliding window (SW) technique is used in the process of offline training the model. According to the research objective of this paper, a set of hyperparameters is preliminarily determined as: epoch = 50, BS = 128, IS = one month's data, TWS = 360 (one hour's data), PWS = SWS = 6 (one minute's data). The SS-ANN consists of six layers: two GRU hidden layers (each with 100 neurons), one SAM layer, one densely hidden layer with linear activation, one input layer, and one output layer.

6. Hyperparameter Optimisation

   The training window can be iteratively trained through the SW technique, allowing the SS-ANN to learn more historical temperature information. Based on the determination of the initial hyperparameters, manual optimization is carried out by adjusting the sizes of TWS, PWS, and SWS, respectively, while keeping other parameters unchanged. Window sizes = {1, 6, 30, 60, 90, 120, 150, 180, 360} represent data volumes of 10 s, 1 min, 5 min, 10 min, 15 min, 20 min, 25 min, 30 min, and 1 h, respectively. Through tenfold cross-validation, the MRE of different windows and sizes on the same test data is obtained, as shown in Fig. 4.3a.

Figure 4.3a shows that as PWS and SWS increase, MREs significantly increase, but after TWS = 6, MREs of different TWS do not significantly increase or decrease. Theoretically, TWS can only support prediction when TWS > PWS, which means that the amount of data for each prediction should be less than the amount of training data, so when TWS is too small, such as TWS = 1 and TWS = 6, TWS cannot support the target prediction. In addition, TWS has a decisive significance for the adjustment of SWS and PWS because it determines the amount of data for each training, and it cannot be too small to flexibly choose PWS. Therefore, TWS is finally determined to be 360, that is, one hour of data.

In principle, a larger SWS will result in less available historical information for each training, and vice versa, so MRE gradually increases with the increase of SWS, as shown in Fig. 4.4a. However, when SWS = 1, its computation is large and requires more computation time, so in this study, SWS is determined to be SWS = 6. In addition, regarding the constant training samples, a longer prediction range will make it more difficult to train, so a smaller PWS can achieve higher prediction accuracy. However, when a battery system experiences temperature anomalies or thermal faults in the next PWS, and when the PWS is small, the time available for drivers and passengers will not be enough to escape danger. Therefore, PWS should be determined flexibly according to the requirements of prediction accuracy and the length of the future prediction period.

Figure 4.4b shows the MRE of different IS (one year of data in 2016, one month of data in February 2016, one week of data in the first week of February 2016, and one day of data on February 7, 2016). When IS increases, MRE has a clear downward trend. When IS = one year of data, MRE is much lower than other IS data. Through calculation, although there are significant differences in the length of training time for different IS, no matter how large the training sample is, the test data is always 3 s, which is much smaller than the sampling interval (10 s). In addition, this is only obtained under the limited computer configuration conditions in this work,

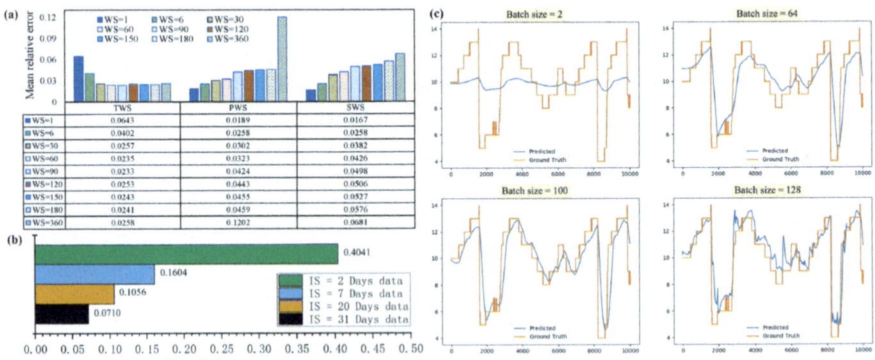

	TWS	PWS	SWS
WS=1	0.0643	0.0189	0.0167
WS=6	0.0402	0.0258	0.0258
WS=30	0.0257	0.0302	0.0382
WS=60	0.0235	0.0323	0.0426
WS=90	0.0233	0.0424	0.0498
WS=120	0.0253	0.0443	0.0506
WS=150	0.0243	0.0455	0.0527
WS=180	0.0241	0.0459	0.0576
WS=360	0.0258	0.1202	0.0681

Fig. 4.4 Hyperparameter optimization process: **a** MRE of different window sizes; **b** MRE of different IS; **c** Prediction results of different batch sizes

so the technology proposed in this section can quickly achieve online temperature prediction through offline models. In order to achieve all-climate prediction of battery temperature, one year of data is the most suitable IS in this section.

When adjusting model parameters, suppressing the overfitting problem of the attention mechanism is key to affecting the model training effect. Among the hyperparameters of the model, the batch size has the most obvious impact on the degree of model overfitting. Figure 4.4c shows the model prediction results when the batch size is 2, 64, 100, and 128. It is observed that when the batch size is 2, the model's prediction result is a straight line around the mean of the true value, and this phenomenon decreases with the increase of batch size. This is because the smaller the batch size, the more training batches per period, and too many training batches will cause serious overfitting problems in the first few epochs, making the model's prediction results biased towards the average value of the training data. Therefore, a batch size of 128 was chosen for subsequent training.

In Fig. 4.4c, the true value is step-shaped, while the predicted value is a continuous curved curve. The original collected temperature is only stored as an integer, the data granularity is large, and there is a significant error and omission of detailed information. They use small granularity data to predict temperature, retaining as much detail as possible.

7. Prediction Results and Discussion

   Figure 4.5a shows the training loss and validation loss before and after hyperparameter optimisation, indicating that the model's loss is small and stable, without overfitting. In addition, Fig. 4.5a also shows that the initial epochs are also sufficient to minimize and stabilize the model's loss. Figures 4.5b and c show the temperature prediction results of the probe based on tenfold cross-validation test data before and after optimization, indicating that the vehicle under study only performed one drive and one charge on February 8, 2017. The MRE before and after optimization were 1.82% and 0.95%, respectively. The temperature prediction value after optimisation in Fig. 4.5b is closer to the actual temperature value than the temperature prediction value before optimisation in Fig. 4.5c. Therefore, the SS-ANN model has better prediction performance when optimizing hyperparameters. In addition, Fig. 4.5b and c also show that the absolute error after optimization is closer to zero, and the prediction results are more accurate and stable than before optimization.

The model's predictive ability is verified by obtaining 10,000 data points in each of the three parts of the data. The validation results are shown in Fig. 4.6. The red dashed line in the figure represents the absolute error between the predicted temperature and the absolute temperature, and the red dashed line represents the Mean Absolute Error, MAE). For each data segment, the corresponding SS-ANN model achieved the best MAE of 0.812C°, 0.395C°, and 0.646C° respectively. Model 1 predicted

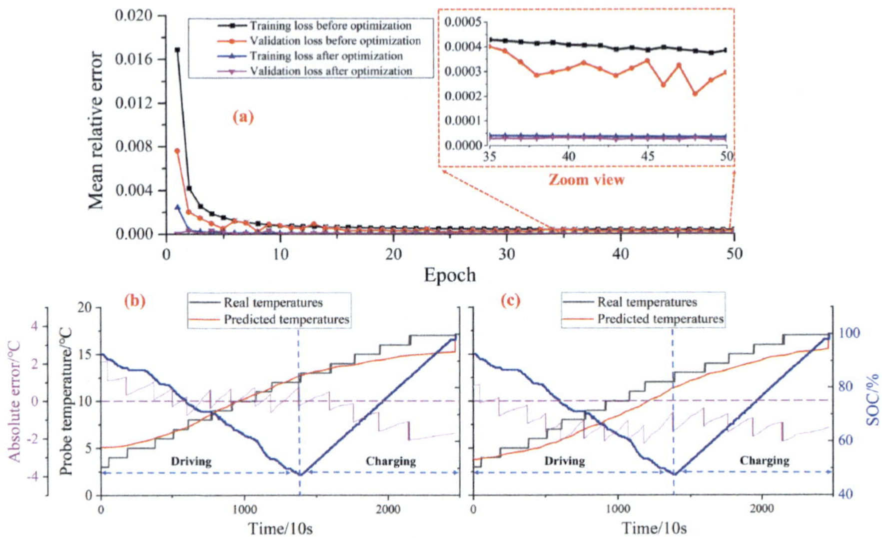

Fig. 4.5 Hyperparameter optimization results: **a** training loss and validation loss; **b** probe temperature prediction results after hyperparameter optimization; **c** probe temperature prediction results before hyperparameter optimization

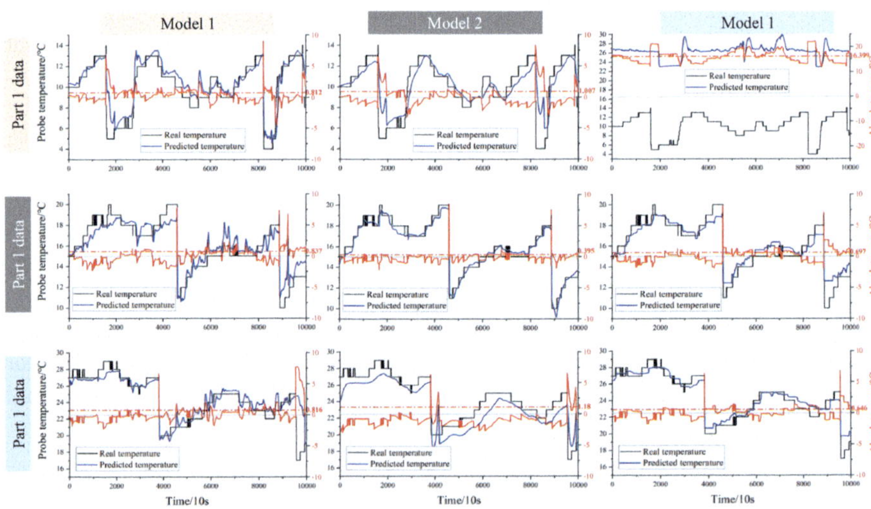

Fig. 4.6 Prediction results of the three SS-ANN models using different parts of the data

temperatures significantly higher than the actual temperatures of the first part of the data, while the MAE of other parts of the data was less than 1C°. This indicates that the partition model has learned the important features of the corresponding data, and has a high prediction accuracy for the relevant parts of the data.

8. Stability and Robustness Verification

   To verify the stability and robustness of the SS-ANN model in different application environments, temperature data from four days in 2017 (April 1, July 1, October 1, January 1) were taken as test data for the four seasons for comparison. Figure 4.7 shows the temperature curves before and after prediction for the four seasons, with MREs of 0.19%, 0.17%, 0.24%, and 0.71%, respectively. Figure 4.7 shows that the SS-ANN model has good general performance in all-climate applications. In addition, the prediction results for spring, summer, and autumn are better than those for winter, and even better than the previously selected test data. This is because the low outdoor temperatures in Beijing in winter are far from the appropriate working temperature range of the battery system, and the working environment is harsh. Compared with other seasons, the prediction of probe temperature values in winter is more challenging.

9. Superiority and Flexibility Verification

   A smaller prediction MRE and a shorter prediction time (PT) will produce better prediction results. To verify the flexibility of this method, we use the relationship between prediction accuracy and PT, and take the prediction sensitivity (PF) as an indicator of prediction performance, with the PF at PWS = 1 as the benchmark, as described in Eq. (4.2). When TWS = 360, SWS = 6, we calculated the PF for different PWSs, as shown in Table 4.1. Table 4.1 shows that the PF values for different PWSs do not change significantly. When PWSs increase, PT does not change significantly, but as PF gradually decreases, the prediction effect becomes worse. Therefore, to achieve good prediction results, we can determine the prediction horizon according to the requirements of PF. The longer the prediction range that meets the PF requirements, the more temperature features are predicted. Once

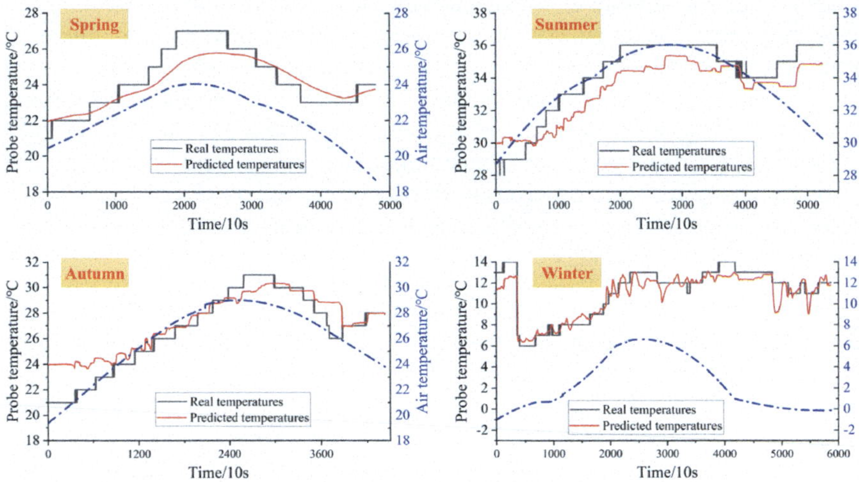

Fig. 4.7 Real and predicted temperature curves for the four seasons

Table 4.1 PF for different PWSs when TWS = 360 and SWS = 6

PWS	MRE	PT(s)	PF
1	0.0189	6.5152	1.0000
6	0.0258	6.5235	0.7335
30	0.0302	6.4645	0.6210
60	0.0323	6.5225	0.5858
90	0.0424	6.5524	0.4483
120	0.0443	6.5332	0.4278
150	0.0455	6.4888	0.4137
180	0.0459	6.4464	0.4074
360	0.1202	6.5586	0.1583

the battery system has a temperature anomaly, we can prevent them from having battery failures, or even thermal runaway, providing more timely protection for the safety of drivers and passengers.

$$PF = \frac{PT}{MRE \times PF(1)} \tag{4.2}$$

where $PF(1)$ is the PF when PWS = 1.

10. Prognostic Effects of Various Temperature Anomalies

    Table 4.2 lists the alarm and warning thresholds for temperature anomalies in the battery system under study. It should be emphasized that this method is also applicable to other types of battery systems, but due to differences in battery performance and type, there are slight differences in thresholds. As shown in Table 4.2, temperature anomalies are divided into three levels according to severity: level one alarm, level two warning, and level three warning. The level three warning state can also be safe, but preventive measures should be taken to prevent further temperature anomalies in the battery. The level two warning state means that the battery system is in a dangerous state and should be stopped for inspection under permissible conditions; the level one alarm state means that the battery system is in a dangerous thermal state, and the driver needs to stop for inspection as soon as possible to avoid more serious accidents and injuries to the driver and passengers.

To verify the model's prognostic effects on various temperature anomalies, we retrieved some temperature anomaly data from the vehicle under study. Figure 4.8a shows the temperature and TRR curves of 16 probes on 5 March 2017, showing that the vehicle started working at 07:10:23 in the morning, and probe 1 issued an OT warning of $T > 40$ °C at 09:11:43 in the morning. Probe 1 showed a different temperature fluctuation from the other probes after the vehicle started running that day, which means that a thermal fault may have occurred in the battery cell or module

Table 4.2 Alarm or warning thresholds for temperature anomalies

Type of alarm or warning	MRE	PT(s)
Excessive temperature difference	40 °C ≤ T < 50 °C	3
	50 °C ≤ T < 65 °C	2
	T ≥ 65 °C	1
Excessive temperature	5 °C ≤ $\triangle Tdmax$ < 8 °C	2
	$\triangle Tdmax$ ≥ 8 °C	1
Excessive temperature rise	$\triangle Tdmax$ ≥ 5 °C/10 s	1

of the battery pack at probe 1. Fortunately, as can be seen from the TRR curve, all probes' TRR are below 5 °C/10 s, all within the safe TRR range.

OTD anomalies are typically caused by one (or more) probe temperature being too high or too low. As shown in Fig. 4.8b on November 3, 2017, all 16 probe temperatures and TD curves, the vehicle started working at 07:55:47 in the morning, and an OTD warning of $\triangle Tdmax$ > 5 °C occurred at 18:14:55 in the afternoon. When all probe temperature readings rise or fall, probe 16 always has the highest temperature, but the temperature from probe 11 gradually becomes lower than the others, which then triggers the OTD warning. This suggests that the battery unit or module near probe 11 may have a thermal fault or other faults. If the OTD warning

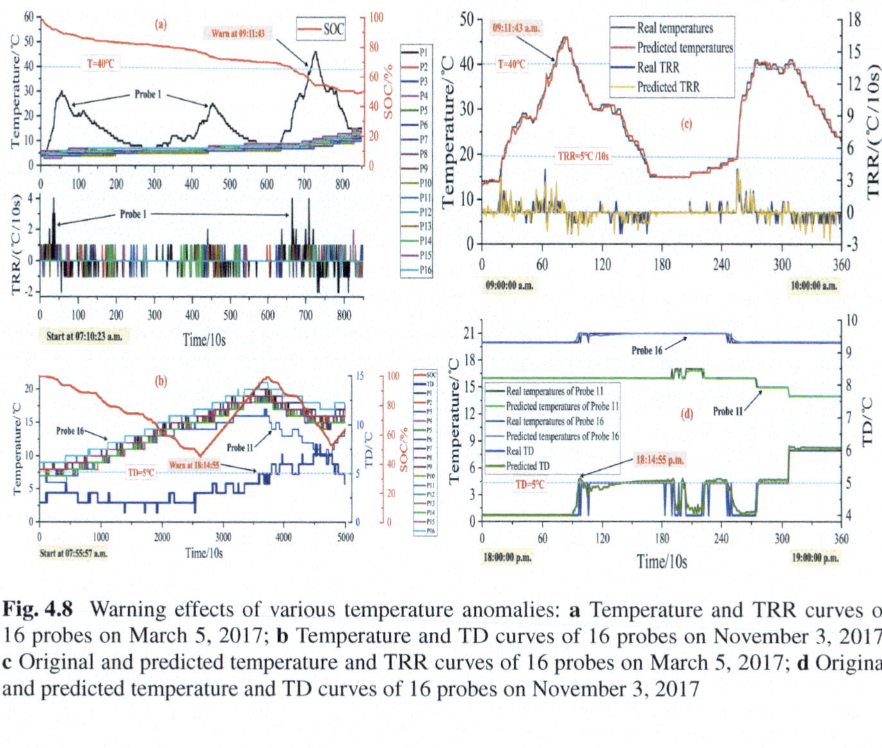

Fig. 4.8 Warning effects of various temperature anomalies: **a** Temperature and TRR curves of 16 probes on March 5, 2017; **b** Temperature and TD curves of 16 probes on November 3, 2017; **c** Original and predicted temperature and TRR curves of 16 probes on March 5, 2017; **d** Original and predicted temperature and TD curves of 16 probes on November 3, 2017

is not predicted and handled in a short time, it may gradually lead to more serious faults, even thermal runaway of the battery system.

We take one hour of operation data with OT anomalies (09:00:00–10:00:00 on March 5, 2017) as the prediction object, with the data from the previous hour as TW. Figure 4.8c shows the temperature and TRR curves before and after prediction using the well-trained SS-ANN model. Figure 4.8c demonstrates excellent prediction performance, with the predicted values of temperature and TRR astonishingly close to the actual values, and the OT anomalies can be predicted timely and accurately.

Referring to the above implementation scheme, we take one hour of data with OTD anomalies (18:00:00–19:00:00 on November 3, 2017) as the prediction object. The basic probe (probe 11 and probe 16) temperature values and TD curves before and after prediction are shown in Fig. 4.8d, which indicates that the prediction effect of the two probe temperatures is very good, and as shown by the predicted TD curve, it shows the accurate prediction effect of OTD anomalies.

Since the determined PWS is 1 min (PWS = 6), the above temperature anomalies can be predicted at least 1 min in advance. Although one minute is short, for less severe temperature anomalies (such as second and third-level warning states), there is a longer time for passengers to react and escape. On the other hand, for temperature anomalies that may lead to battery thermal runaway (such as a first-level warning state), one minute will also be sufficient, which is very valuable for drivers and passengers to quickly escape from danger. In addition, according to the accuracy requirements in Fig. 4.3a, PWS can be flexibly adjusted. Therefore, the proposed technique is universal, and PWS can be flexibly selected according to the required prediction accuracy and range, making it suitable for all types of vehicles.

Based on the above analysis, we further developed a multi-stage research plan for online thermal fault prediction of battery systems in real-world electric vehicles, as shown in Fig. 4.9. With the data expansion of the new energy vehicle big data platform, more actual vehicle data of the same model and different models, as well as vehicles that have had battery faults or thermal runaway accidents, a universal model can be obtained for full-state, all-weather application to actual vehicles. This study uses the trained offline model for online multi-step temperature prediction, so the computation of temperature prediction is small. More importantly, both the new energy vehicle big data platform and the edge computing device integrated into the battery management system can achieve fast and accurate multi-step forward temperature prediction, making the proposed method easy to implement in real-world electric vehicles.

4.2.2 Voltage Prediction

To achieve the required voltage and capacity, a set of batteries are often connected in series or parallel. Due to the ageing process or overuse in actual operation, each component or related accessory may experience various failures. If these failures are not controlled, they will severely affect the safety of the battery and may even lead

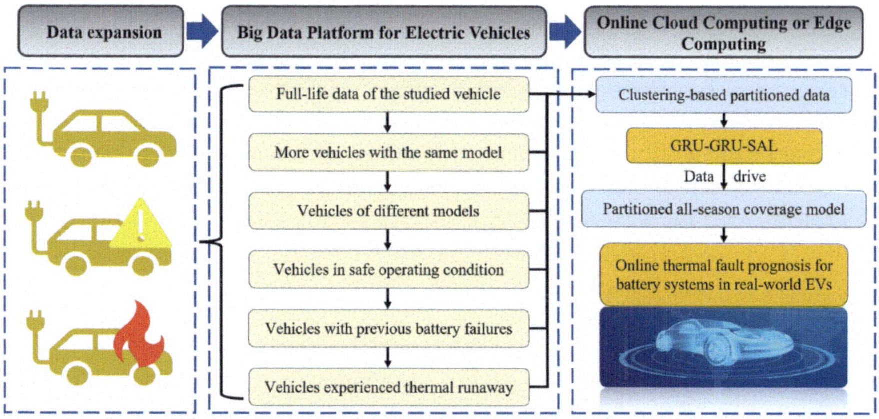

Fig. 4.9 Multi-stage research plan for online thermal fault warning of real vehicle battery systems

to thermal runaway of the battery system under certain extreme conditions. Many studies have shown that battery failures can lead to various voltage abnormalities. Typical voltage abnormalities can be categorized into overvoltage, undervoltage, overvoltage rate of change, and poor voltage consistency. For example, overvoltage may indicate that the battery system has to overcharge or charging protection circuit failure; Undervoltage represents that the battery system may have over-discharge or internal short circuit failure. In addition, voltage abnormalities often mean that more serious failures may be encountered. Therefore, timely and accurate prediction of voltage abnormalities during vehicle operation is of great significance in preventing voltage failures. This also means that some human intervention is needed to identify potential problems brought about by voltage abnormalities, such as real-time monitoring and accurate prediction of battery voltage, to ensure the safe operation of the vehicle.

1. Current Research

   As is well known, many battery failures can cause voltage abnormalities or be caused by overvoltage, undervoltage, or uneven voltage distribution. Usually, explicit voltage abnormalities may indicate that a more serious internal battery failure has occurred or will form, such as internal short circuits and electrode structure failures. The frequent occurrence of thermal runaway accidents in electric vehicles in recent years has attracted widespread attention to battery failure prediction, but there is little research on voltage abnormality prediction. Previous research on voltage abnormalities in battery management systems was limited to research on voltage sensor failures, and there is a lack of in-depth analysis of the connection between voltage abnormalities and other battery performance.

   However, the voltage fault diagnosis method based on improved Shannon entropy can only amplify a smaller voltage abnormality according to the current voltage; it cannot accurately predict future voltage.

Machine learning algorithms are deeply rooted in our lives. Its applications span almost all areas of artificial intelligence; the progress of many of its applications can be attributed to a small category of artificial neural networks—RNN. This type of neural network includes LSTM, which gives recursive neural networks the ability to continue self-learning. In recent years, the application research of neural networks in the field of electric vehicles or battery management systems has gradually increased; they mostly focus on the estimation of battery SOC, SOP, SOE or SOH.

2. LSTM Neural Network

The LSTM network is a type of RNNs. RNNs are a type of artificial neural network for sequence dataset pattern recognition; it tries to simulate time or sequence-dependent behavior, such as language, stock prices, power demand, etc. In addition to these applications, RNNs are mainly used for any time series data. However, standard RNNs cannot ignore the shortcomings of long-range dependence, and the gradient either explodes or disappears during backpropagation. LSTM has achieved good success in capturing long-term dependencies in sequences, so it has been widely used in these research fields.

When we apply LSTM to battery voltage prediction, the typical dataset for training the network is: $DS = \{(IU_1, OU_{1*}), (IU_2, OU_{2*}),\ldots, (IU_n, OU_{n*})\}$. In the formula, OU_k is the actual voltage value at time step k, IU_k is the input at the same time step. The network schematic of the LSTM battery voltage predictor is shown in Fig. 4.10. i_t, o_t, f_t and c^t are the input gate, output gate, forget gate and storage unit respectively. During the training process, the input, output and forget gates allow LSTM to forget or write new information into the storage unit.

The established battery voltage predictor LSTM network shows that unit blocks have replaced the standard neural network layers. At the end of each forward pass, the difference between the predicted voltage of the LSTM and the actual voltage can be calculated by the following loss function.

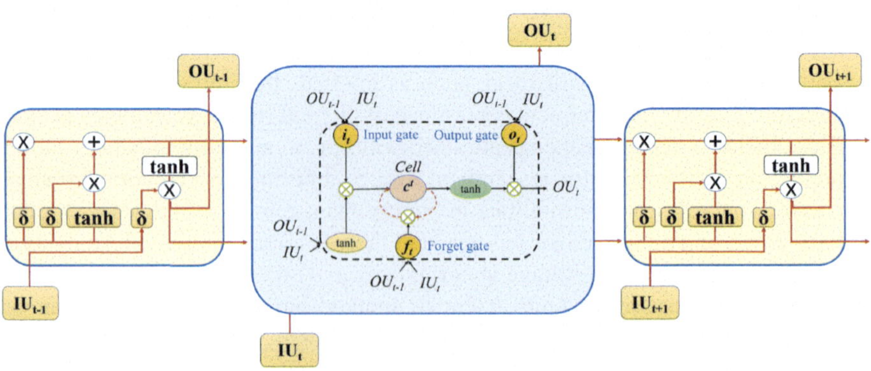

Fig. 4.10 Schematic of the LSTM battery voltage predictor network

$$Loss = \sum_{t=0}^{N} \frac{1}{2}\left(VOL_t - VOL_t^*\right)^2 \tag{4.3}$$

where N is the sequence length, VOL_t and $VOL_t{}^*$ are the predicted voltage value and actual voltage value at time step t.

When the training dataset is input into the LSTM network, the forward pass begins and ends when a voltage prediction is generated at each time step, while calculating the total loss. However, when configuring the learning process with Keras-LSTM, the optimizer is one of the most critical components in determining the compilation effect by updating the network's weights W and biases b. The most widely used optimizers include SGD, Adagrad, Adam, Adamelta, RMSprop, Nadam, Adamax, etc. Among them, the most commonly used optimizer is Adam. Therefore, we use Adam as the optimization method, updating W and b according to the gradient of the loss function. The network parameters (weights and biases) of Adam are updated as follows:

$$m_t = \mu * m_{t-1} + (1 - \mu) * g_t \tag{4.4}$$

$$q_t = v * q_{t-1} + (1 - v) * g_t^2 \tag{4.5}$$

$$\widehat{m}_t = \frac{m_t}{1 - \mu^t} \tag{4.6}$$

$$\widehat{q}_t = \frac{q_t}{1 - v^t} \tag{4.7}$$

$$W_{t+1} = W_t - \frac{\widehat{m}_t}{\sqrt{\widehat{q}_t + \epsilon}} * \alpha \tag{4.8}$$

where m_t and q_t are respectively the first and second-moment estimates of the gradient, which can be considered as the expected predictions of $E|g_t|$, $E|g_t{}^2|$; $\widehat{m}_t$ and $\widehat{q}_t$ are corrections of m_t and q_t respectively. It can be seen that this direct gradient estimation can be dynamically adjusted according to the gradient. Through $-\frac{\widehat{m}_t}{\sqrt{\widehat{q}_{t+\epsilon}}}$ a clear dynamic constraint on the learning rate can be formed. Through backpropagation, the network can self-learn its weights, greatly improving the traditional voltage prediction method, which requires manual construction on the battery model, which is time-consuming and laborious.

When using a finite dataset to optimise LSTM network learning, gradient descent will be an iterative process. Therefore, updating weights with a single pass or a cycle is insufficient. Network weights will change with the increase of cycles, and the prediction curve will go from underfitting to optimal and then to overfitting. Some literature has reported techniques related to preventing neural network overfitting. In previous techniques, L1 and L2 regularisation are the two most commonly used methods. However, both methods add extra terms to the cost function to regularise learning weights. L1 regularisation is the sum of all network weights, and L2 regularisation is the sum of squares of all weights. To solve this problem, Srivastava et al. proposed a more effective dropout technique, i.e., randomly dropping neurons from the network during training. But this dropout method also has a significant problem, that is, only the randomly retained neurons form the network; each processed network is less sensitive to the weights of specific neurons. On the contrary, this will lead to a weaker network to predict the output voltage.

In response to the above problems, this paper will propose a pre-dropout technique to prevent LSTM overfitting. By analysing the combination of weather-vehicle-driver, the most suitable parameters are specified as input. This technique will be detailed in the data preprocessing section. Figure 4.11 shows the neural network models with and without dropout. As shown in Fig. 4.11c, the pre-dropout technique proposed through clustering can not only effectively select beneficial parameters but also maximize the retention of neurons and connections beneficial to prediction results. Therefore, using this pre-dropout technique can achieve better training and prediction effects.

3. Voltage Prediction Based on LSTM Neural Network

   Based on the above introduction, the developed LSTM voltage prediction model mainly includes four parts: a many-to-one LSTM structure, a network parameter optimizer using the Adam method, a multi-input parameter (MIP) technique separating charging/driving states, and a pre-dropout technique to prevent LSTM overfitting. Compared with traditional parameter prediction methods and deep learning network structures, the LSTM model proposed in this paper has the following advantages:

   (1) Firstly, traditional deep learning network training algorithms based on batch gradient descent or stochastic gradient descent converge too slowly and cannot converge to the correct network weights, making them unsuitable for online prediction of battery system voltage. LSTM is a rare variant of

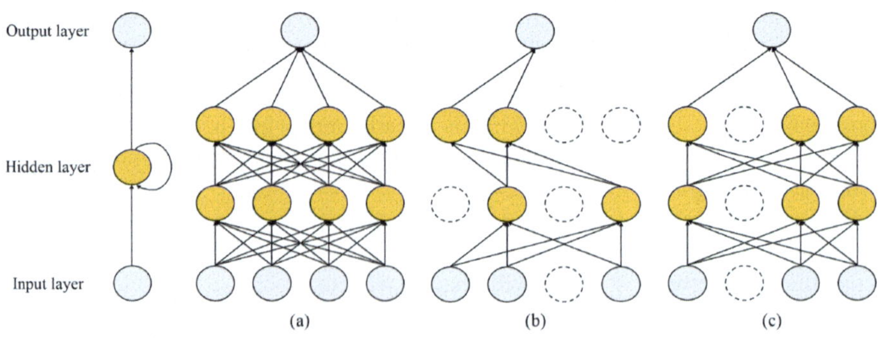

Fig. 4.11 Neural network models with and without dropout

RNN, inheriting most of the features of RNN models. It effectively solves the problem of gradient vanishing caused by the gradual decrease in the gradient backpropagation process; it has a more efficient processing effect than the standard RNN. Therefore, LSTM is very suitable for dealing with problems highly related to time series, such as the problem studied in this paper.

(2) Secondly, LSTM is a better RNN architecture for storing and accessing information than the standard RNN. LSTM has recently given state-of-the-art results in various sequence processing tasks. Therefore, it provides us with new ideas and opportunities to reveal the capabilities of LSTM, accurate voltage prediction, without using any battery models, filters, or inference systems.

(3) Thirdly, due to the inherent flaws of RNN in dealing with long-term memory, LSTM filters the previous states based on RNN. Unlike RNN, which only considers the most recent state, the cell state of LSTM determines which states should be retained and which should be forgotten. This allows for the selection of which states have a greater impact on the current situation, rather than simply choosing the most recent state. Therefore, LSTM has better long-range data processing effects, which is more conducive to research on parameter prediction based on time series.

(4) Fourthly, existing literature on LSTM-oriented parameter prediction mainly uses single parameters or multiple parameters for training, without conducting correlation analysis. Therefore, overfitting is an inevitable problem, leading to poor learning effects; the characteristics of the training parameters limit the prediction accuracy. This paper will introduce a many-to-one LSTM architecture, using pre-dropout technology to prevent LSTM overfitting, and combining correlation analysis between various parameters.

Typically, LSTM has five types of mappings, such as one-to-one, one-to-many, many-to-one, many-to-many (m–m), and many-to-many (m–n). In this plan, we will perform multi-step forward voltage prediction and various window size optimizations, so we use many-to-one LSTM to build our desired model. When the LSTM prediction model is completed, we can further study voltage anomaly prediction based on accurate prediction of future battery voltage. The flowchart of battery system voltage fault prediction based on LSTM is shown in Fig. 4.12.

4. Data Description and Preprocessing

In the actual operation of electric vehicles, due to the variability and unpredictability of unknown factors such as road conditions and driving behavior, the battery voltage will also experience severe random fluctuations. The vehicle data studied is collected from the new energy vehicle big data platform. The monitoring management mechanism of the new energy vehicle big data platform is shown in Fig. 4.13. The construction goal of this platform is to monitor and manage various types of public service vehicles in real-time, and to provide feedback to the vehicle or driver in a timely manner, enabling more efficient and

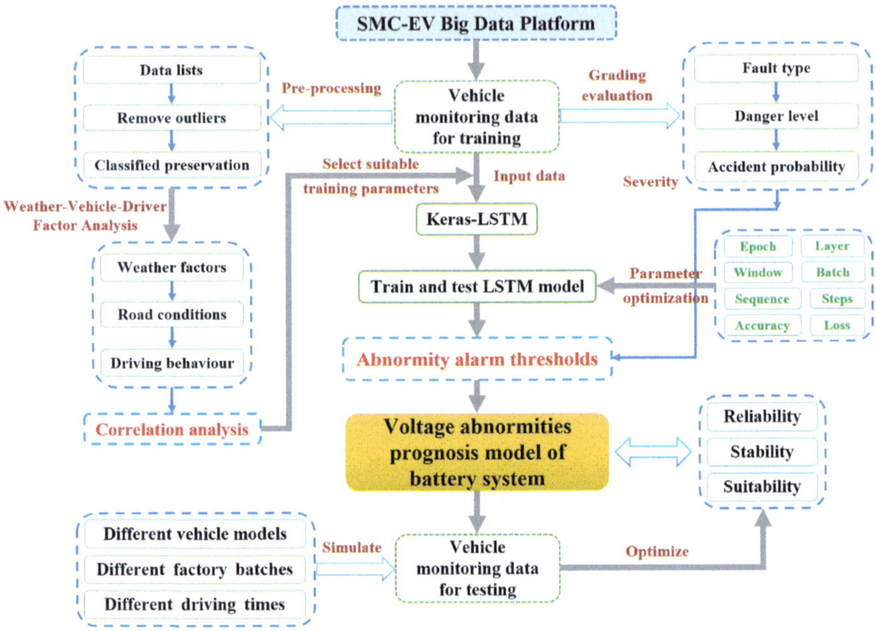

Fig. 4.12 Flowchart of battery system voltage fault prediction based on LSTM

safer operation; it provides information technology support and data services for electric vehicle enterprises and government departments. The center was recently upgraded to the National Electric Vehicle Monitoring Management Centre (China), which provides round-the-clock monitoring services for about 1 million electric vehicles of various types, mainly including taxis, buses, sanitation vehicles and other public vehicles. Therefore, more vehicle operation data will be collected, providing a broader application prospect for machine learning technologies such as the LSTM network proposed in this paper.

Fig. 4.13 Monitoring management mechanism of the new energy vehicle big data platform

Table 4.3 Voltage anomaly and alarm threshold

Abnormalities and alarm categories	Classification rule		Anomaly and alarm levels
Over-voltage or under-voltage of a single battery	Overtension	3.6n≤Upack3.9n	2
		Upack≥3.9n	1
	Undertension	2.5n< Upack≤3.0n	3
		2.0n < Upack≤2.5n	2
		Upack≤2.0n	1
Rapid total voltage variation	Overtension	3.6n≤Ucell< 3.9n	2
		Ucell≥3.9n	1
	Undertension	2.5 < Ucell≤3.0	3
		2.0< Ucell≤2.5	2
		Ucell≤2.0	1
Rapid cell voltage variation	Steep voltage rise	ΔUpack≥0.4n/10	1
	Steep voltage decline	ΔUpack≤–0.4n/10	1
	Steep voltage rise	ΔUcell≥0.4n/10	1
	Steep voltage decline	ΔUcell≤–0.4n/10	1

The vehicle studied in this work is an electric taxi assembled with an 18,650 lithium-ion battery pack. The main parameters of the studied vehicle are shown in Table 4.3. The battery pack consists of 100 individual batteries, with a total energy of 29.1 kWh and a rated voltage of 326 V. The data sampling interval of this vehicle is 10 s. The layout of the battery pack of the studied vehicle is shown in Fig. 4.14. From the figure, we can see that 80 individual batteries (from battery 1 to battery 80) are arranged horizontally; the remaining 20 individual batteries (81–100) are arranged vertically. For example, battery 1 is at the top, battery 4 is at the bottom; batteries 81–100 are arranged from back to front in sequence.

However, with 100 individual cells, training and building 100 models would take a significant amount of time; and in future real-time voltage predictions, invoking too many models at once will greatly reduce prediction efficiency. To address this issue, this paper connects the voltage data of 100 batteries in sequence to form a new set of battery voltage data; then trains these latest voltage data as inputs to obtain a universal battery voltage prediction model. This approach has two advantages: firstly, it simplifies the final voltage model into a module model and a battery model; secondly, the integrated battery model contains the voltage information of all individual cells. Therefore, a good model can achieve the effect of synchronized prediction of all battery voltages.

Because meteorological factors, vehicle factors, and driver behavior will not affect the voltage characteristics or fluctuations of the battery system when the vehicle is driving or charging. For a more comprehensive analysis, a weather-vehicle-driver analysis needs to be conducted when preprocessing the input dataset, thus considering all three factors simultaneously for battery voltage prediction.

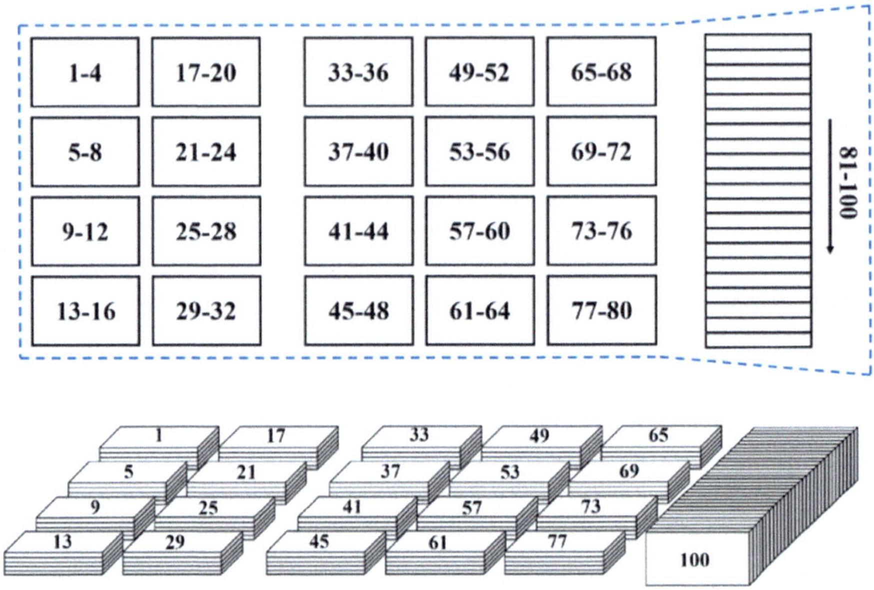

Fig. 4.14 Layout of the battery pack of the studied vehicle

Considering the meteorological factors that may affect battery voltage, six weather parameters were added, collected, and analyzed from the weather website (https://www.wunderground.com); they are humidity, precipitation, atmospheric pressure, air temperature, visibility, and wind speed. However, the sampling interval of historical meteorological data is 3 h, which is too long and inconsistent with the 10 s sampling interval of the new energy vehicle big data platform. To solve this problem, we use Lagrange interpolation to interpolate the weather parameters every 10 s. Then, the interpolated weather data is merged with the vehicle data according to the time alignment.

When it comes to batteries and vehicles, factors related to battery voltage, probe temperature, SOC, current, and remaining useful life (RUL) are closely related. However, it is difficult to monitor or predict the RUL value of the battery in real time accurately. There is currently no very effective method. It is well known that the RUL of a battery is negatively correlated with the total mileage of the vehicle, i.e., the longer the total mileage, the shorter the RUL of the battery. Therefore, in this work, the total mileage of the vehicle available from the new energy vehicle big data platform will be used to represent the RUL of the battery used for preprocessing data.

Undoubtedly, the impact of driver driving behavior on electric vehicles and power batteries can be negligible, but current research in this field mainly focuses on the impact of driving behavior on vehicle economy; there is currently no relevant research considering the impact of driving behavior on battery safety and durability. Due to the data advantage of the new energy vehicle big data platform, this paper will consider the relevant parameters of driver driving behaviour when preprocessing data. Parameters reflecting driver driving behavior include brake pedal stroke value, vehicle speed, and vehicle acceleration. However, the new energy vehicle big data platform cannot record acceleration in real-time. In addition, due to the long sampling interval of 10 s, it is impossible to estimate acceleration by deriving speed. Therefore, to obtain acceleration data, some other methods need to be considered to calculate or represent acceleration. The relationship between acceleration and motor speed, rated power, wheel radius, and other parameters is shown in Eq. (4.9).

$$\begin{cases} T_{motor} = \frac{9550*P}{n} \\ T_{wheel} = T_{motor} * i \\ a = \frac{F}{m} = \frac{T_{wheel}}{R*m} \end{cases} \Rightarrow a = \frac{9550 * P * i}{R * m * n} \tag{4.9}$$

In the above formula, T_{motor} represents motor torque, P represents motor rated power, n represents motor speed, T_{wheel} represents wheel torque, i represents the transmission efficiency of the motor to the wheel, F is the vehicle driving force, m is the vehicle's curb weight, R is the tyre radius. From Eqs. (4.9), it can be seen that except for n and a, all other parameters are known constants, from which it can be deduced that acceleration is inversely proportional to motor speed. But because the transmission efficiency i and other various parameters inevitably have errors, there will inevitably be errors in the process of calculating acceleration. Therefore, to avoid the occurrence and impact of errors, this study will directly represent acceleration a with motor speed n.

After the above weather-vehicle-driver analysis, the final selection of battery voltage, probe temperature, brake pedal stroke value, motor speed, vehicle speed, battery pack voltage, current, SOC, humidity, precipitation, atmospheric pressure, air temperature, visibility, wind speed, mileage and other 15 parameters as the training parameters for the LSTM model.

Taking a day's vehicle operation data as an example, some typical parameter curves for February 8, 2017, are shown in Fig. 4.15. As can be seen from the enlarged view of Fig. 4.15, the voltage and current curves are inversely proportional. When the current increases, the driving power provided by the motor also increases, so the speed of the vehicle also increases. In addition, when charging is about to be completed, the battery pack voltage will slightly decrease, and the current will change from constant negative (cross-constant current charging) to zero. However, it is difficult to see the precise relationship between each parameter and the battery voltage through these curves. Furthermore, some parameters have too strong a relationship with the voltage (such as current), or some parameters have too weak a relationship with the voltage (such as various weather parameters), making it unsuitable to train all parameters together.

However, when we first trained the LSTM model with all 15 parameters as inputs, the results showed that no matter which of these 15 parameters was the output, the training would quickly stop due to overfitting. Analysis suggests that this is because there is still some particularly strong or weak correlation between these parameters,

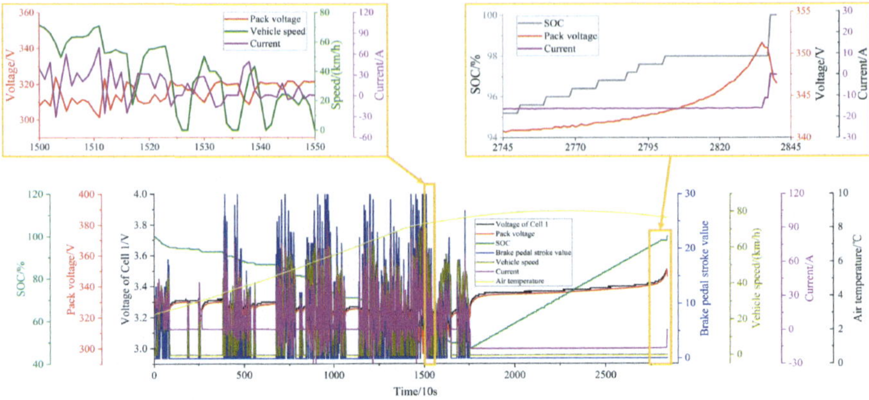

Fig. 4.15 Typical parameter curves for part of February 8, 2017

which can interfere with each other during training; this can also lead to overfitting before training begins, making the final training effect worse. Therefore, it is not feasible to use all parameters as inputs, and it is very necessary to perform correlation analysis on all input parameters before training.

In order to perform correlation analysis, select suitable input parameters, and eliminate harmful parameters, this paper uses the Pearson Correlation Coefficient (PCC), which is widely used in scientific research. The PCC was developed by Karl Pearson in the 1880s based on the concept of correlation proposed by Francis Galton. The phase change curves between the selected 15 parameters are shown in Fig. 4.16, where all phase change curves are taken in absolute value for simplified analysis.

The absolute value of the PCC between two parameters can be used to judge the correlation between the two parameters: 0.8–1.0 is extremely strong correlation, 0.6–0.8 is strong correlation, 0.4–0.6 is moderate correlation, 0.2–0.4 is weak correlation, 0–0.2 is extremely weak or no correlation. Figure 4.16 shows the PCC list between all

No correlation	Weak correlation	Moderate correlation	Strong correlation	Extremely strong correlation

	Air temperature	Barometric pressure	Brake pedal stroke value	Current	Humidity	Mileage	Motor speed	Pack voltage	Precipitation	SOC	Temperature of probe 1	Vehicle speed	Visibility	Voltage of Cell 1	Wind speed
Air temperature	1	0.8757	0.0355	0.0198	0.1247	0.1415	0.0382	0.0219	0.0534	0.0674	0.9275	0.0377	0.0387	0.04	0.0129
Barometric pressure	0.8757	1	0.0421	0.0023	0.2063	0.0751	0.024	0.0504	0.1142	0.0849	0.838	0.0237	0.1058	0.067	0.0132
Brake pedal stroke value	0.0355	0.0421	1	0.8364	0.0372	0.0073	0.6815	0.6377	0.0105	0.0689	0.0775	0.6683	0.0141	0.6228	0.0137
Current	0.0198	0.0023	0.8364	1	0.031	0.0047	0.6143	0.8054	0.0009	0.0039	0.0613	0.6018	0.0041	0.7931	0.0099
Humidity	0.1247	0.2063	0.0372	0.031	1	0.4116	0.0367	0.0521	0.2822	0.035	0.19	0.0369	0.3083	0.0571	0.4404
Mileage	0.1415	0.0751	0.0073	0.0047	0.4116	1	0.0003	0.0202	0.0421	0.0185	0.1026	0.0001	0.0432	0.0146	0.3136
Motor speed	0.0382	0.024	0.6815	0.6143	0.0367	0.0003	1	0.6185	0.0088	0.0572	0.0153	0.999	0.0114	0.6226	0.0202
Pack voltage	0.0219	0.0504	0.6377	0.8054	0.0521	0.0202	0.6185	1	0.0065	0.3405	0.1428	0.6127	0.0059	0.9942	0.0185
Precipitation	0.0534	0.1142	0.0105	0.0009	0.2822	0.0421	0.0088	0.0065	1	0.0156	0.0466	0.0091	0.6157	0.0079	0.0104
SOC	0.0674	0.0849	0.0689	0.0039	0.035	0.0185	0.0572	0.3405	0.0156	1	0.0645	0.0574	0.0059	0.3453	0.0285
Temperature of probe 1	0.9275	0.838	0.0775	0.0613	0.19	0.1026	0.0153	0.1428	0.0466	0.0645	1	0.0154	0.0428	0.1622	0.0136
Vehicle speed	0.0377	0.0237	0.6683	0.6018	0.0369	0.0001	0.999	0.6127	0.0091	0.0574	0.0154	1	0.0117	0.6172	0.0202
Visibility	0.0387	0.1058	0.0141	0.0041	0.3083	0.0432	0.0114	0.0059	0.6157	0.0059	0.0428	0.0117	1	0.007	0.0302
Voltage of Cell 1	0.04	0.067	0.6228	0.7931	0.0571	0.0146	0.6226	0.9942	0.0079	0.3453	0.1622	0.6172	0.007	1	0.0201
Wind speed	0.0129	0.0132	0.0137	0.0099	0.4404	0.3136	0.0202	0.0185	0.0104	0.0285	0.0136	0.0202	0.0302	0.0201	1

Fig. 4.16 Values of correlation coefficients between all potential input parameters

considered parameters, where the voltage of Battery 1 represents the battery voltage, and the temperature of Probe 1 represents the probe temperature. By eliminating irrelevant and extremely strongly correlated parameters, parameters that meet the training requirements can be obtained. As can be seen from Fig. 4.16, the parameters matching the battery voltage are brake pedal stroke value, current, motor speed, SOC and vehicle speed; the parameters matching SOC are pack voltage and battery voltage; no parameters can match the probe temperature. Finally, based on the analysis of the correlation between all existing matching parameters, only 4 input parameters can be used to train the LSTM model, namely battery voltage/battery pack voltage, SOC, brake pedal stroke value and vehicle speed.

The LSTM models for battery voltage and battery pack voltage can be trained and established separately. The battery voltage curve for 2016 can be obtained, as shown in Fig. 4.17. As shown in Fig. 4.17, the battery voltage/battery pack voltage changes with the seasons; the environmental temperature in spring and summer is higher, closer to the suitable working temperature of the battery system, so the voltage working range in these three seasons is much smaller than in autumn and winter; in late autumn and winter, due to the lower outdoor temperature, electric vehicles

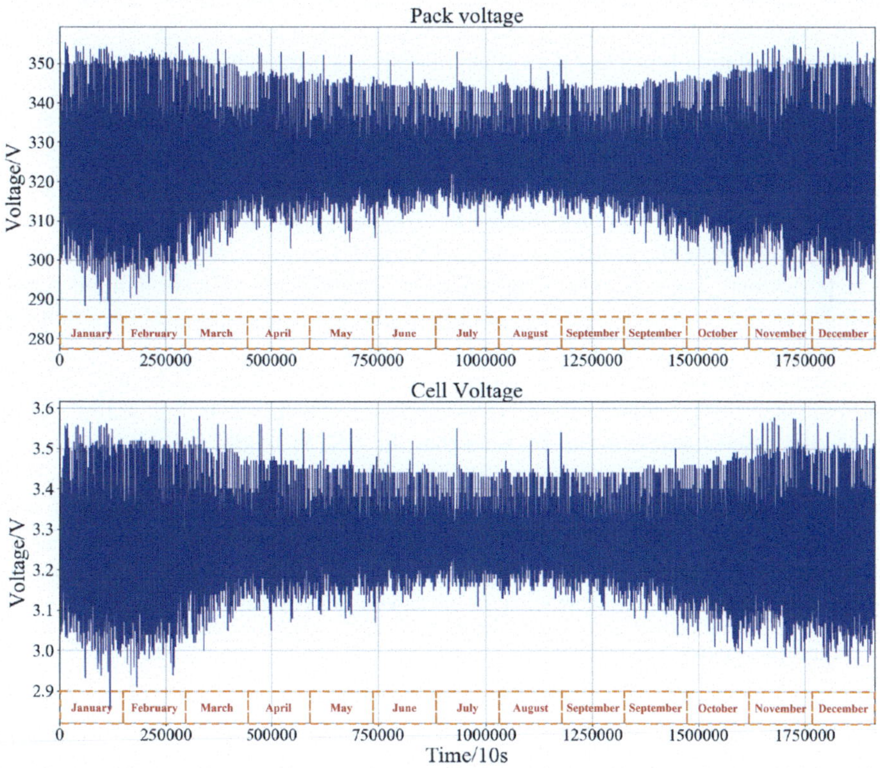

Fig. 4.17 Battery voltage curve for 2016

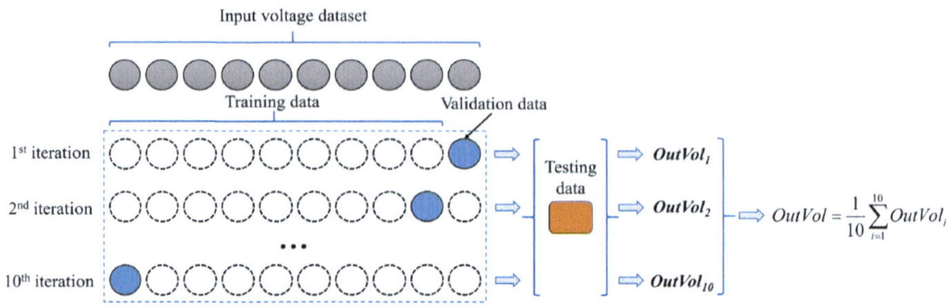

Fig. 4.18 Schematic diagram of tenfold cross-validation

need the power battery to provide higher power and greater energy consumption. In order to analyze the voltage prediction effect and robustness in different seasons by training these input parameters, we will specifically discuss the predicted voltage characteristics in different seasons in the prediction results and discussion section.

Datasets in machine learning algorithms are usually divided into three categories: training data, validation data, and test data. Training data is used to calculate the gradient update weights; validation data can determine some hyperparameters to avoid overfitting during training (for example, determining the size of the epoch based on validation loss); test data after training can accurately evaluate the quality of training.

As mentioned earlier, the vehicle operation data for one month in February 2016 was determined as the initial IS. In order to conveniently compare the effects after training with different ISs (1 year of data, 1 month of data, 1 week of data, 1 day of data), the data for one day on February 8, 2017, was used as the test data. In order to verify and ensure the stability of the LSTM model, tenfold cross-validation needs to be performed. The schematic diagram of the tenfold cross-validation in this paper is shown in Fig. 4.18. The training data and validation data respectively account for 90% and 10% of the entire IS.

Based on the pre-set hyperparameters, the prediction results of battery voltage and package voltage after tenfold cross-validation are shown in Fig. 4.19a and b. From these two figures, it can be seen that the LSTM model has achieved good prediction results, with the predicted voltage values being very close to the actual voltage values. The MREs of the battery voltage and package voltage are 0.88% and 0.89% respectively. In addition, it can be seen from these two figures that the vehicle under study only drove and charged once on 8th February 2017, and the voltage curve shows different characteristics under driving and charging conditions.

5. Hyperparameter Selection

   During the training process of the LSTM model, many parameters need to be set and optimized; training these parameters will take a long time, especially when the model has a large dataset. However, it is impossible to optimize the parameters in a short time. Therefore, a set of operable hyperparameters needs to be established empirically first, and these parameters are then optimized step by step to obtain the optimized model.

In fact, in the machine learning process, when the data is too large, it needs to be broken down into smaller sizes, and terms such as epochs and batch size (BS) are required parameters. An epoch means that the entire dataset is passed forward and backward through the LSTM neural network once; BS represents the number of training examples in a single batch.

Based on the research objective of this paper, a set of hyperparameters was initially determined: initial epochs = 50, BS = 128, IS = 1 month of data, TWS = 360 (1 h of data), PWS = 6 (1 min of data), SWS = 60 (10 min of data). The LSTM network constructed in this paper consists of 6 layers: 3 LSTM hidden layers, each with 100 neurons, 1 input layer, 1 densely hidden layer with linear activation, and 1 output layer.

6. Confirming Hyperparameters

   BS is an important parameter in machine learning. On the one hand, for large datasets like the one in this study, it is not feasible to load all the data at once due to the large amount of data and the memory limitations of old laptops. On the other hand, if BS is blindly increased, the computer's memory will not support it, and the time spent on running epochs will greatly increase. Therefore, the number of BS should be determined based on the prediction efficiency of the LSTM model. In this study, the research range of BS is {8, 16, 32, 64, 128, 256, 512, 1024}. To verify the prediction accuracy of battery voltage and package voltage, the MRE of test data was calculated according to different BS, as shown in Fig. 4.20. As can be seen from Fig. 4.20, when BS = 16, the MRE is the lowest and the test effect is the best. Therefore, BS = 16 is used to predict the battery voltage of the battery system under study.

   Based on the initial determination of hyperparameters, the method of tenfold cross-validation was used for manual optimization to determine TWS, PWS, and SWS. On the basis of the original hyperparameters, keeping other parameters unchanged, the sizes of TWS, PWS and SWS were adjusted respectively, and the MRE was compared; window sizes = {6, 30, 60, 90, 120, 150, 180, 360} represent data volumes of 1 min, 5 min, 10 min, 15 min, 20 min, 25 min, 30 min and 1 h respectively, and the MRE of test data for different windows and window sizes is shown in Fig. 4.21. Since the initial TWS = 60, the smallest TWS is TWS = 60.

   As can be seen from Fig. 4.21, with the change of TWS, there is no obvious increase or decrease in MRE. As long as TWS > PWS, i.e., the amount of data predicted each time is less than the amount of data trained, TWS can make objective predictions. In addition, TWS determines the amount of data trained each time, so TWS has a decisive influence on the adjustment range of SWS and PWS; for more different predictions in the future, TWS can be a manageable size. Therefore, TWS = 360 is determined.

   By applying SW technology, each time SW is refreshed, the training window can be iteratively trained, allowing the LSTM network to learn more historical voltage information. As can be seen from Fig. 4.21, as SWS increases, the MRE of battery voltage and package voltage gradually increases. This is because a larger SWS will result in less available historical information for each training session. In principle, when SWS = 1, all historical information is available. However, a smaller SWS means a longer training time for each session. As can be seen from Fig. 4.21, the MRE at SWS = 6 is very close to the MRE at SWS = 1. Therefore, this study determines SWS = 6.

As can be seen from Fig. 4.21, as PWS increases, the prediction accuracy of the test data decreases; the MRE at PWS = 1 is greater than the MRE at PWS = The MRE of 6. It can be inferred that from a constant training sample, a longer prediction horizon will suffer more significant learning difficulties. In general, the smallest PWS should be chosen to obtain the best prediction accuracy. However, when the prediction range is too small, once the battery system experiences a voltage abnormality, the available time for the driver to take action will be too short. Therefore, the PWS should be determined jointly based on the requirements for prediction accuracy and available emergency repair time. To achieve higher prediction accuracy, determine PWS = 6.

Based on the optimized hyperparameters, the prediction results for battery voltage and battery pack voltage are shown in Fig. 4.22a and b. The MREs for battery voltage and battery pack voltage are 0.64% and 0.63% respectively. Compared with the prediction results before optimization in Fig. 4.19, it can be seen that the LSTM model's prediction results are better, and the predicted voltage values are closer to the actual voltage values.

7. The Effect of Pre-Dropout in Preventing Overfitting

When training is excessive or there is insufficient training data, the LSTM model's training process is prone to overfitting. The results before and after using the improved pre-dropout method for hyperparameter optimization to prevent overfitting are shown in Fig. 4.23. From Fig. 4.23, it can be seen that the pre-dropout technique effectively prevents overfitting during training. The intuitive manifestation of overfitting can also be seen from the figure: as training progresses, the loss of the training data gradually decreases, but when the validation data cannot cooperate with the training network, the loss of the validation data will increase.

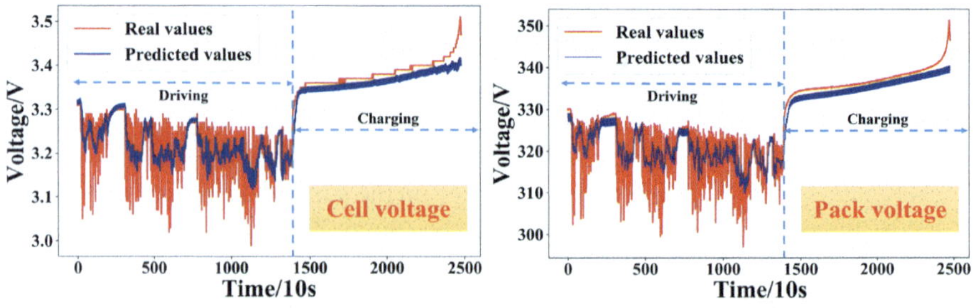

Fig. 4.19 Prediction results of battery voltage and package voltage after tenfold cross-validation

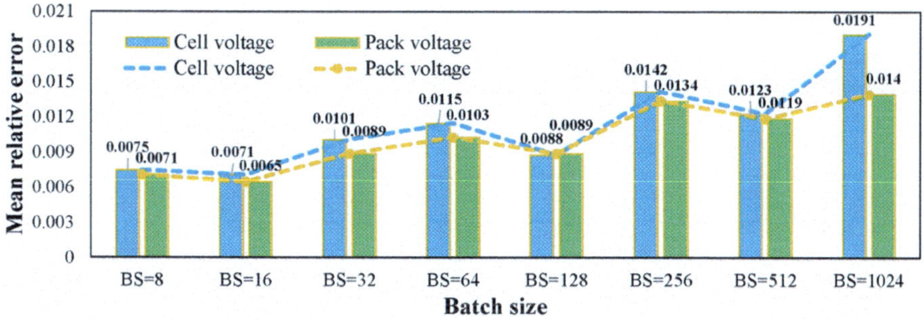

Fig. 4.20 MRE of test data calculated with different BS

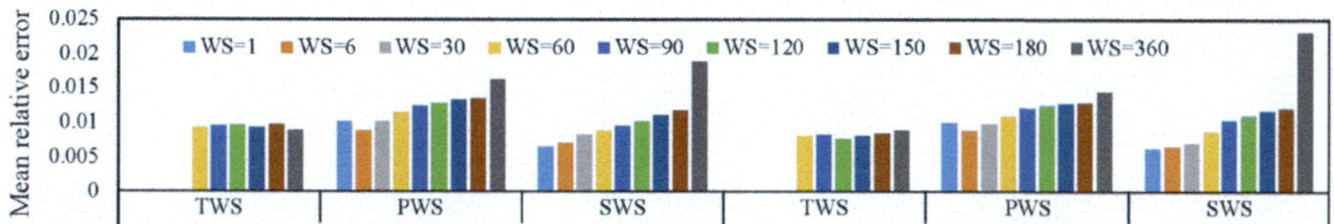

	TWS	PWS	SWS	TWS	PWS	SWS
	Cell voltage			Pack voltage		
WS=1		0.0102	0.0065		0.01	0.0063
WS=6		0.0088	0.0071		0.0089	0.0066
WS=30		0.0102	0.0082		0.0098	0.0071
WS=60	0.0093	0.0115	0.0088	0.0081	0.011	0.0089
WS=90	0.0095	0.0124	0.0095	0.0083	0.0121	0.0104
WS=120	0.0096	0.0128	0.0102	0.0077	0.0125	0.0111
WS=150	0.0093	0.0133	0.0111	0.0081	0.0128	0.0118
WS=180	0.0097	0.0135	0.0118	0.0085	0.0129	0.0122
WS=360	0.0088	0.0162	0.0189	0.0089	0.0145	0.0232

Fig. 4.21 MRE of test data for different windows and window sizes

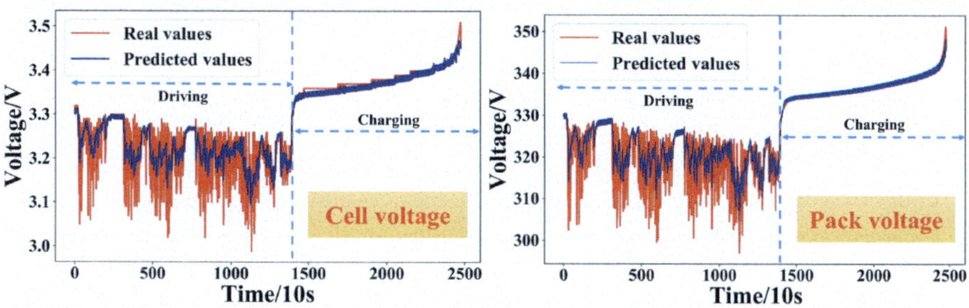

Fig. 4.22 Prediction results for battery voltage and battery pack voltage based on optimized hyperparameters

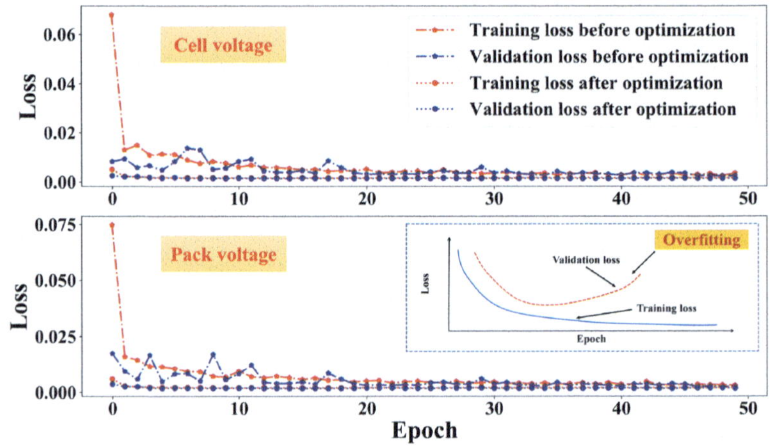

Fig. 4.23 Results before and after using the improved pre-dropout method for hyperparameter optimization to prevent overfitting

In addition, Fig. 4.23 also shows that the number of initial periods is sufficient to minimize and stabilize the model loss.

8. Dual Model Cooperation Prediction Strategy

As can be seen from Fig. 4.20, the vehicle under study exhibits different voltage characteristics during the charging and driving stages. There could be three reasons for this: firstly, as the prediction progresses, the timeliness of the model becomes worse. Secondly, the battery parameters themselves have different characteristics during the charging and driving stages, causing this phenomenon. Thirdly, some input parameters such as brake pedal stroke value and vehicle speed only function during driving and are zero during charging. Therefore, training the LSTM model with four input parameters during charging may have a negative impact on the prediction results.

However, the first two reasons are largely irreversible. The third reason is due to the different characteristics of various parameters of the battery system under driving/charging conditions, so the models and voltage predictions under these two conditions also need to be distinguished. To solve this problem and eliminate the influence of irrelevant parameters during model training, this paper proposes a Dual Model Cooperation (DMC) prediction strategy; during charging, the input parameters are reduced from four to two (voltage and SOC). Through multiple tests and validations of these two models, the DMC prediction strategy intelligently applies two LSTM models under driving/charging conditions, which can not only reduce the computational load but also maximize the prediction accuracy.

9. Verification of Effectiveness, Stability, and Robustness

Based on the determined hyperparameters above, we will compare the prediction effects of 4 different ISs (1 year of data in 2016, 1 month of data in February 2016, 1 week of data in the first week of February 2017, and 1 day of data on February 7, 2017), with the test data being 1 day of data on February 8, 2017. Figure 4.24 shows the MRE comparison chart after tenfold cross-validation of different IS test data. In Fig. 4.24, the smaller the IS, the more obvious the increasing trend; there is a significant difference in MRE between different ISs, but the difference between IS = 1 year of data and IS = 1 month of data is small. As mentioned earlier, the initial data for IS is February 2016. In principle, the larger the IS, the better the model prediction effect. However, as used in this paper, for calculating such a large dataset in a short time, computer performance is a limited configuration. But for a new energy vehicle big data platform, a well-configured GPU server can build an ideal LSTM model. In addition, the technology proposed in this paper achieves online voltage prediction through offline training of the model; the length of the model training time will not cause any interference to the prediction speed and accuracy. Therefore, the larger the IS, the stronger the model's learning and prediction capabilities, and this paper is more suitable for using 1 year of data as the IS.

In summary of the above analysis and conclusions, a set of hyperparameters can finally be determined as: IS = one year of data, TWS = 360, PWS = 6, SWS = 6, epochs = 50, BS = 16.

Within the LSTM architecture, there are five types of model architectures. A many-to-one (4.1) model architecture is used for multi-step forward voltage prediction of the battery system. To verify the superiority of the applied model structure, the MRE of the one-to-one and many-to-one (4.1) model structures were compared. The one-to-one LSTM model has only one input and one output, that is, the LSTM model only learns information about the target parameter itself. Under the premise of the same hyperparameters, the MREs of the battery voltage and battery pack voltage of the one-to-one LSTM model are 0.81% and 0.79% respectively, both higher than the 0.58% and 0.56% of the many-to-one (4.1) model. Therefore, the many-to-one (4.1) model has better prediction performance.

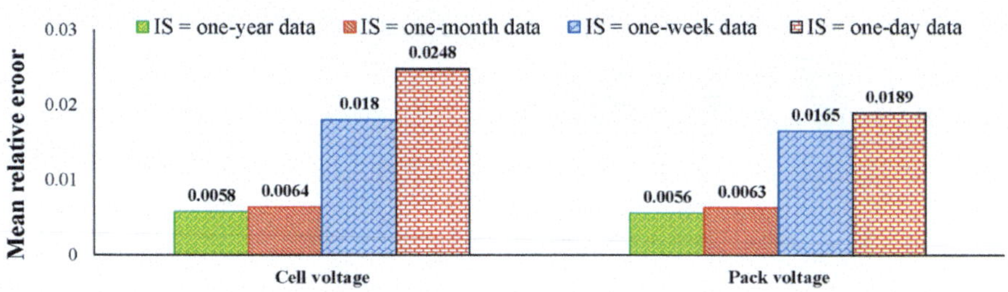

Fig. 4.24 MRE comparison chart after tenfold cross-validation of different IS test data

In fact, for the same raw material, the same batch of single cells, performance parameters such as open circuit voltage, capacity, internal resistance, and cycle life all follow a normal distribution with finite dispersion. That is to say, all our efforts to improve consistency are to change the aggregation of these parameters, which can be represented by the standard deviation (SD). In the normal distribution of battery voltage, the number of batteries with voltage values within 3σ of the average voltage accounts for 90% of the total number of batteries. The better the consistency of the battery voltage, the smaller the SD, and the more concentrated the voltage distribution. This paper evaluates the level of battery voltage consistency with a threshold of 0.05; when the SD of each battery voltage reaches >0.05, it indicates that the consistency of the current battery voltage is very poor, and vice versa. This can also be explained as, batteries within a range of 0.15 V of the average voltage need to account for more than 99% of all batteries.

The voltage curve and SD of 100 battery cells on the day of test data collection are shown in Fig. 4.25. As can be seen from Fig. 4.25, the consistency of each cell is good, with SD not exceeding 0.05. In addition, it can be seen from the figure that these batteries have much worse consistency during the driving phase than during the charging phase; especially during the stable charging phase, all battery voltages are almost the same, and only when close to full charge does the SD increase. This is easy to understand, as the vehicle is always in a non-complex, stable operating state during the charging phase, so the battery system operates relatively stably and evenly; during the driving process, due to the uncertainty of road conditions and weather, as well as the complexity of driver behavior, the operation of the battery system is quite complex, and its consistency is difficult to guarantee.

In order to verify the effectiveness and stability of the established LSTM battery model in predicting all battery voltages, we used the same test data to test the other 99 batteries separately, obtaining the MRE of all 100 batteries, as shown in Fig. 4.26. As can be seen from Fig. 4.26, although the same model was used, the prediction accuracy of all battery voltages is very small, the values are very close, and the SD is only about 0.0001. Therefore, the LSTM battery model

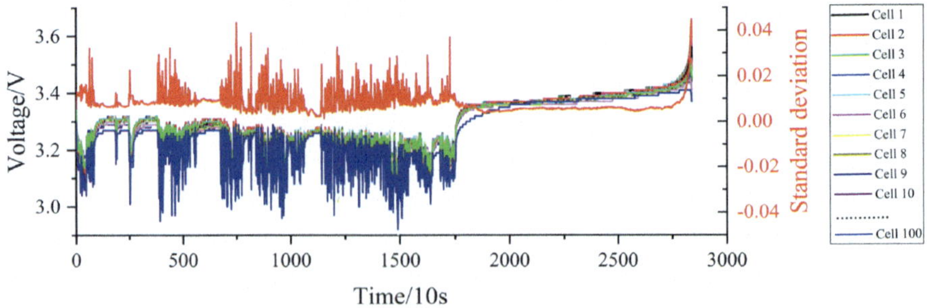

Fig. 4.25 Voltage curve and SD of 100 cells on the day of test data collection

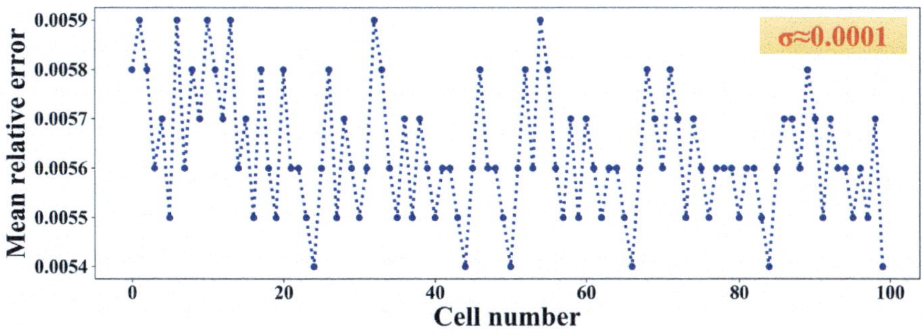

Fig. 4.26 MRE of all 100 batteries

proposed in this paper can meet the purpose of predicting all battery voltages while ensuring high prediction accuracy and stability.

In order to verify the robustness of the LSTM model, we used the data from April 1, July 1, October 1, and January 1, 2017 as test data for the four seasons. The voltage curves before and after prediction for the four seasons are shown in Fig. 4.27, with MREs of 0.37%, 0.31%, 0.42%, and 0.62% respectively. From this chart, we can see that the LSTM model proposed in this paper has good predictive performance in all four seasons, especially in spring, summer, and autumn, which is much better than the test data we previously selected. As is well known, the external environment varies greatly in different seasons, and batteries, due to their performance defects, will exhibit different operating characteristics in the four seasons; due to the low outdoor temperature in Beijing in winter, which is far from the suitable working temperature range of the battery system, the voltage fluctuation during vehicle operation is more severe, with a range of fluctuation much greater than the other three seasons. Therefore, the battery voltage in winter is more difficult to predict than in the other three seasons. In summary, the LSTM model proposed in this paper has strong robustness and all-weather applicability. If the training sample is large enough and the training parameters are continuously optimized, more accurate prediction results can be obtained.

10. Voltage Abnormality Prediction

The threshold and warning threshold for voltage abnormality alarms in the battery system under study are shown in Table 4.4. It should be emphasised that this method is also applicable to other battery systems; however, due to differences in battery types and parameters, the alarm thresholds may vary slightly.

As can be seen from the table, voltage abnormalities are divided into three levels by setting thresholds, namely level one alarm, level two warning, and level three warning. The third level warning status is also a safe status, but it is necessary to prevent the battery voltage from deteriorating; the second level warning status indicates that the battery is in a dangerous state, and it is required

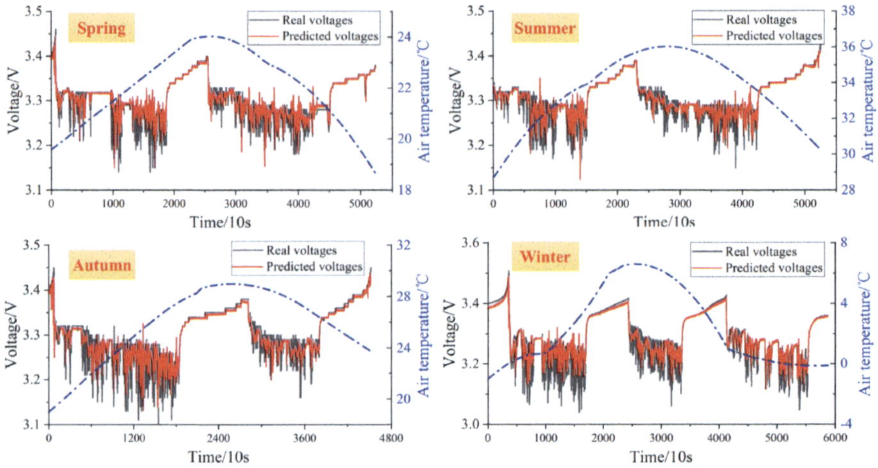

Fig. 4.27 Voltage curves before and after prediction for the four seasons

Table 4.4 Unimportant parameter coefficients

Coefficient	Probe temperature	Brake pedal travel value	Motor speed	Vehicle speed	Air pressure	Air temperature
SOC	0.0645	0.0689	0.0572	0.0574	0.0849	0.0674
Probe temperature	1	0.0775	0.0153	0.0154	0.838	0.9275
Brake pedal travel value	0.0775	1	0.6815	0.6683	0.0421	0.0355
Motor speed	0.0153	0.6815	1	0.999	0.024	0.0382
Vehicle speed	0.0154	0.6683	0.999	1	0.0237	0.0377
Air pressure	0.838	0.0421	0.024	0.0237	1	0.8757
Air temperature	0.9275	0.0355	0.0382	0.0377	0.8757	1

to stop the car for inspection under permissible conditions; when a level one alarm is triggered, the driver needs to stop or wait for professionals to arrive at a safe place to avoid accidents that may harm the driver and passengers.

As mentioned in the introduction, common voltage abnormalities can be summarised as overvoltage, undervoltage, overvoltage, and poor battery voltage consistency. To verify the effectiveness of the battery system voltage abnormality prediction, voltage abnormality test data of the research vehicle was retrieved from the new energy vehicle big data platform. According to the finally selected hyperparameters, the voltage curve of the research vehicle before and after the prediction on September 15, 2017, is shown in Fig. 4.28. As can be seen from Fig. 4.28, the vehicle started working at 09:27:10 in the morning and stopped working at 19:29:30 in the evening. Figure 4.28 also shows that the battery system only performed one drive and one charge on this day. Battery 100 had an undervoltage warning of $U_{\text{battery}} < 3.0$ V at 15:56:20 in the early morning; Battery 19 had an overvoltage warning of $U_{\text{battery}} > 3.6$ V at 19:28:10 in the evening. Since the vehicle started running that day, Battery 100 showed different fluctuation characteristics from other batteries, became an outlier, and maintained a lower voltage than other batteries. In addition, in the short time when the battery system was almost fully charged, all battery voltages rose rapidly, but only Battery 19 had overvoltage at the end of charging, which may be due to the inconsistency between different batteries causing the smaller capacity Battery 19 to be overcharged.

As shown in Fig. 4.28, because TWS = 360, only the data of the first hour of the day can be used as the initial WS during the test, and all subsequent data can be gradually predicted according to the set SWS and PWS. By partially magnifying the overvoltage and undervoltage areas, we can see that all voltages, including the encapsulated voltage, and the abnormal Batteries 19 and 100 can be accurately predicted. This figure also shows that the predicted voltage is very close to the actual voltage. Theoretically, the more stable the single battery voltage characteristics, the higher the prediction accuracy. This also ensures that

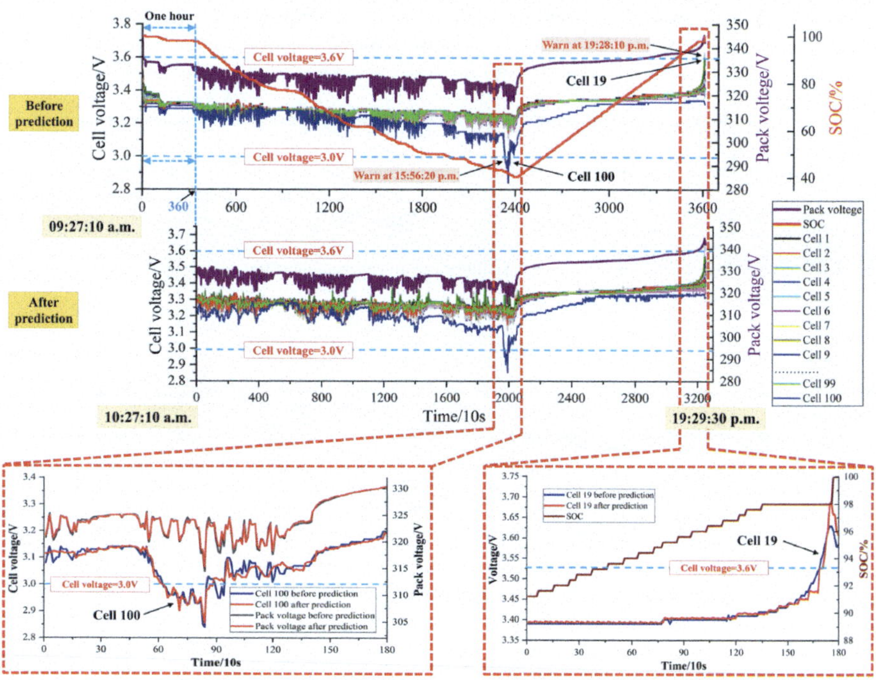

Fig. 4.28 Voltage curve of the studied power battery system before and after prediction

the proposed method has good feasibility and reliability, as it can accurately predict abnormal voltages, and normal battery voltages will not be misreported.

Figure 4.29 shows the VCR curve before and after prediction. Since there has been no VCR abnormality in the current battery system, the VCR values are generally very small ($\Delta U_{\text{battery}}$ always satisfies $|\Delta U_{\text{battery}}| \leq 0.4$ V/10 s), the predicted value of VCR is slightly different from the real value, but it is close at the position where the VCR value is larger, ensuring that once a super large VCR occurs, it can be predicted in time.

Figure 4.30 shows the SD curve before and after prediction, indicating that the SD is greater than 0.05 at the end of charging. As can be seen from Fig. 4.30, the predicted value of SD is very close to the real value, and the high consistency difference of battery voltage SD can be predicted accurately and timely. At the same time, by real-time monitoring of battery voltage, we can find the battery that causes this consistency difference, such as Battery 19 in this study.

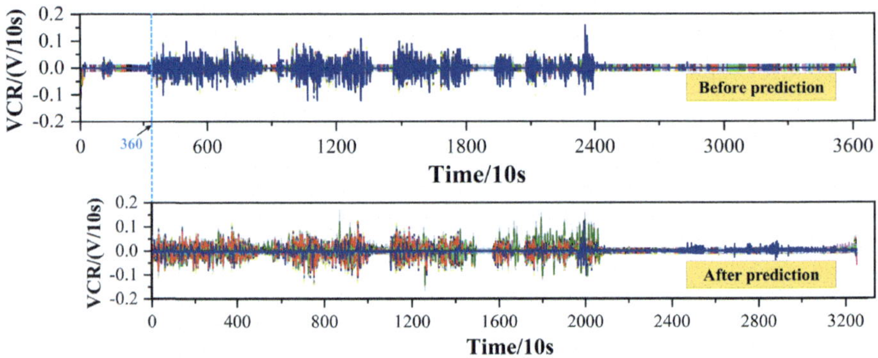

Fig. 4.29 Prediction effect of single cell voltage temperature rise rate

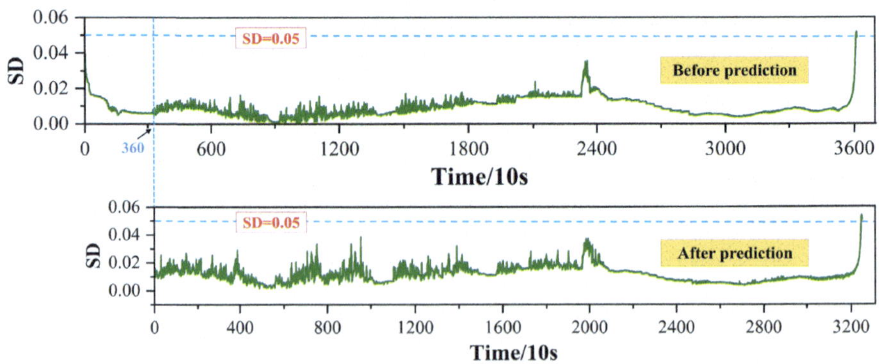

Fig. 4.30 Prediction effect of single cell voltage standard deviation

In addition, to verify the stability of the model, we obtained more new energy vehicle big data platform data at different times in 2017. These methods all achieved good prediction results and were able to successfully predict different degrees of voltage abnormalities within a PWS. Since the PWS finally selected in this study is 1 min (PWS = 6), Therefore, the aforementioned voltage anomaly can be predicted at least 1 min in advance. Admittedly, one minute is very short, but for severe temperature anomalies that could lead to battery thermal runaway (such as a level one alarm state), one minute is sufficient and valuable, allowing drivers and passengers to evacuate in time. Moreover, according to this study, PWSs can also vary flexibly, but as the PWS increases, the prediction accuracy will decrease, as shown in Fig. 4.21. Therefore, when establishing a model for another type of vehicle, it is necessary to flexibly choose the PWS according to the required prediction accuracy and visibility. In principle, regardless of the range of the PWS, the proposed LSTM model can predict voltage anomalies by accurately predicting changes in battery voltage in future PWSs. However, the purpose of this study is to predict battery voltage, serving the prediction of various battery faults caused by voltage anomalies, so pursuing a wider range of smaller PWSs may not yield ideal results.

4.2.3 SOC Prediction

1. Current Research Status

   SOC estimation methods can be roughly divided into four categories: lookup table method, ampere-hour integration method, model-based method, and data-driven method. The first two methods estimate SOC based on the working characteristics of lithium-ion batteries. The lookup table method is the most basic SOC estimation method, which determines the relationship between SOC and some other measurable battery parameters, such as open circuit voltage (OCV) or alternating current impedance (AC). However, since the accurate measurement of OCV or AC impedance requires the battery to be left idle for a long time to keep the parameters stable, this method is not suitable for real vehicle applications. The ampere-hour integration method integrates the current over time, which can easily estimate the real-time SOC of lithium batteries. But this method requires the initial SOC to be known and ignores the error of the current sensor. In addition, due to aging effects, the capacity of lithium-ion batteries needs to be recalibrated over time. Therefore, this method is only suitable for laboratories with stable measurement environments. Despite the limitations of these two methods at the methodological level, the main SOC estimation methods used in commercial electric vehicles are a combination of the lookup table method and the ampere-hour integration method, because these two methods are simple, require less computational resources, and can be met by existing on-board computing resources.

   Due to the inherent flaws of this method, errors are inevitable, so more general model-based methods are widely explored. Common models include electrochemical models and equivalent circuit models. The electrochemical model is established based on various coupling mechanisms such as chemical reactions and heat transfer inside the lithium-ion battery. However, in practical applications, many model parameters could be more measurable, and complex partial differential equations are difficult to calculate, making this method only suitable for laboratory use. The equivalent circuit model is established by combining circuit elements such as resistance and capacitance, considering the effects of battery aging, temperature, and other factors, and conducting empirical analysis. A variety of online SOC estimation algorithms have been developed by combining the battery model with different iterative calculation methods such as Kalman filters, nonlinear observers, sliding mode observers, and H-infinity observers. In particular, the equivalent circuit model is easily expressed as a state space model with SOC as a state, so the application of Kalman filtering is the most widespread, such as extended Kalman filtering (EKF), unscented Kalman filtering (UKF), and cubature Kalman filtering (CKF). Although model-based methods are relatively more suitable for on-board applications because they can simulate battery dynamics online and correct errors, their accuracy largely depends on the modeling experience of researchers.

In order to eliminate the influence of subjective factors, researchers are trying to explore methods that are more suitable for actual vehicle applications. With the significant progress of artificial intelligence technology, data-driven methods show significant advantages in SOC estimation, such as no need for models, high accuracy, and strong robustness. Unlike model expressions based on specific mechanisms, data-driven methods use machine learning or deep learning techniques to train and establish black box models with historical data. Then, the future SOC value is predicted based on the trained model. Compared with different data-driven methods, such as RNN and LSTM, the gated recurrent unit is receiving increasing attention due to its significant performance in SOC prediction modeling.

2. Structure of the Multi-Step SOC Prediction Model

The GRU-RNN model is a feedback recursive neural network with a memory function, used to solve the gradient vanishing and gradient explosion problems of the classic RNN. Compared with LSTM, this method simplifies the structure by reducing the number of storage units without losing performance, thereby significantly improving training efficiency. In this study, a multi-step SOC prediction model based on GRU-RNN was constructed, including an input bin module and a calculation module, as shown in Fig. 4.31. The double dropout method consists of two parts, including multi-layer correlation analysis (first dropout) and random dropout (second dropout). The input binning module uses the first dropout to derive the input data type required for training. The calculation module consists of an input layer, three hidden layers, and an output layer. The second dropout is included in the calculation module, and the double dropout process is described in Sect. 4.3.

After the 15 types of original running data are processed in the input bin module, they are organized into a training dataset containing 5 types of data

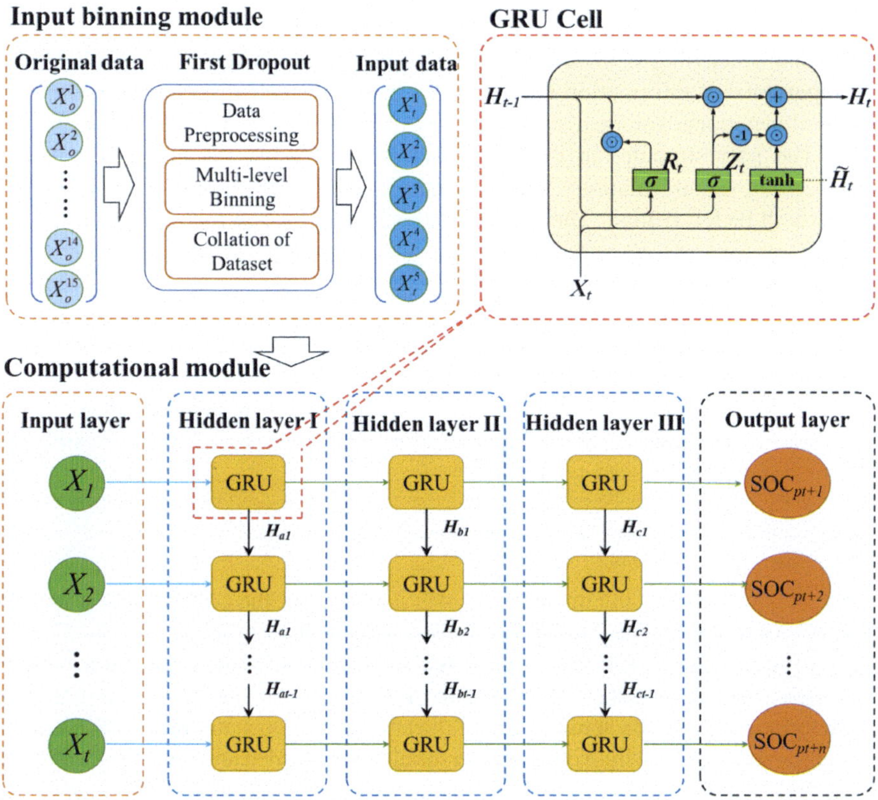

Fig. 4.31 Multi-step prediction GRU model

including SOC, pack voltage, vehicle speed, probe temperature, and brake pedal stroke value through the first dropout, and input into the input layer of the calculation module. Section 3.2 introduces the detailed extraction process of the first dropout. X_t represents the input data at different moments in time. At a certain moment (for example, the first row of the calculation module), the input data X_1 is passed horizontally between the three hidden layers, with a single GRU in the diagram representing 100 parallel GRU units. For multiple time matrices (such as the first and second rows of the calculation module), in addition to the horizontal transmission of input data in the corresponding GRU layer, the GRU of the previous matrix will pass the state information (H_{at-1}, H_{pt-1}, H_{ct-1} representing the state information of different hidden layers) to the GRU in a single hidden layer of the next matrix. At the final moment, the last GRU channel of the third hidden layer outputs the prediction of future SOC (SOC_{pt+n}) to the output layer.

The basic GRU consists of a reset gate, an update gate, and a temporary output, and can be seen as a single neuron. In its operation, it receives information from the outside world and from the previous neuron, and refines and updates the received information through its internal information processing mechanism, then passes it on to the next neuron. The basic principle of information processing is to selectively retain and forget information based on the importance of information features. This process is called forward propagation of information, explained as follows.

The gate structure in the GRU unit is responsible for the selection and screening of information. The reset gate controls the amount of information that will be left behind. The specific expression is as shown in Eq. (4.1):

$$R_t = \sigma(W_{xr}X_t + W_{hr}h_{t-1} + b_r) \tag{4.10}$$

where R_t represents the reset gate, W_{xr} and W_{hr} are the weight matrices corresponding to the input data X_t at time t and the hidden state information h_{t-1} from the previous neuron, b_r is the bias of the weight matrix in the reset gate.

The update gate controls the amount of past information to be considered, which can be expressed as Eq. (4.2):

$$Z_t = \sigma(W_{xz}X_t + W_{hz}h_{t-1} + b_z) \tag{4.11}$$

where Z_t represents the update gate, W_{xz} and W_{hz} are the corresponding weight matrices in the update gate, b_z is the bias of the weight matrix.

In addition to the two gates $_t$, there is also a candidate hidden state information, which can be regarded as the temporary new information at the current moment. Before outputting the actual value of the current hidden state information h_t, the neuron will decide which information should be discarded and which information should be retained by multiplying the corresponding elements in R_t and Z_t. R_t determines how much previous information needs to be retained from h_{t-1}, Z_t determines how much previous information needs to be discarded. If the values of R_t and Z_t are 1, choose the corresponding elements in h_{t-1} and h_t, or discard the element when R_t and Z_t are 0.

$$\tilde{h}_t = \tanh[W_{xh}X_t + W_{hh}(R_t \circ h_{t-1}) + b_h] \tag{4.12}$$

$$\begin{aligned} h\sigma(x) &= \frac{1}{1 - e^{-x}} \\ \tanh(x) &= \frac{e^x - e^{-x}}{e^x + e^{-x}} \end{aligned} \tag{4.13}$$

Here, it represents the multiplication of each element, $_t$ represents the candidate hidden state information, h_t represents the current hidden state information, W_{xh} and W_{hh} represent the corresponding weight matrices, b_h is the bias in the candidate hidden state information. σ (x) and *tanh* (x) represent *sigmoid* and *tanh* activation functions, their expressions can be seen as Eq. (4.14).

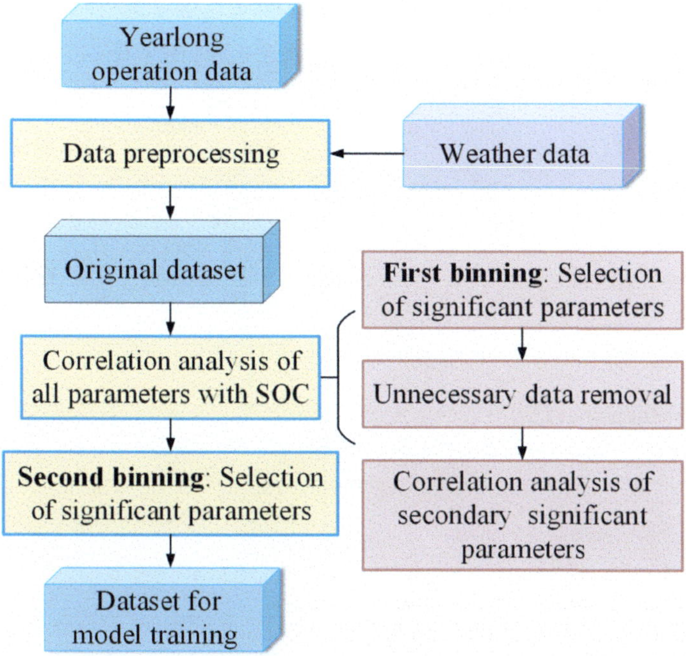

Fig. 4.32 Multi-level binning of original data

$$
\begin{aligned}
\sigma(x) &= \frac{1}{1+e^{-x}} \\
\tanh(x) &= \frac{e^{x}-e^{-x}}{e^{x}+e^{-x}}
\end{aligned}
\tag{4.14}
$$

3. Double Dropout to Prevent Overfitting

Due to the large volume of actual operational data, which is far more complex than experimental data, it is difficult to obtain satisfactory prediction results by directly training the model with it. Therefore, this paper proposes a new double dropout method to optimize model training. The first dropout is a multi-level correlation analysis based on the Pearson correlation analysis method, the specific flowchart is shown in Fig. 4.32. After all parameters are subjected to a primary correlation analysis with SOC, the three main parameters obtained are SOC, battery pack voltage, and battery 1 voltage (with corresponding coefficients of 1, 0.341, and 0.345, respectively). The battery pack voltage is the sum of the voltages of individual battery packs based on the series and parallel relationships within the battery pack, so the battery pack voltage and battery pack voltage can be considered as one parameter. The correlation coefficients of the remaining parameters are all below 0.1, as shown in Fig. 4.33. From the correlation analysis results, it can be seen that the correlation coefficient of the current is only 0.004, but

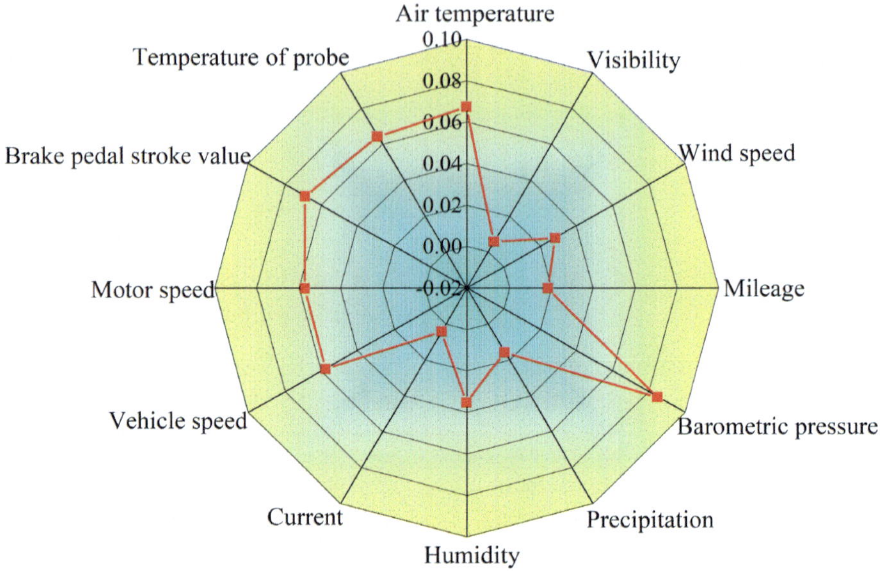

Fig. 4.33 Correlation coefficients of different parameters

the current parameter is widely used in experimental research as an input, which indicates that there is a significant difference in the selection of input parameters for SOC prediction using real vehicle data. A secondary correlation analysis is performed on the remaining parameters, and six parameters with coefficients greater than 0.05 (probe temperature, brake pedal stroke value, motor speed, vehicle speed, air pressure, and air temperature) are selected for further analysis, as shown in Table 4.5. Among the six available parameters, the coefficients of probe temperature, air pressure, and air temperature are all close to 1, about 0.7 of the other parameters. Therefore, the parameters are divided into environmental parameters and vehicle operation parameters. Considering the high correlation of environmental parameters, only the probe temperature (the probe temperature is a direct measurement parameter of the battery system) was selected, and the correlation between vehicle operation parameters is around 0.7, so the brake pedal stroke value and vehicle speed were selected. Therefore, the inputs used for model training include SOC, pack voltage, vehicle speed, probe temperature, and brake pedal stroke value.

The second dropout is constructed based on random dropout, which includes the state information transfer process between the adjacent layers of the computing module (except for the third hidden layer and the output layer), as shown in Fig. 4.34. Taking the random dropout between the input layer and the first hidden layer as an example: at time 1, the input data X_1 is stored in the dropout layer, and the dropout layer is built into the GRU logic in advance during

Table 4.5 Coefficients of unimportant parameters

Coefficient	SOC	Probe temperature	Brake pedal stroke value	Motor speed	Air pressure	Air temperature
SOC	0.0645	0.0689	0.0572	0.0574	0.0849	0.0674
Probe temperature	1	0.0775	0.0153	0.0154	0.838	0.9275
Brake pedal stroke value	0.0775	1	0.6815	0.6683	0.0421	0.0355
Motor speed	0.0153	0.6815	1	0.999	0.024	0.0382
Vehicle speed	0.0154	0.6683	0.999	1	0.0237	0.0377
Air pressure	0.838	0.0421	0.024	0.0237	1	0.8757
Air temperature	0.9275	0.0355	0.0382	0.0377	0.8757	1

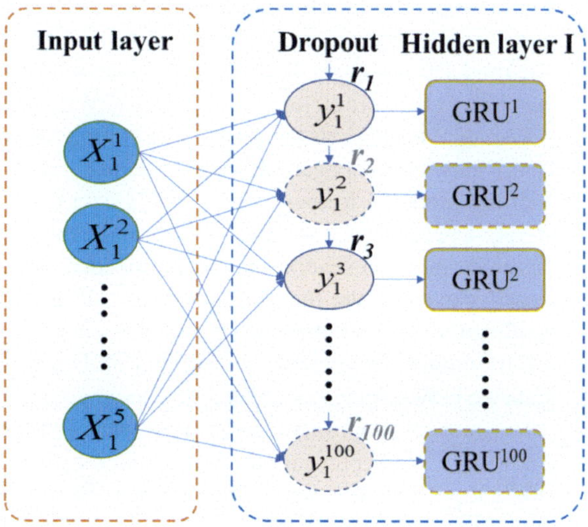

Fig. 4.34 Second dropout analysis

programming, instead of directly entering the first hidden layer. The specific implementation process is as follows.

Step one: Set the node retention probability r_i. For the input information stored in the dropout layer, a Bernoulli distribution with a probability of p is used to randomly generate r_i, its value is 0, r_i the number of is the same as the number of nodes. After multiple attempts, when p is 0.2, the model performance is optimal.

$$r_i \sim Bernoulli(p) \tag{4.15}$$

Step two: Multiply the corresponding r_i with the node connected after r_i, to implement the dropout of the GRU unit in the network structure, which means that only 80% of the nodes participate in the calculation and transmission of information. It can be represented as Eq. (4.7).

$$y_s^i = \mathrm{Y} - r_i * y_t^i \tag{4.16}$$

where Y represents all the information stored in the dropout layer at time t, $y^i{}_s$ represents the retained information, $y^i{}_t$ represents the discarded information.

Step three: After the first two steps, the network density of the model is diluted to a certain extent, and the remaining information can be used to complete the model training.

4. Determination of Initial Hyperparameters

Generally speaking, the larger the input sample (IS), the better the prediction effect, because more features can be learned. However, due to the limitation of computational resources, the operation data for a year is too large to achieve high computational efficiency. Therefore, the optimization of hyperparameters is a guarantee for accurate and efficient multi-step SOC prediction after building the model framework. Four hyperparameters were analyzed, including epochs, batch size (BS), prediction window size (PWS), and training window size (TWS).

An epoch refers to the process of completing one forward computation and one backward propagation with the entire dataset. In other words, this means that the model is trained with a year's data once. Since gradient descent is an iterative process, one training cannot make the network weights converge to the optimal value, so several epochs are needed. The model's operation data for a year far exceeds the load of computational resources, which is not suitable for application in real battery systems, and needs to be divided into several batches for training. BS represents the amount of data in each batch. Achieving the best balance between memory efficiency and memory capacity is the key to choosing BS, which will optimize the performance and training speed of the model. In addition, PWS represents the prediction length of future SOC fluctuations, and TWS represents the size of the sample. Based on empirical analysis, a set of initial hyperparameters was selected: IS = 1 month of data, epoch = 50, BS = 32, PWS = 30 (SOC prediction for 5 min), TWS = 360 (1 h of data). In addition, this paper uses the mean relative error (MRE) to evaluate the results, which can be expressed as Eq. (4.17).

$$\mathrm{MRE} = \frac{1}{n}\sum_{t=1}^{n}\frac{\left|\mathrm{SOC}_{pt} - \mathrm{SOC}_{rt}\right|}{\mathrm{SOC}_{rt}} \times 100\% \tag{4.17}$$

where n is the number of SOC samples, $\mathrm{SOC}_{\mathrm{pt}}$ is the predicted value of the battery system, and $\mathrm{SOC}_{\mathrm{rt}}$ is the actual value of the battery system. To ensure that all data features can be learned, a sliding window is used for sampling, and the sliding window size (SWS) is 60 (10 min of data). The sliding window in sampling refers to the interval between two starting points being less than the length of a sample. The schematic diagram of different windows is shown in Fig. 4.35.

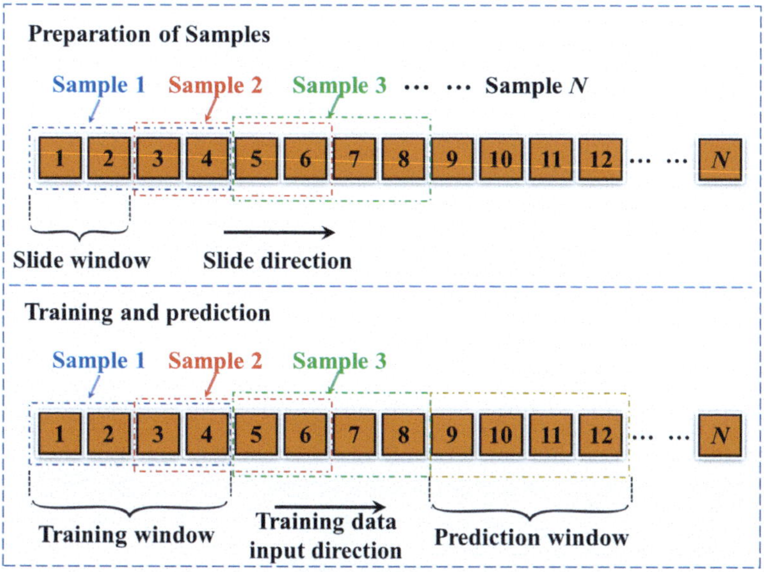

Fig. 4.35 Schematic diagram of different window hyperparameters

To verify the reliability of the initial hyperparameters, one month's data was used for training, and one day's data from the following year was used for testing. The prediction results and absolute errors based on the initial hyperparameters are shown in Fig. 4.36. The prediction error is basically controlled within 5%, but in places where the data fluctuates greatly due to insufficient model training and sudden changes in data features, the error can reach 5–10%, indicating that the model performs well on the test data and provides a good basis for further optimization.

5. Results and Discussion

Firstly, the model still uses a month's worth of data as the training dataset, and another day's data as the test dataset for training. The test dataset is randomly selected and not used for model training to ensure the distribution difference between the two datasets. The training process is shown in Fig. 4.37. The training loss (TRL) and validation loss (VAL) are recorded, and the model shows rapid convergence using the Adam optimizer. Due to the different distribution of the test dataset, VAL cannot reach the same level as TRL. At the same time, as the number of epochs increases, TRL gradually decreases and tends to stabilise, indicating that the model has been trained on the training dataset. However, when the number of epochs exceeds 30, the trend of VAL and TRL is opposite, with the maximum value of VAL at 0.00113 at 30 epochs. VAL represents the generalisation of the model during training, so we finally choose 30 epochs.

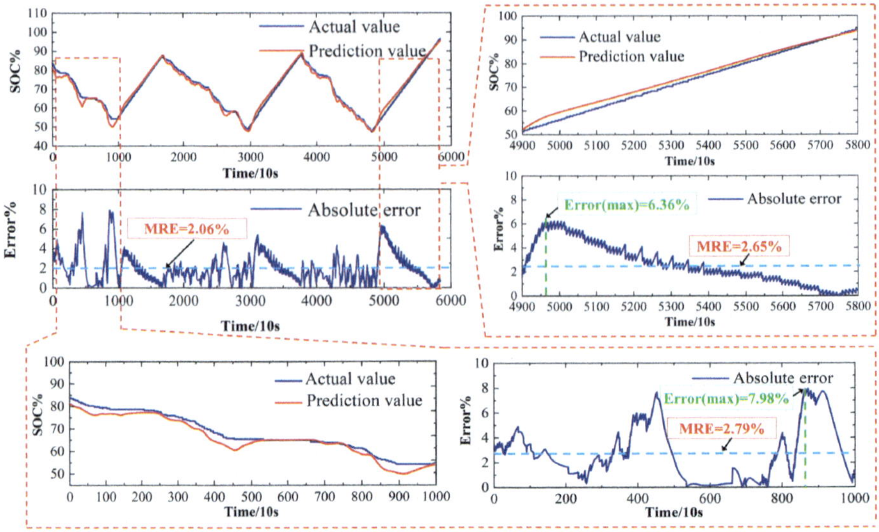

Fig. 4.36 Predictive performance of initial hyperparameters

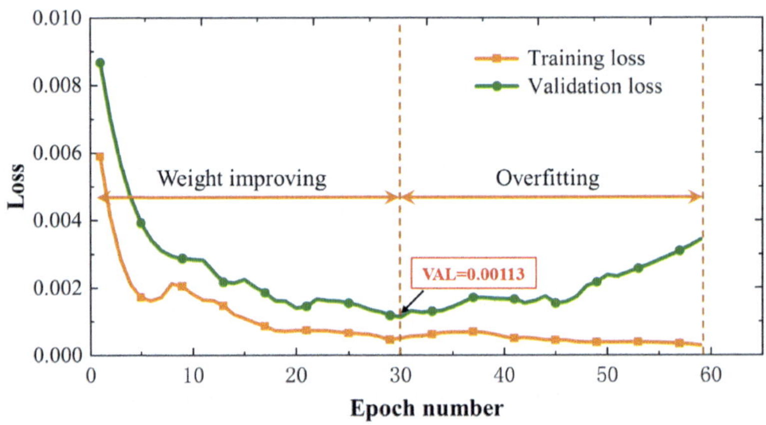

Fig. 4.37 Determining Epoch

As mentioned in Sect. 4.1, not all data is input into training every time, and the choice of BS is crucial for achieving the best balance between computational efficiency and storage capacity. Although a larger batch size can improve the utilization of computational resources and process data faster, an excessively large batch size often leads to the network converging to a local optimum. Similarly, a batch size that is too small can also lead to inefficient training. Therefore, when choosing the batch size, the number and complexity of the dataset samples should be fully considered. Referring to the experience of network parameter

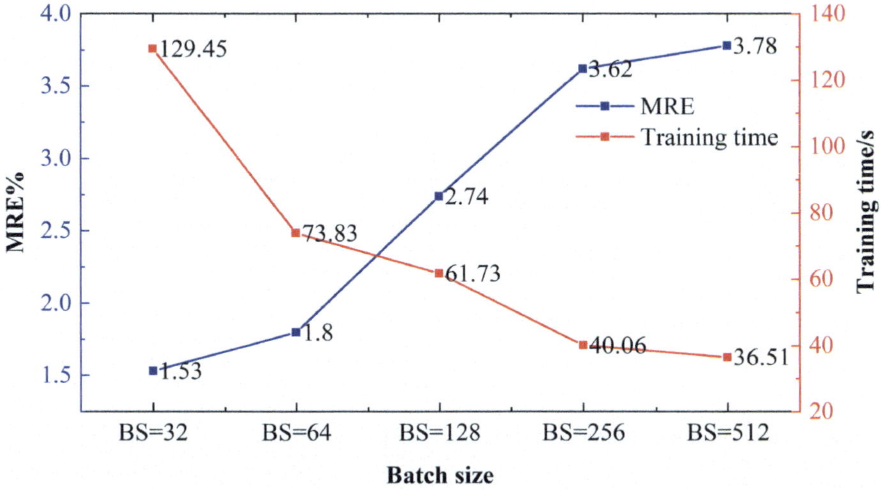

Fig. 4.38 MRE and training efficiency under different BSs

optimization, an analysis of different BSs ({32, 64, 128, 256, 512}) was carried out for 30 epochs and a 5-min prediction time. Figure 4.38 shows the MRE and corresponding training time under different BSs. It can be seen that the prediction accuracy of the model is negatively correlated with BS, but the increase in BS significantly improves training efficiency. During training, each BS feeds back a loss to the model's optimization, i.e., the average loss of all samples in the BS. The smaller the BS, the more accurate the feedback, and the prediction performance of the model will significantly improve. However, a smaller BS means that the number of processes will also increase, thus reducing efficiency. Considering the need for training efficiency and accuracy, the final BS is determined to be 64 (MRE = 1.8%, training time = 72.83 s).

TWS is selected to divide the vehicle operation data into samples. Considering that electric taxis run for about 5~8 h a day, a 5 min multi-step prediction method is used to analyze different TWSs ({30, 60, 90, 120, 180, 270, 360, 480, 600}, representing 5 min, 10 min, 15 min, 20 min, 30 min, 45 min, 1 h, 80 min, 100 min of data respectively). The prediction accuracy and training time of the model under different TWSs are shown in Fig. 4.39. As can be seen from Fig. 4.39, when TWS is 480, the prediction accuracy of the model is the highest (MRE = 1.74%). At the same time, as TWS increases, the training time of the model gradually increases. Considering that when TWS increases from 360 to 480, the improvement of MRE is only 0.06%, but the training time increases by 25.6%, the optimal TWS determined in this paper is 360.

The choice of PWS is based on the premise of TWS being 360. In prediction, the trained model is based on the most recent input TWS for feature fitting, so the upper limit of PWS selection is 360. This paper analyses different PWSs ({6, 30, 60, 120, 240, 360}, representing 1 min, 5 min, 10 min, 20 min, 40 min, and 1 h of

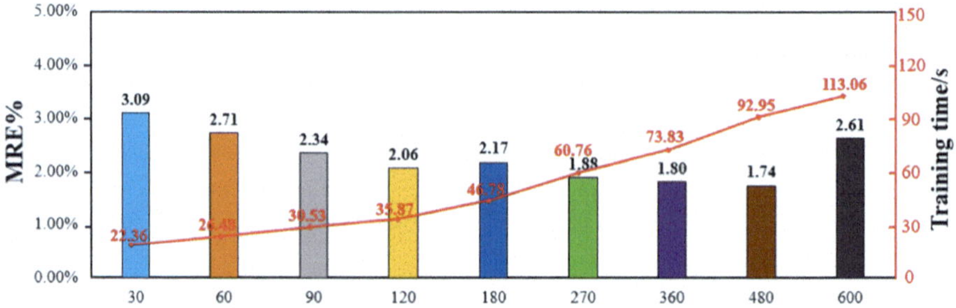

Fig. 4.39 MRE and training efficiency under different TWSs

multi-step prediction, respectively). MRE and prediction cost time are shown in Fig. 4.40. Since there are sufficient computational resources, the prediction time stays mostly the same with the increase of PWS. When PWS is 360, the change in MRE is significant, reaching 9.69%. This is because the larger the PWS, the more features are fitted under limited input conditions, and the GRU network has a limited memory length for historical data features. The further the distance in the time series output, the more uncertainty it will cause in model prediction. Considering that the impact of prediction time is not significant, the final choice of PWS is 30 (best accuracy, MRE = 1.8%).

As can be seen from the above sections, the optimised model has achieved satisfactory prediction performance, but there is still a problem that a month's data does not have enough features. When the time period is extended to a quarter or a whole year, factors such as environment and traffic conditions change significantly, bringing significant changes to the dataset. For example, the charging

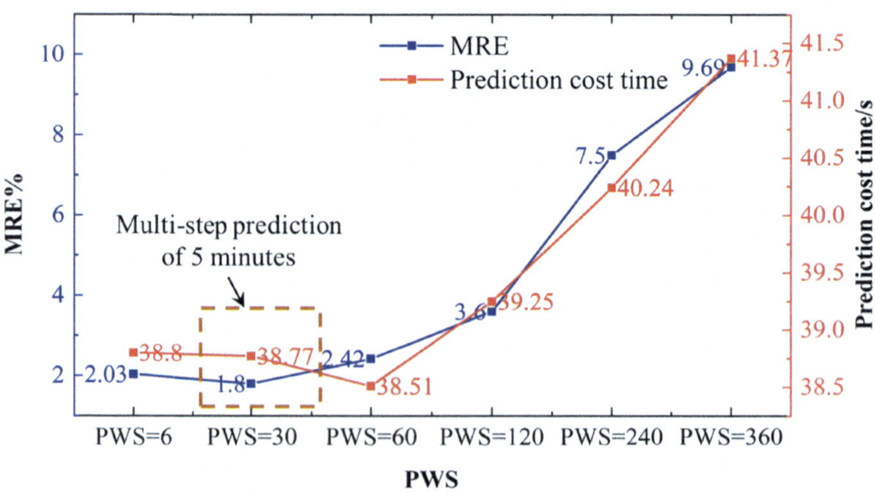

Fig. 4.40 MRE and prediction cost time under different PWSs

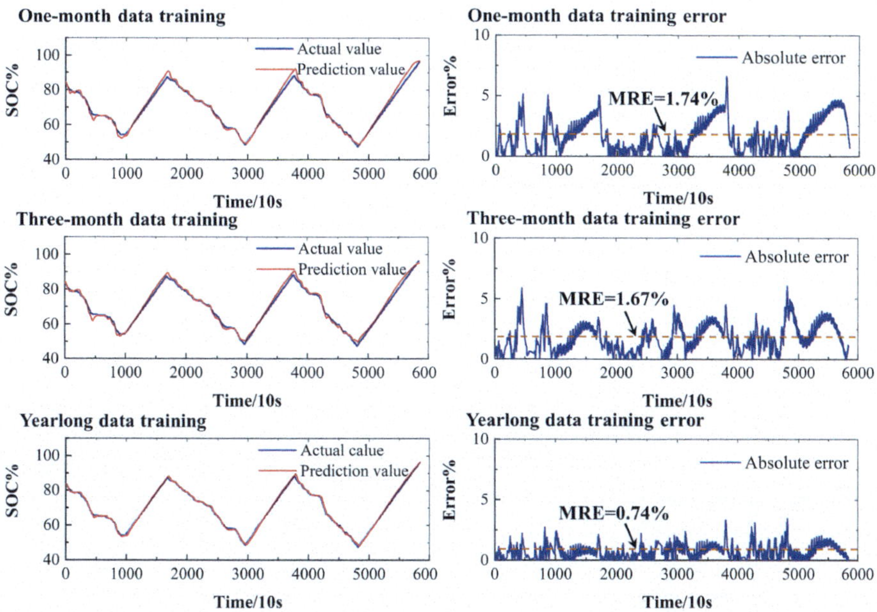

Fig. 4.41 Prediction performance under different intervals

and discharging performance of batteries decreases in winter, and the frequency of charging and discharging increases. Therefore, this section will analyze the impact of different ISs (1 month's data, 3 months' data, and 1 year's data) on multi-step prediction.

The prediction results and acoustic emission results of different ISs are shown in Fig. 4.41. Based on the same 1 month's data, MRE reaches 1.74%, which is 0.32% lower than the initial hyperparameters, as shown in Fig. 4.41a. At the same time, the prediction performance of the model does not improve significantly, AE fluctuates significantly between 0 and 7.5%, and MRE only decreases by 0.07%. However, when the model is trained with a year's data, significant improvements are obtained. AE is basically controlled below 2.5%, and MRE reaches 0.74%, indicating that sufficient training samples significantly improve the prediction performance of the model. Based on the above analysis, a set of optimal hyperparameters (Epoch = 30, BS = 64, TWS = 360, PWS = 30, IS = year data) is determined, as shown in Table 4.6.

6. Validating the Stability, Robustness, and Superiority of the Model

The above analysis shows that the predictive performance of the multi-step prediction model trained with interannual data is the best. Considering the complexity of the actual operating conditions of the vehicle, six days of data for the next year are randomly selected to test the stability and robustness of the model, as shown in Fig. 4.42. From the actual SOC curves of the six days, there are significant differences in the actual operating time and charging and discharging conditions under the six conditions. Especially in Case 3, you can see a long time of variable speed driving, and the number of charging and discharging cycles is more than the other five cases. The MREs under each test are 0.69%, 0.78%, 0.86%, 0.75%, 0.79% and 0.74%, respectively, indicating that the model has good SOC multi-step prediction accuracy.

Figure 4.43 presents a comparison of the predictive performance of the dual dropout GRU proposed in this paper with three common methods (LSTM, GRU, and linear regression). The three common methods have all been validated through experimental datasets. To eliminate the impact of different datasets, all models use the same IS (one year of data) for training and testing with the same one day of data. It can be found that among the four methods, the predictive performance of the dual dropout GRU method is the best. Although the other three methods can predict the trend of SOC changes, they cannot avoid large errors. Therefore, the proposed method has the potential for practical application in electric vehicles.

4.2.4 SOP Prediction

1. Introduction to SOP Evaluation Method

   SOP is commonly used in the field of battery state estimation to represent the maximum charging or discharging power that the battery can reach in the future. Charging and discharging power and peak power are two other expressions of SOP. SOP estimation and prediction is one of the core functions of BMS. The mainstream SOP estimation methods are: Hybrid Pulse Power Characteristic (HPPC) method, prediction method based on SOC constraints, prediction method based on voltage constraints, and multi-constraint dynamic method.

2. HPPC Method

   The HPPC method is a calculation method proposed by the Idaho National Engineering and Environmental Laboratory in the United States. The HPPC method is used to determine the dynamic power capability of the battery under hybrid

Table 4.6 Summary of hyperparameters and prediction performance

Hyperparameter	Parameter value			Determination
Epoch	[1, 2, 3, ..., 30, ..., 57, 58, 59]			30
BS	[32, 64, 128, 256, 512]			64
TWS	[30, 60, 90, 120, 180, 270, 360, 480, 600]			360
PWS	[6, 30, 60, 120, 240, 360]			30
IS	One month	Three months	One year	One year
Best MRE	0.74%			

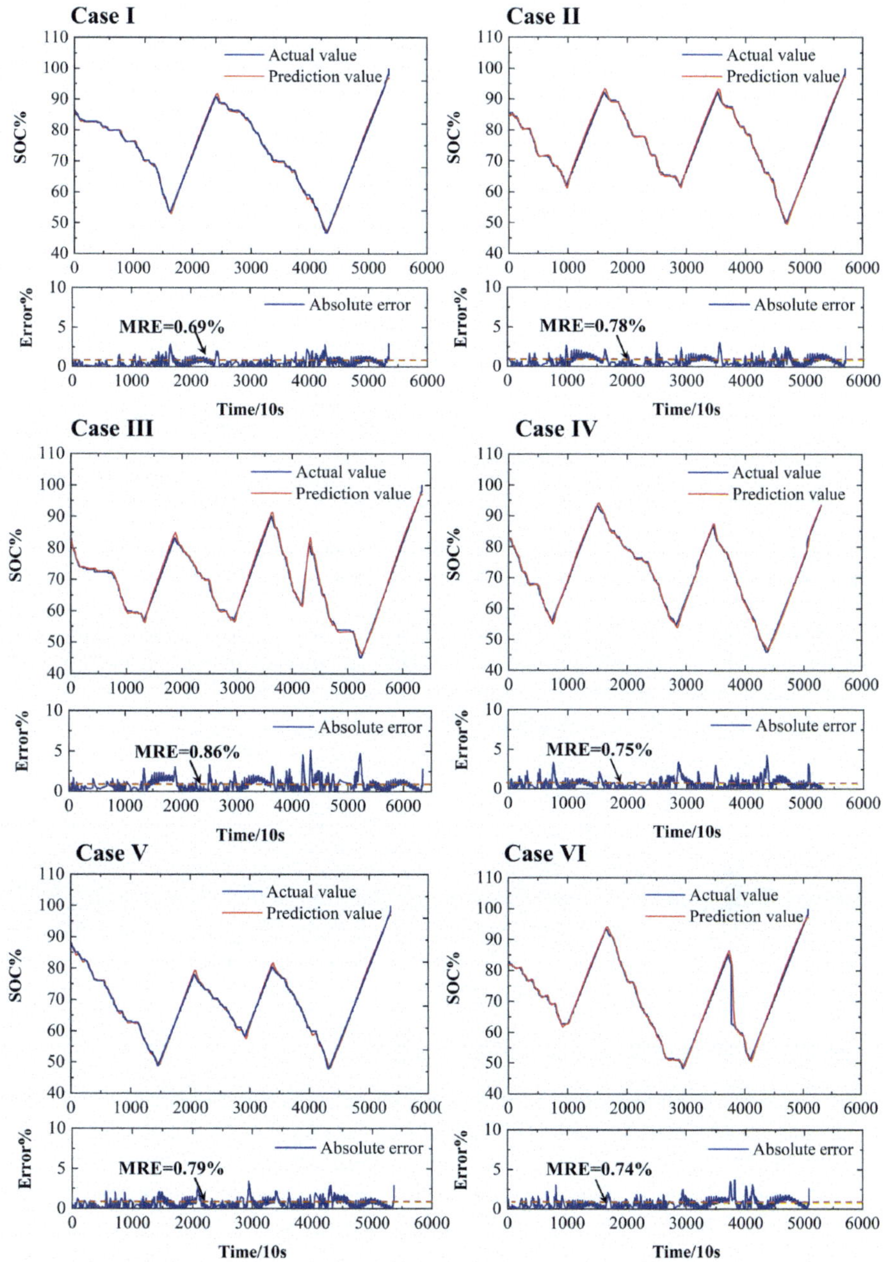

Fig. 4.42 Verification of stability and robustness under different conditions

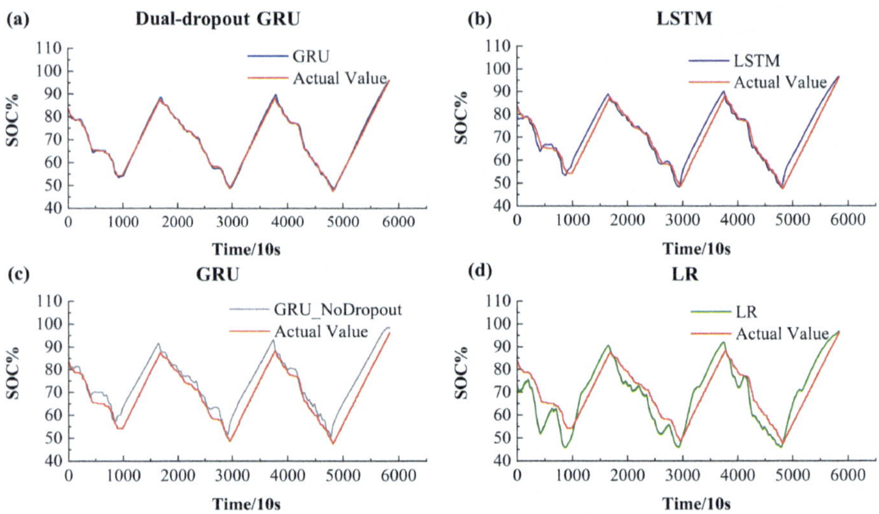

Fig. 4.43 Predictive performance of different methods: **a** prediction based on dual dropout GRU; **b** prediction based on LSTM; **c** prediction based on GRU; **d** prediction based on LR

pulse conditions. The test process of the HPPC method measures the battery's 10 s discharge power and 10 s charging power for every 10% of SOC. In the HPPC method, the formula for calculating SOP is as follows.

$$\begin{cases} SOP_{\text{chg}} = U_{\max} \cdot (\frac{OCV - U_{\max}}{R_{\text{chg}}}) \\ SOP_{\text{dis}} = U_{\min} \cdot (\frac{OCV - U_{\min}}{R_{\text{dis}}}) \end{cases} \tag{4.18}$$

where, SOP_{chg} and SOP_{dis} are the battery's charging SOP and discharging SOP; $U_{\max}$ and $U_{\min}$ are the cut-off voltages for battery charging and discharging respectively; R_{chg} and R_{dis} are the internal resistances of the battery during charging and discharging respectively.

3. Prediction Method Based on SOC Constraints

In the HPPC method, a single cell's peak charging current and peak discharging current are obtained through HPPC, and then the charging SOP and discharging SOP of the battery are calculated. In the prediction method based on SOC constraints, the peak charging and discharging currents of the power battery are obtained based on the maximum and minimum SOC constraints during the use of the power battery, and then the charging SOP and discharging SOP of the battery are calculated. In the prediction method based on SOC constraints, the formula for calculating SOP is as follows.

$$\begin{cases} SOP_{\text{chg}} = U_{\max} \cdot (\frac{z_m(t) - z_{\max}}{\eta_i \Delta t / C_{\max}}) \\ SOP_{\text{dis}} = U_{\min} \cdot (\frac{z_m(t) - z_{\min}}{\eta_i \Delta t / C_{\max}}) \end{cases} \tag{4.19}$$

Where, η_i is the charging and discharging efficiency; $z_{\max}$ and $z_{\min}$ are the upper and lower constraints of SOC respectively; $z_{\mathrm{m}}(t)$ is the SOC of single cell m at time t.

4. Prediction Method Based on Voltage Constraints

The HPPC method can only predict the instantaneous peak current in a short time interval, but the battery will face overcharging or over-discharging problems in a long time interval, and the prediction of peak current in a long time interval is equally important. In order to achieve the prediction of peak current in a long time interval, researchers have proposed a peak current prediction method based on voltage constraints. This method is used to predict the SOP of the battery.

Specifically, the improved Rint model is used to calculate the terminal voltage of a single cell, and the formula is as follows.

$$U(t+\Delta t) = OCV_m[z_m(t+\Delta t)] - Ri_m(t) \tag{4.20}$$

where, R is the internal resistance of the monomer during charging and discharging; $i_m(t)$ is the SOC of the monomer m at the moment t.

Since the peak current cannot be directly solved, the function needs to be Taylor expanded, and the formula is as follows.

$$\begin{aligned} OCV_m[z_m(t+\Delta t)] &= OCV_m\left[z_m(t) - i_m(t)\frac{\eta_i \Delta t}{C_{\max}}\right] \\ &= OCV_m[z_m(t)] - i_m(t)\frac{\eta_i \Delta t}{C_{\max}} \frac{\partial OCV(z)}{\partial z}\bigg|_{z=z_m(t)} + R[z_m(t), i_m(t)\frac{\eta_i \Delta t}{C_{\max}}] \end{aligned} \tag{4.21}$$

where, $R(\cdot)$ is the remainder of the first-order Taylor expansion, which is approximately equal to 0 and is omitted; the derivative of OCV $\frac{\partial OCV(z)}{\partial z}$ can be obtained from the OCV–SOC curve. As shown in Fig. 4.44.

The peak current prediction formula is as follows.

$$\begin{aligned} i_{\max}^{dis} &= OCV[z_m(t)] - \frac{U_{\min}\eta_i \Delta t C_{\max}^{\partial}}{OCV(z)\partial z}\bigg|_{z_m(t)} + R_{dis} \\ i_{\min}^{chg} &= OCV[z_k(t)] - \frac{U_{\max}\eta_i \Delta t C_{\max}}{\partial OCV(z)\partial z}\bigg|_{z_m(t)} \end{aligned} \tag{4.22}$$

$$\begin{cases} rU_t = OCV - U_D - R_i \\ \dot{U}_D = \frac{i}{C_D} - \frac{U_D}{C_D R_D} \end{cases}$$

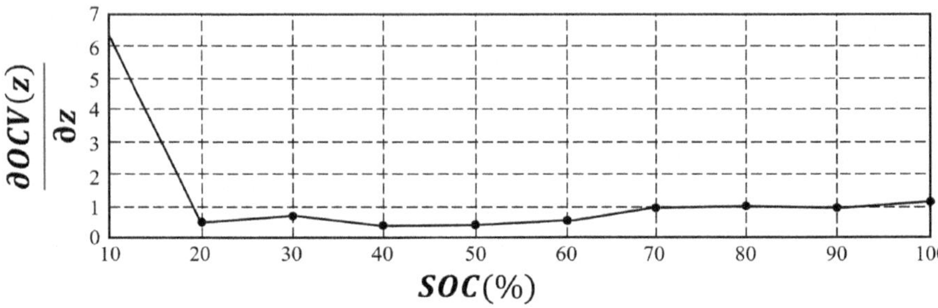

Fig. 4.44 OCV–SOC curve

Then, the peak charging current and peak discharging current can be substituted into Eq. (4.18–4.2) to obtain the charging SOP and discharging SOP.

5. Multi-Constraint Dynamic Method

The Thevenin model uses an RC network to describe the dynamic voltage characteristics of power batteries, and has the advantage of accurately simulating phenomena such as polarisation and hysteresis of power batteries. The Thevenin model is shown in Fig. 4.45.

The state space equation of the Thevenin model is shown in the following equation.

$$\begin{cases} U_t = OCV - U_D - Ri \\ \dot{U}_D = \frac{i}{C_D} - \frac{U_D}{C_D R_D} \end{cases} \tag{4.23}$$

The formula for the single-cell terminal voltage can then be obtained as follows.

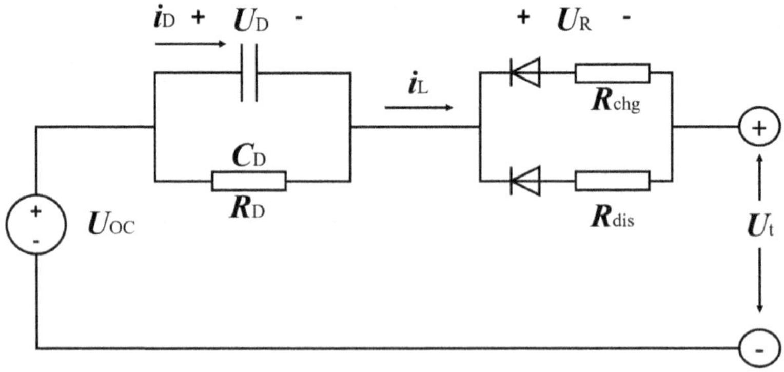

Fig. 4.45 Thevenin model

$$U_D(t+\Delta t) = e^{\frac{-\Delta t}{\tau}} U_{D,m}(t) + R_D\left(1 - e^{\frac{-\Delta t}{\tau}}\right) \underset{m}{i}(t) \tag{4.24}$$

where, $\tau = R_D C_D$.

Then, after substituting into Eq. (4.18–4.3), the peak current calculation formula is obtained as follows.

$$\begin{cases} i_{\max}^{dis,MCD} = \dfrac{OCV[z_m(t)] - U_{D,m}(t)e^{\frac{-\Delta t}{\tau}} - U_{\min}}{\frac{\eta_i \Delta t}{C_{\max}} \left.\frac{\partial OCV(z)}{\partial z}\right|_{zm(t)} + R_D\left(1 - e^{\frac{-\Delta t}{\tau}}\right) + R_{dis}} \\ i_{\min}^{chg,MCD} = \dfrac{OCV[z_k(t)] - U_{D,m}(t)e^{\frac{-\Delta t}{\tau}} - U_{\max}}{\frac{\eta_i \Delta t}{C_{\max}} \left.\frac{\partial OCV(z)}{\partial z}\right|_{zm(t)} + R_D\left(1 - e^{\frac{-\Delta t}{\tau}}\right) + R_{chg}} \end{cases} \tag{4.25}$$

The current of the power battery cell is limited in real situations, and is constrained by the following formula.

$$\begin{cases} i_{\max}^{dis} = min\left(i_{\max}, \min_{m} i_{\max}^{dis,MCD}\right) \\ i_{\min}^{chg} == max\left(i_{\min}, \max_{m} i_{\min}^{chg,MCD}\right) \end{cases} \tag{4.26}$$

Afterward, the peak charging current and peak discharging current can be substituted into Eq. (4.18–4.2) to calculate the charging SOP and discharging SOP.

4.2.5 SOE Prediction

1. Introduction to SOE Evaluation Method

The definition of a battery's SOE is the percentage of the remaining energy of the battery to the rated energy under rated conditions. It refers to the energy of the battery discharging to the cut-off voltage under the current state and standard conditions.

Currently, SOE estimation methods are mainly divided into estimation-based methods and data-driven methods. Common data-driven methods include BP neural networks, fuzzy control, etc. The concept of SOE was proposed by Kelli Mamadou. The battery's current energy situation is usually judged by SOC, but the change in SOC can only reflect the change in current and cannot describe the change in voltage.

2. Power Integration Method

The power accumulation method is similar to the ampere accumulation method in SOC estimation. It is a method of integrating the battery power according to the definition. By integrating the product of battery current and voltage, the change in energy is calculated to obtain the current SOE value of the battery. Like ampere-watt integration, watt-watt integration also has disadvantages. In the initial stage of battery charging and discharging, the correct initial SOE value is important to maintain high estimation accuracy and dynamic estimation ability for watt integration. However, it is difficult to obtain the initial value of SOE, and the accuracy of the initial value directly affects the estimation result. As the charging and discharging proceed, the process noise gradually accumulates, and the deviation between the SOE estimation result and the true value gradually increases. Therefore, the initial value of SOE must be correctly obtained for the accumulated energy, the accumulated energy cannot be used for a long time, and it cannot be used alone.

3. Open Circuit Voltage Method

   The open-circuit voltage method requires a long period of battery idle time. After the battery status stabilises, the relationship between the open-circuit voltage and the SOE is determined, establishing an OCV–SOE relationship curve, and the SOE under the current state is calculated. However, during the driving process of electric vehicles, due to complex working conditions such as acceleration, climbing, energy braking recovery, etc., the working current changes greatly, and the open-circuit voltage is difficult to stabilize in a short time. The stabilization time is closely related to factors such as current size and temperature. Therefore, this method has a great uncertainty when estimating SOE, and it is impossible for electric vehicles to stand for a long time during driving, so it is not suitable for online estimation of SOE. The open-circuit voltage method is suitable for the research and maintenance of experimental batteries and can be combined with the watt-hour accumulation method to calculate the initial value of SOE.

4. Data-Driven SOE Prediction Method

   In the data-driven approach, the battery is viewed as a black box system. Based on the input external vibrations and the corresponding output, the fitting ability of the data-driven approach is used for learning. Most of the current data-driven SOE prediction methods use other data for prediction. A multi-dimensional lookup table of maximum available energy and remaining available energy is constructed, and the current SOC, SOH, temperature, and current are interpolated to obtain the current maximum available energy and remaining available energy. SOE is obtained based on the remaining available energy/maximum available energy. However, the establishment of a multi-dimensional lookup table is based on a large amount of experimental data, and this method depends on the accurate estimation of battery SOC and SOH. In addition, this paper needs to discuss the applicability of electric vehicles under drastic changes in dynamic working conditions. To simulate the dynamic voltage characteristics of the battery, a battery model based on wavelet neural networks has been proposed. In order to suppress the measurement noise of current and voltage, some researchers use particle filter estimators to estimate the energy state of the battery.

4.3 SOS Joint Estimation

Real-time monitoring and accurate prediction of various states of the power battery system are the basis for power failure diagnosis and risk prediction, which are crucial for ensuring the safe and reliable operation of electric vehicles. The various state parameters of the power battery system have non-linear and strong time-varying characteristics, so it is difficult to accurately model and effectively reproduce them through conventional physical experiments and mathematical models. This chapter will use the LSTM neural network to complete the online multi-step joint prediction of key states such as voltage, temperature, and SOC of the vehicle power battery system based on the big data of actual vehicle operation, considering complex influencing factors such as road conditions, weather, and driving behavior, and at the same time achieve online accurate estimation of the SOH of the vehicle power battery system.

4.3.1 LSTM-Based Multi-State Joint Prediction Model

1. Data Preprocessing

   The data sampling interval is 10 s. Vehicle A is a pure electric taxi equipped with a high specific energy lithium iron phosphate power battery system. Its main performance parameter configuration is shown in Table 4.7. The power battery system consists of 100 single batteries, and there are 16 temperature probes distributed inside and outside the battery box to monitor the temperature of the battery pack. To simplify the analysis, the single cell voltage in the following text will be represented by the single cell voltage of cell 1, and the probe temperature will be represented by the temperature of probe 1.

The running data of vehicle A for the whole year of 2016 is used as the input set, and part of the running data for 2017 is used as the test set for research. In order to meet the research requirements, the original vehicle operation data needs to go through a series of preprocessing processes, such as unifying the formats of various data items, eliminating outliers, and supplementing missing values, etc. In addition, to conduct a more comprehensive analysis and more accurate prediction, this paper proposes an analysis method for weather, vehicle, and driver, which considers factors such as weather, vehicle, and driver behavior in addition to the characteristics of the battery system itself.

Table 4.7 Vehicle A performance parameter configuration

Curb weight	Wheelbase	Top speed	Motor rated power	Motor torque	Maximum range
1350 kg	2500 mm	125 km/h	20 kW	144N·m	200 km

By utilising web scraping technology, weather data that could potentially affect vehicle operation was exported from the weather website https://www.wunderground.com. Ultimately, six meteorological parameters were added for potential input parameter analysis: humidity, precipitation, atmospheric pressure, temperature, visibility, and wind speed. The original meteorological data had a time interval of 3 h, which was inconsistent with the 10 s sampling interval of the new energy vehicle big data platform. Therefore, the Lagrange interpolation method was used to interpolate each meteorological parameter at 10 s intervals, and then the interpolated weather data was merged and stored with the vehicle data according to the time alignment principle. Meteorological parameters usually fluctuate very little in the short term and have a small impact on battery performance, so the interpolated meteorological parameters can meet the requirements of this study. Figure 4.46 shows the preprocessing flowchart of vehicle data and meteorological data, and the temperature curve after interpolation on January 1, 2016, is shown in Fig. 4.47.

For vehicle operating factors related to battery voltage, temperature, and SOC, the remaining useful life (RUL) of the power battery is a parameter that needs to be considered. At present, for a vehicle in actual operation and the existing RUL estimation algorithm, it is almost impossible to monitor the real RUL of the power battery in real-time accurately. However, the RUL of the pure electric vehicle power battery system is negatively correlated with the vehicle mileage, that is, the longer the vehicle mileage, the smaller the power battery RUL. Many pure electric vehicles' full life cycle is also expressed in total mileage. Therefore, in the parameter analysis

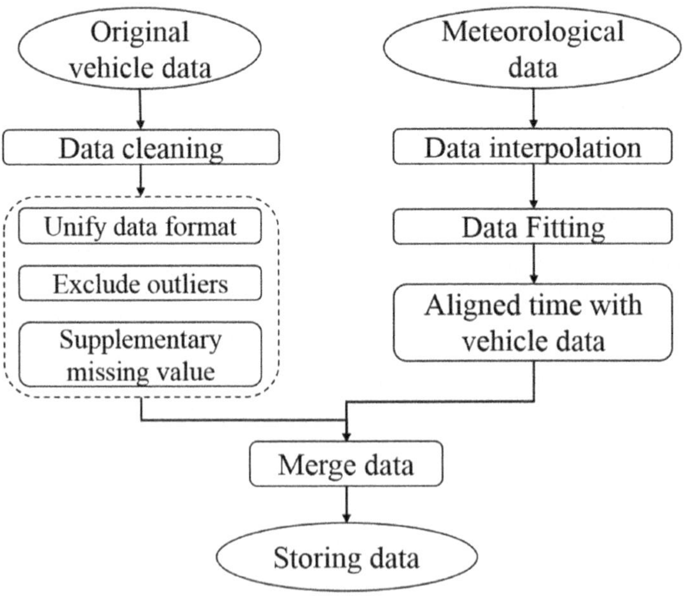

Fig. 4.46 Preprocessing flowchart of vehicle data and meteorological data

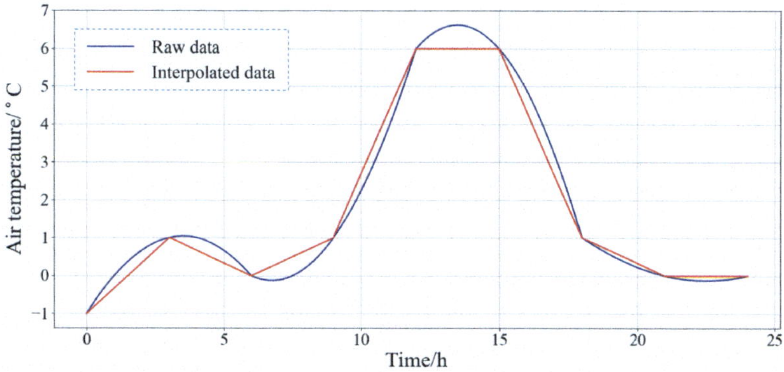

Fig. 4.47 Temperature curve before and after Lagrange interpolation

below, the vehicle mileage directly measurable from the new energy vehicle big data platform will be used to indirectly represent the battery RUL status.

The driver's driving behavior can have a significant impact on the safety and durability of the real vehicle power battery system, but the new energy vehicle big data platform currently cannot monitor and record the vehicle's driving acceleration in real-time. For the method of estimating acceleration based on vehicle speed, the 10 s sampling interval is also too large. Therefore, to obtain the vehicle acceleration status, it must be represented or calculated through other parameters that can be measured in real-time. To solve this problem, first analyze the correlation between acceleration and motor speed, rated motor power, wheel radius, and other parameters, as shown below:

$$T_{motor} = \frac{9550 * P}{n} \tag{4.27}$$

$$T_{wheel} = T_{motor} * i \tag{4.28}$$

$$a = \frac{F}{m} = \frac{T_{wheel}}{R * m} \tag{4.29}$$

Therefore, the acceleration a can be calculated as follows:

$$a = \frac{9550 * P * i}{R * m * n} \tag{4.30}$$

where, T_{motor} is the motor torque, P is the rated power of the motor, n is the motor speed, T_{wheel} is the wheel torque, i is the transmission efficiency from the motor to the wheel, F is the driving force of the vehicle, m is the vehicle's curb weight, and R is the tire radius.

In Eq. (4.30), except for the motor speed n and acceleration a, all other parameters are known/unknown constants, so acceleration and motor speed are inversely proportional to each other. Since the transmission efficiency i. The calculation or estimation of parameters will inevitably contain some errors, and the acceleration calculated through Eq. (4.30) will also contain certain errors. To avoid computational errors and computational consumption, this study will directly use the motor speed to indirectly represent the vehicle acceleration state.

Based on the above "weather-vehicle-driver" approach to analyze each potential input parameter, the preliminary determination of the 15 parameters used for training the LSTM model are: battery voltage, probe temperature, SOC, brake pedal stroke value, motor speed, vehicle speed, total voltage, current, humidity, precipitation, air pressure, air temperature, visibility, wind speed and mileage. The feature curves of all parameters of the vehicle under study for the whole year of 2016 are shown in Fig. 4.48.

2. Multi-Parameter Joint Prediction Based on LSTM

   The basic principle of the LSTM model has been introduced in Chap. 2. LSTM neural network. When using LSTM to predict the real vehicle power battery system status, there are still three difficulties: ① The speed of convergence to the optimal network weight of the traditional deep learning network based on batch gradient descent or stochastic gradient descent is very slow, so it is not suitable for online prediction of the real vehicle battery system status. ② How to effectively

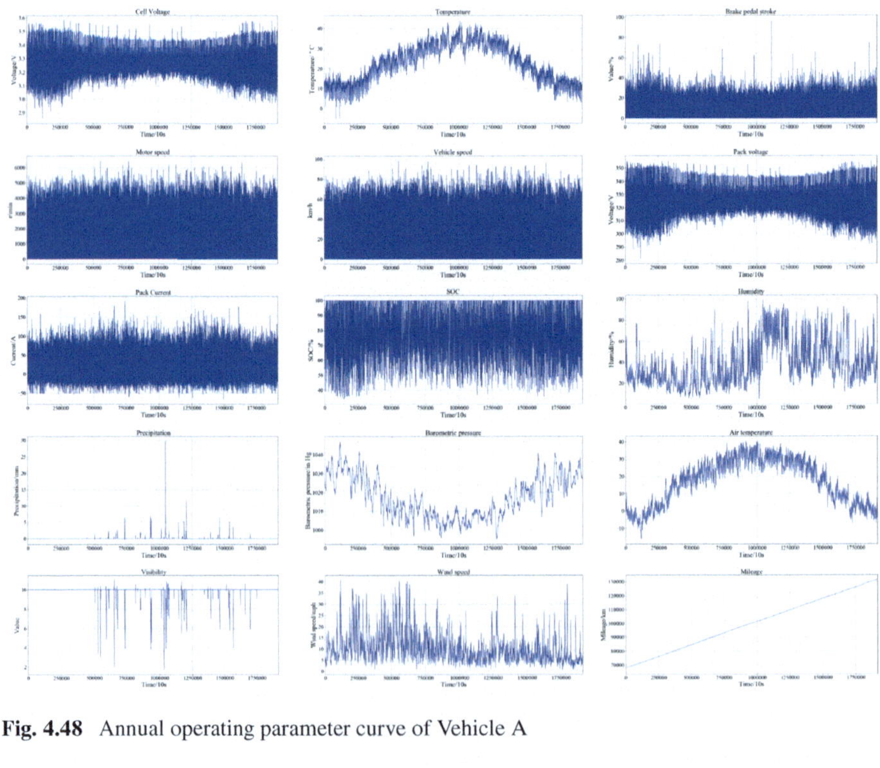

Fig. 4.48 Annual operating parameter curve of Vehicle A

capture the nonlinear coupling relationship between the predicted parameters and the complex and variable road conditions, weather and driver driving behavior, and how to distinguish between the complex and variable driving state of the vehicle and the relatively stable charging state, is still a major challenge for the full-weather full-state application of the LSTM model. ③ For the LSTM deep neural network, overfitting in training will degrade the learning process and prediction results.

To solve the above problems, the LSTM model constructed in this study mainly includes four parts: an LSTM(m–n) network architecture, a network parameter optimizer based on the Adam method, a "Pre-dropout" technique to prevent LSTM overfitting, and a "Dual-model-cooperation" (DMC) prediction strategy that can adaptively distinguish between charging/driving states.

4.3.2 Hyperparameter Optimization and Prediction Result Analysis

1. Preventing Overfitting Effects

   The generalization ability of the LSTM model during training will be affected by many factors, among which overfitting is one of the most common reasons, which is also the result of the mismatch between the model's learning ability and the complexity of the training data. There may be many reasons for the occurrence of overfitting, such as the selection of samples or the selection method is wrong, which will lead to the selected good training data set is not enough to represent the predetermined classification rules, too many input or output parameters will lead to the model complexity is too high, etc. The solutions to overfitting mainly include: regularisation (L1 and L2); data augmentation, that is, increasing the training data samples; Dropout technology; Early stopping. A considerable number of papers have been published on the research of various technologies to prevent neural network overfitting. However, both L1 and L2 regularisation methods add some extra terms to the original function to regularise the learning weights, where L1 regularisation adds the sum of all network weights and L2 regularisation adds the sum of the squares of the weights. The specific details of the Dropout technique are introduced in Sect. 4.1.2 (2).

2. Determination of LSTM Network Parameters

   In the training process of the LSTM model, it is not that the more input parameters, the better. When all strongly related parameters and weakly related parameters are added to the training, it will produce counterproductive training effects and increase the probability of overfitting. When all the above 15 parameters are used as input parameters to train the LSTM model, it is found that no matter which parameter is used as output, the training will end soon due to overfitting. This is because many of these parameters have particularly strong or particularly weak correlations, so it is very necessary to carry out correlation analysis between parameters before modeling to select the most suitable input parameters and eliminate disadvantageous parameters. In statistics, the Pearson correlation coefficient developed by Karl Pearson in the 1880s from an idea proposed by Francis Galton has been widely used in scientific research. The value of the Pearson correlation coefficient is between 1 and 1, where 1/1 is a positive/negative linear correlation relationship and 0 represents a nonlinear relationship. The PCC correlation coefficients between all 15 considered parameters are shown in Fig. 4.49 (to simplify the analysis, all PCC correlation coefficients have been processed in absolute value).

Then, after analyzing the correlation between all matchable parameters, only 5 input parameters can participate in training the LSTM model, which are single cell voltage, probe temperature, SOC, brake pedal opening and vehicle speed.

3. Determination of Different Types of Window Sizes

   Before training the LSTM model, many model parameters need to be set and optimized. However, models with large datasets require more time to train a set of hyperparameters, making it difficult to optimise the model's hyperparameters in a short time. Therefore, a set of hyperparameters needs to be preset based on past experience and then gradually optimized.

In the machine learning process, the training dataset needs to be divided into smaller sizes and provided to the computer one by one, and the weights of the neural network should be updated at the end of each step to suit the given dataset. Therefore, parameters like Epoch and Batch size (abbreviated as BS) are needed. Various window sizes also have a significant impact on the prediction effect because they can determine the number of samples input and output each time. In addition, to enable the training process to learn and remember more historical information, this paper applies the sliding window technique. Figure 4.50 is a schematic diagram of preparing LSTM training samples using the sliding window technique.

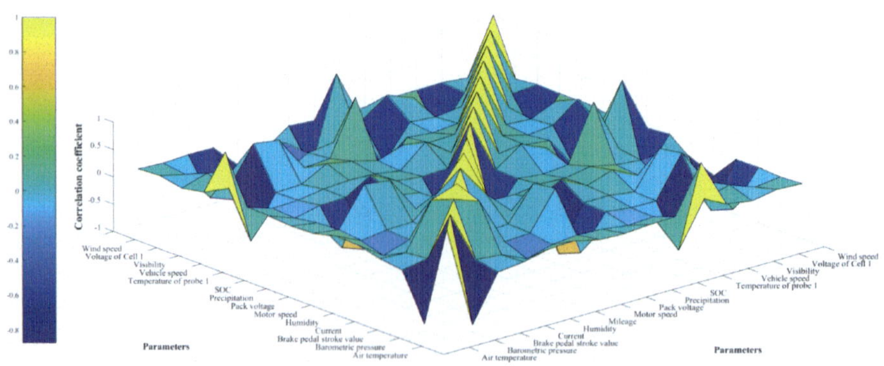

Fig. 4.49 Correlation coefficients between all potential input parameters

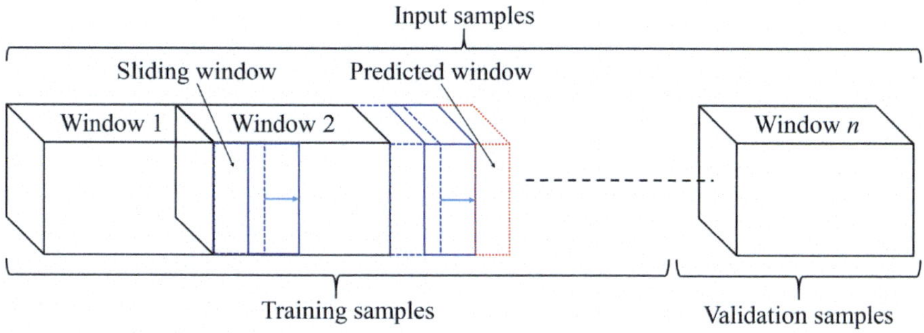

Fig. 4.50 Preparation of LSTM training samples using sliding window

Based on the above analysis, a preliminary set of hyperparameters is determined: Epoch = 50, BS = 128, the training set is one month's running data, the training window size (Training window size, TWS) is 360 (because the sampling interval is 10 s, so 360 represents one hour's data), the predicted window size (Predicted window size, PWS) is 6 (one minute's data), SWS = 60 (10 min' data). The constructed LSTM network consists of six layers: an input layer, three hidden layers (each with 100 neurons), a fully connected hidden layer with linear activation, and an output layer.

To facilitate the comparison of prediction effects of different training datasets (data for the whole year of 2016, data for one month in 2016, data for one week in 2016) in the following text, this paper uses the running data of vehicle A for one month in February 2016 as the initial training dataset, and the data for February 8, 2017, as the test data. To evaluate the predictive performance of the model, the Mean relative error (MRE) and Root mean square error (RSME) are the two most commonly used error evaluation methods. In this study, the formulas for MRE and RSME can be expressed as:

$$MRE = \frac{1}{n}\sum_{i=1}^{n}\frac{\left|\widehat{P_i} - P_i\right|}{P_i} \times 100\% \tag{4.31}$$

$$RSME = \sqrt{\frac{1}{n}\sum_{i=1}^{n}\left(\widehat{P_i} - P_i\right)^2} \tag{4.32}$$

where n is the number of training samples or test samples, ^Pi and Pi represent the predicted value and the actual value, respectively.

In addition, to obtain a comprehensive validation effect, "cross-validation" is one of the most commonly used validation methods in modeling applications, sometimes referred to as "rotation estimation", which is a practical method in statistics to divide data samples into smaller subsets, proposed by Seymour Geisser in 1993. Common forms of cross-validation include Holdout, K-fold cross-validation and Retention, among which K-fold cross-validation is the most widely used method. This method can repeatedly use randomly generated sub-samples for training and validation, and the results of each training will be validated once. To ensure the effectiveness and stability of the LSTM model, all training processes in this study have undergone tenfold cross-validation, the principle diagram of which is shown in Fig. 4.51; the training data and validation data account for 90% and 10% of the entire training dataset, respectively. In addition, MRE is used to evaluate the prediction results of the model on the test data after each cross-validation, and the average MRE of ten cross-validations is taken as the final prediction MRE of each set of hyperparameters.

Based on the preliminary selection of hyperparameters and tenfold cross-validation, the prediction results for cell voltage, probe temperature, and SOC are shown in Fig. 4.52, with their MREs being 1.15%, 5.56% and 3.55%, respectively. The figure shows that the predicted values of the three target parameters are very close to the actual values. In addition, Fig. 4.52c indicates that Car A only drove and charged once on 8th February 2017.

The correct choice of BS is to find the best balance between memory efficiency and memory capacity, as shown in Fig. 4.53a. Full batch (blue) training is suitable for smaller training datasets, its determined direction can better represent the overall sample, thus more accurately pointing towards the direction of the extremum, but it is not suitable for larger batch datasets. Mini-batch (green) training refers to choosing a moderate BS value, inputting the batch size data into the deep learning network, and

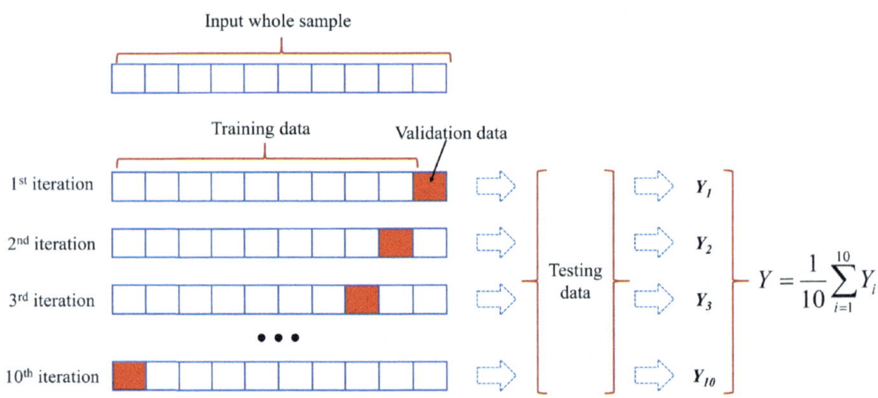

Fig. 4.51 Schematic diagram of tenfold cross-validation

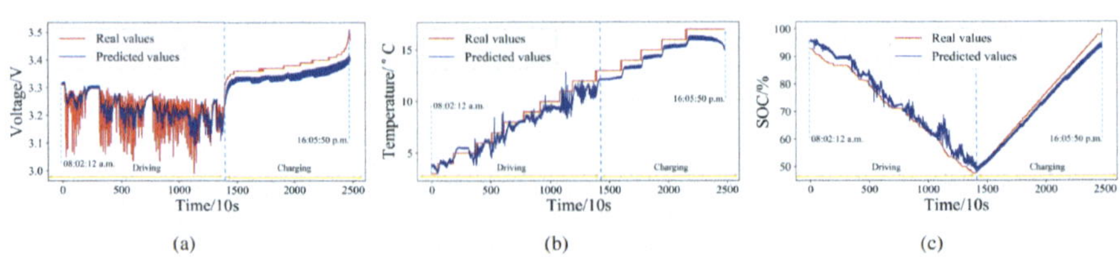

Fig. 4.52 Multi-parameter prediction results based on preliminarily selected hyperparameters. **a** cell voltage; **b** probe temperature; **c** SOC

then calculating the average loss function of all samples in this batch. Random batch (red, BS = 1) refers to each training adjusting according to the gradient direction of each sample; at this time, the global effect is not considered, so it isn't easy to achieve good convergence results in this case. Based on a large amount of empirical research, this study chose BS sizes in the range of {8, 16, 32, 64, 128, 256, 512, 1024} for comparison. The MRE on the test data based on different BSs is calculated to verify the prediction accuracy of the three target states, as shown in Fig. 4.53b. As can be seen from Fig. 4.53b, the MRE is lowest when BS = 16, indicating the best prediction performance.

To verify the prediction effect of the LSTM network at different prediction window sizes PWS, such as PWS = {6, 30, 60, 90, 120, 150, 180, 360}, they represent the predicted values within the next 1 min, 5 min, 10 min, 15 min, 20 min, 25 min, 30 min and 1 h respectively. The prediction MREs of different PWSs are compared and verified, as shown in Fig. 4.54. As PWS increases, the MRE values of the three target states also gradually increase, with the temperature MRE rising the steepest and the voltage MRE changing the most gently. Because TWS remains constant, as the prediction window gets larger and the prediction time gets longer, the learning difficulty becomes greater and the prediction accuracy becomes lower. Smaller PWSs should have higher prediction accuracy, however, if a severe battery failure or safety risk is predicted to occur within the next PWS, a too small PWS means that the escape and rescue time left for the driver is very short. Therefore, in this study, the prediction window is set to PWS = 30, that is, the prediction time window is 5 min.

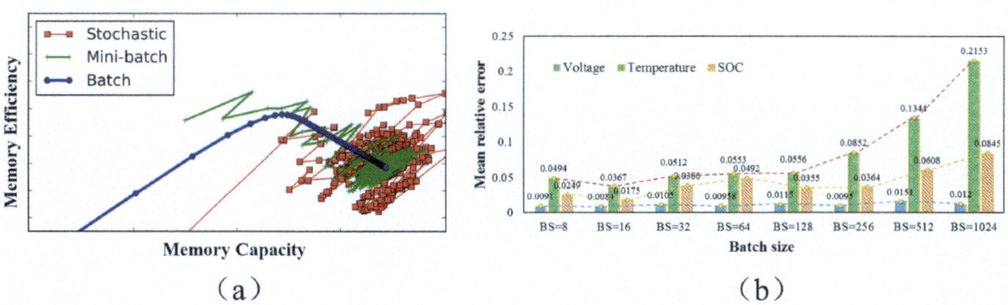

Fig. 4.53 BS values: **a** Relationship between memory efficiency and memory capacity for different BSs; **b** Prediction accuracy based on different BSs

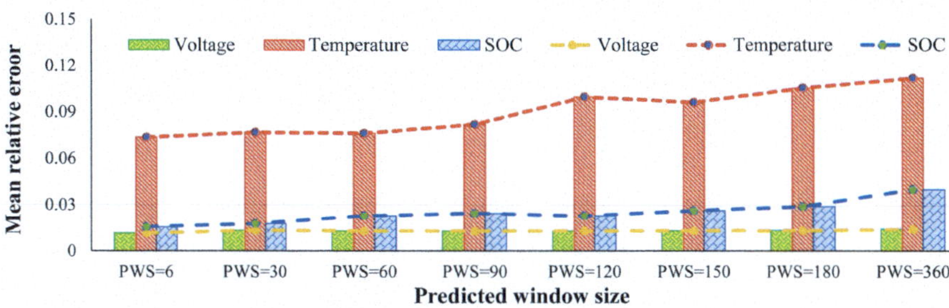

Fig. 4.54 Prediction results of different prediction window sizes

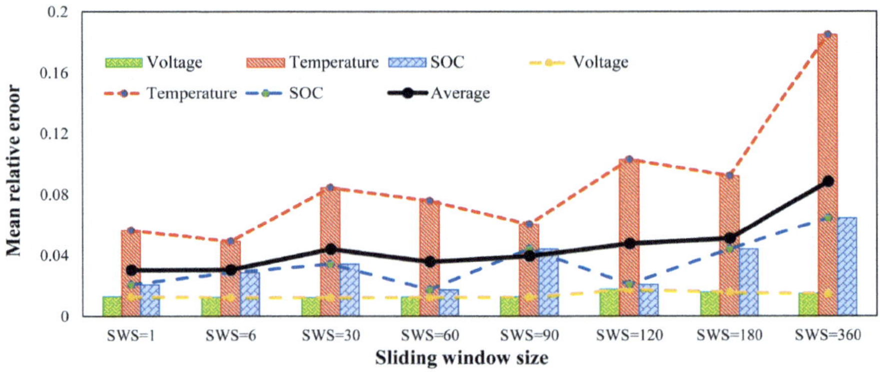

Fig. 4.55 Prediction results of different sliding window sizes

To verify the prediction effect of different sliding windows, SWS = {1, 6, 30, 60, 90, 120, 180, 360} are used for comparison analysis, that is, 10 s, 1 min, 5 min, 10 min, 15 min, 20 min, 30 min and 1 h are used as sliding steps respectively, where SWS = 360 = TWS means that the training window does not iterate or slide. The prediction MRE based on different SWSs is shown in Fig. 4.55. The prediction MREs of the three battery states have a trend of gradually increasing with the increase of SWS, among which the MREs of temperature and SOC fluctuate greatly, and the change of voltage MRE is not obvious. Ideally, the smallest SWS (SWS = 1) can obtain almost all available historical information, but a smaller SWS also means higher computer configuration requirements and longer training time. Figure 4.55 shows that the error at SWS = 6 is very close to that at SWS = 1, which means that the learning ability at SWS = 6 is almost equivalent to that at SWS = 1.

4. Comparison of Different Training Datasets

   Using the data of one day on 8th February 2017 as the test data, the prediction effects of different training datasets (data for the year 2016, data for February 2016, data for the first week of February 2016) are compared. The prediction MREs of different training datasets are shown in Fig. 4.56. As the training dataset increases, the prediction MREs of the three target states significantly decrease, especially when the training dataset is the smallest one-week data, and the model prediction accuracy is much worse.

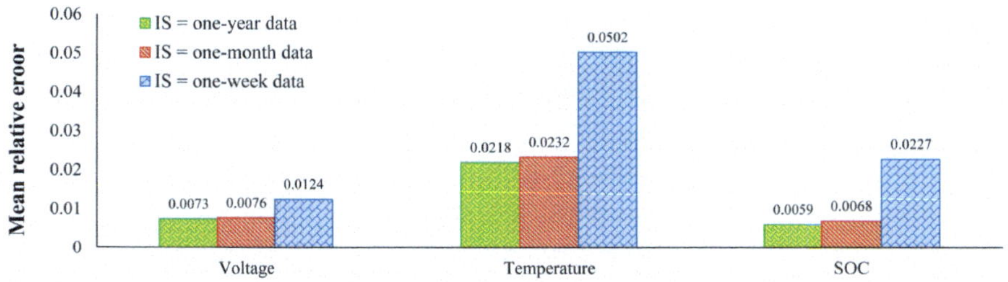

Fig. 4.56 Comparison of prediction results from different training datasets

4.3.3 Results Verification and Analysis

1. DMC Prediction Strategy

   Figure 4.57 shows the prediction errors of the many-to-many (5–3) model and the many-to-many (3–3) model tested on charging data and driving data, respectively. LSTM (3–3) uses voltage, temperature, and SOC as output and input. Through Fig. 4.57, it is found that the two model structures have different prediction effects on charging data and driving data. When the vehicle is driving, the prediction accuracy of the LSTM (5–3) model is significantly better than that of the LSTM (3–3) model because it considers the driver's driving behaviour. Conversely, the predictive performance of the LSTM (3–3) model during the charging phase is slightly better than that of the LSTM (5–3) model because when the vehicle is charging, the speed and brake pedal stroke value is constant at zero, so for the LSTM (5–3) model, the actual input parameters used are only three target parameters, so using five input parameters for training at this time will adversely affect the prediction results of the charging data.

In order to ensure the prediction accuracy under both charging and driving states, this study proposes a DMC prediction strategy, which allows the LSTM (5–3) and LSTM (3–3) models to work during the driving and charging phases, respectively.

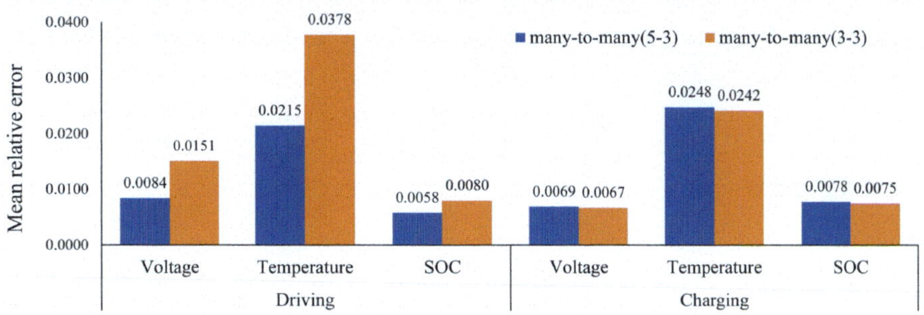

Fig. 4.57 Comparison of the prediction effects of the LSTM (5–3) model and the LSTM (5–3) model during charging and driving phases

Table 4.8 Vehicle state analysis based on current and speed

Current and speed status	Vehicle state
Current $I > 0$ and speed $V = 0$	Stopped during driving
Current $I > 0$ and speed $V > 0$	Driving
Current $I > 0$ and speed $V < 0$	Reversing
Current $I < 0$ and speed $V = 0$	Charging
Current $I < 0$ and speed $V > 0$	Regenerative braking
Current $I < 0$ and speed $V < 0$	Will not occur
Current $I = 0$	Vehicle not started

The working steps of the prediction strategy are as follows: first, the vehicle state is judged according to the state of the battery pack current I and the vehicle speed V, as shown in Table 4.8. When the vehicle is charging, $I > 0$; when the vehicle is driving, $I < 0$. From Table 4.8, when $I < 0$ and $V = 0$, the vehicle is judged to be in a charging state, and all other conditions except $I = 0$ are considered to be driving states. When the vehicle is in a complex and changing driving state, the LSTM (5–3) model is used for prediction; when the vehicle is in a stable charging state, it automatically switches to the LSTM (3–3) model for prediction.

In addition, under the existing computer configuration and selected hyperparameters, through cross-validation, when the training dataset is one year data, one month data, and one week data, the average time required to train the LSTM model is 69.38 h, 16.10 h, and 4.26 h, respectively, but the test time for the same test data is about 7 s. Therefore, although the offline training of the LSTM model takes a long time, when the hyperparameters are determined and the model is well trained offline, the LSTM model only needs a very short time to perform online prediction, and it is not constrained by the size of the training dataset. When performing online predictions on actual running vehicles, the test data is just the data of the previous hour, not the one-day data in this study, and the time for online prediction will be shorter. Therefore, the proposed LSTM model can achieve synchronous online rapid prediction of battery voltage, temperature, and SOC. Based on the selected hyperparameters, when the training dataset is one year data, and the prediction window is PWS = 30, the prediction MRE of the three target states are 0.7%, 2.02%, and 0.58%, respectively. When the prediction window is PWS = 1, the prediction MRE of the

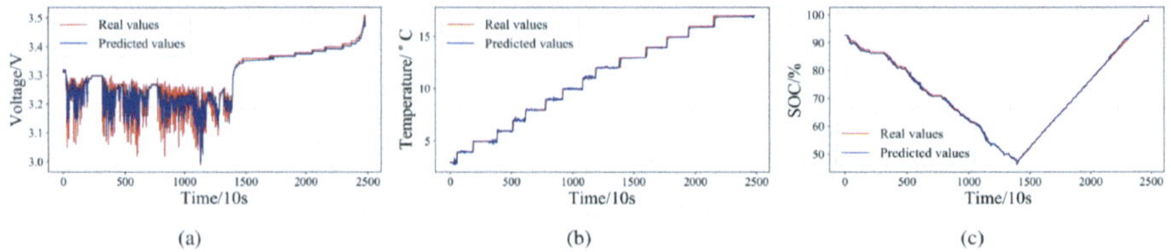

Fig. 4.58 Predicted results based on the final hyperparameters: **a** single-cell voltage; **b** probe temperature; **c** SOC

three target states are 0.58%, 0.82%, and 0.48%, respectively. The prediction results are shown in Fig. 4.58. From the figure, it can be seen that the predicted values after hyperparameter optimisation are closer to the actual values than before optimisation, especially since the temperature and SOC almost coincide with the real values.

2. Validity and Flexibility Verification

   Comparing the prediction results of the LSTM model when PWS = 1 with other prediction algorithms shown in Table 4.9, the results show that the LSTM model has superior prediction performance than other methods. More importantly, this method can be applied to all-weather and all-state applications in actual running vehicles, accurately predicting various battery states without being affected by complex operating environments. Therefore, this method has broad application prospects in the joint prediction and safety control management of multiple states of power battery systems.

The goal of this study is to jointly predict the changes in various battery states over a future period of time, so overly pursuing a small PWS may not necessarily yield ideal results. As shown in Fig. 4.59, a smaller PWS represents short-term prediction, which can achieve higher prediction accuracy, but the available time

Table 4.9 Comparison of prediction effects of various battery parameter prediction studies

<table>
<tr><th colspan="2">Prediction method</th><th>Error</th><th>Test environment</th><th>Prediction target</th><th>Prediction steps</th></tr>
<tr><td colspan="2">Simplified physical model</td><td>MRE < 5%</td><td>−10, 0, 20, and 45 °C</td><td rowspan="2">Voltage</td><td rowspan="7">Single step</td></tr>
<tr><td colspan="2">KLMS-X filter</td><td>RMSE < 0.329</td><td>Constant speed driving (3.5 km/h)</td></tr>
<tr><td colspan="2">Two-step prediction method</td><td>MRE < 3.05%</td><td>20, 30, and 45 °C (room temperature)</td><td rowspan="2">Temperature</td></tr>
<tr><td colspan="2">Kalman filter</td><td>RMSE < 0.269</td><td>60A charge and discharge</td></tr>
<tr><td colspan="2">Support vector machine</td><td>MRE < 2%</td><td>25 °C (room temperature)</td><td rowspan="3">SOC</td></tr>
<tr><td colspan="2">LSTM-RNN model</td><td>MRE < 0.6%</td><td>0, 10, and 25 °C (room temperature)</td></tr>
<tr><td colspan="2">Fuzzy neural network</td><td>MRE < 0.9%</td><td>25 °C(Room Temperature)</td></tr>
<tr><td rowspan="3">LSTM model of this study</td><td>Voltage</td><td>MRE < 0.58%
RMSE < 0.037</td><td rowspan="3">All-weather All-state real vehicle Application</td><td rowspan="3">Multiple parameters</td><td rowspan="3">Adjustable multi-step</td></tr>
<tr><td>Temperature</td><td>MRE < 0.82%
RMSE < 0.1513</td></tr>
<tr><td>SOC</td><td>MRE < 0.48%</td></tr>
</table>

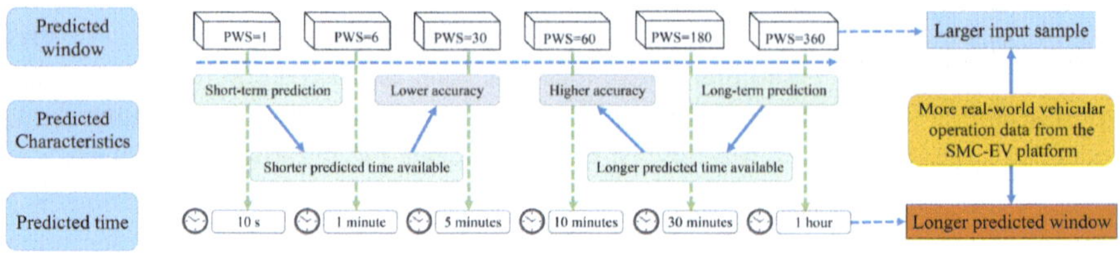

Fig. 4.59 Relationship between prediction window size and prediction accuracy

to take emergency measures when any safety risk is predicted is shorter; a larger PWS represents long-term prediction, its prediction accuracy is lower, but it can provide more emergency handling time for predicted safety risks. Therefore, how to choose PWS also depends on the requirements of PWS and prediction accuracy. In order to improve prediction accuracy and extend the prediction window, it is necessary to continuously expand training data and update and optimise the model's hyperparameters.

It is well known that a smaller prediction MRE and a shorter predicted time (Predicted time, abbreviated as PT) will produce better prediction results. To verify the sensitivity of the LSTM model PWS, the relationship between the predicted MRE and PT is calculated, and the predictive sensitivity (Predictive sensitivity, abbreviated as PS) is used as the final performance indicator for determining PWS. The predictive sensitivity at PWS = 6 is used as the baseline, as shown in Eq. (4.9). Where PS(6) is the predictive sensitivity at PWS = 6. The calculated predictive sensitivities for different PWS are shown in Table 4.6. The results show that as PWS increases, the predicted time *PT* does not change significantly, but the predictive sensitivity gradually decreases, and the prediction effect gradually worsens. Therefore, the size of PWS can be determined by setting a predictive sensitivity threshold.

$$PS = \frac{PT}{MRE * PS(6)} \tag{4.33}$$

3. Verification of Stability and Robustness

   In order to verify the stability and robustness of the LSTM model under different operating environments, the data of April 1, July 1, October 1, and January 1, 2017, were used as the test data for the four seasons for comparison, and their predicted MREs were compared, as shown in Table 4.10. The results show that the LSTM model has good predictive performance in all four seasonal environments, demonstrating excellent all-weather application effects. Among them, because Beijing has a lower outdoor temperature in winter and the operating environment is relatively harsh, the predictive performance in winter is slightly worse than in the other four seasons.

In addition, four operating condition cases are listed to verify the prediction effect of the model under different driving conditions. Figure 4.60 shows the one-hour

Table 4.10 Comparison of predicted MREs in different seasons

Date\parameter	Temperature	Voltage	Temperature	SOC
1st April 2017	5~22 °C	0.0051	0.0212	0.0042
1st June 2017	24~36 °C	0.0039	0.0064	0.0104
1st October 2017	13~26 °C	0.007	0.0121	0.0088
1st January 2017	−3~6 °C	0.0067	0.0325	0.0092

Table 4.11 Comparison of predicted MRE under different driving conditions

Condition\parameter	Voltage	Temperature	SOC
Condition 1	0.0057	0.0278	0.0051
Condition 2	0.0055	0.0252	0.0048
Condition 3	0.0047	0.0098	0.0041
Condition 4	0.0045	0.0101	0.0042

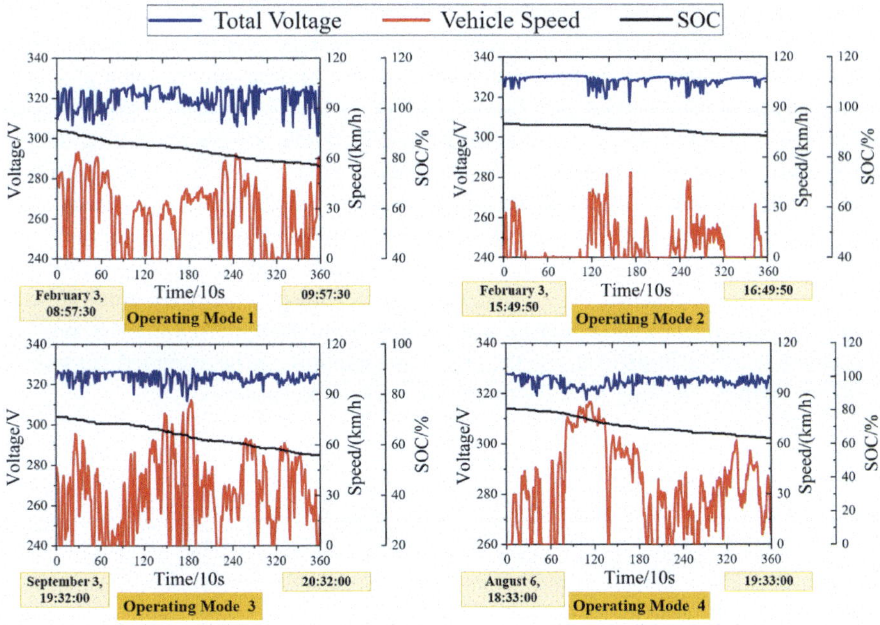

Fig. 4.60 One-hour driving characteristics under different driving conditions

driving data characteristics of the vehicle under study in conditions 1 to 4. Condition 1 is when the vehicle is driving at less than 60 Operating at normal speeds of km/h; Condition 2 indicates the vehicle operating under low speed and frequent idling conditions; Conditions 3 and 4 respectively show two sets of high-speed driving conditions exceeding 80 km/h, the predicted MRE of 1-h test data under these four conditions are listed in Table 4.11. The results show that the proposed LSTM model has stable predictive performance under different driving conditions.

4.4 Summary

This chapter provides a detailed introduction to the prediction methods of power battery system temperature, voltage, SOC, SOP, SOE, and introduces the principles and verification process of LSTM, GRU, SAM and other methods, and discusses in detail the parameter tuning and verification effects of each method.

Based on the LSTM neural network and real vehicle operation big data, a real vehicle power battery system voltage, temperature and SOC multi-state online multi-step joint prediction model is proposed, which overcomes the problem of poor parameter identification and modelling effect of existing battery models facing real vehicle operation data, and can effectively capture the non-linear coupling relationship between battery state and various complex alternating stresses, achieving multi-step prediction of multiple states of real vehicle power battery system in the future. The "weather-vehicle-driver" analysis method takes into account environmental stress, ageing stress, dynamic load and driving behaviour and other multi-stress redundancy conditions, Pre-dropout technology can effectively prevent overfitting during model training, DMC prediction strategy can adaptively switch charging/driving state, effectively improving model training efficiency and prediction accuracy. Test results show that the single-step prediction (next 10 s) MRE of voltage, temperature and SOC reached 0.58%, 0.82% and 0.51% respectively, and the 30-step prediction (next 5 min) MRE reached 0.7%, 2.02% and 0.58% respectively. By continuously expanding training data and updating and optimising model hyperparameters, prediction accuracy can be improved, and the prediction window can be extended, providing more emergency handling time for potential safety risk warnings.

Chapter 5
Research on Fault Diagnosis of Power Battery System

5.1 Overview

With the increasing prominence of energy and environmental issues, the global energy structure is undergoing a huge transformation. As the energy device of new energy vehicles, the safety and stability of power batteries directly affect the safety of vehicles and drivers. Therefore, fault type classification and fault location positioning through power battery system fault diagnosis are important methods to improve the safety of electric vehicles. For power battery system faults, extensive research has been carried out based on various observers, filters and machine learning methods, but existing fault diagnosis methods often have disadvantages such as slow calculation speed, low real-time diagnosis accuracy, and poor robustness. To overcome the defects of traditional threshold method diagnosis indicators being single and early warning time being short, this chapter will deeply study the typical fault mechanism and fault diagnosis method of power batteries, based on real vehicle operation big data, explore the diagnosis method and diagnosis strategy of early hidden battery faults within the conventional safety threshold range, and realize the qualitative classification of power battery fault features and quantitative evaluation of fault levels (Fig. 5.1).

J. Hong, *Big Data Analysis and Application of Power Battery Operation Safety*, Key Technologies on New Energy Vehicles, https://doi.org/10.1007/978-981-96-4399-8_5

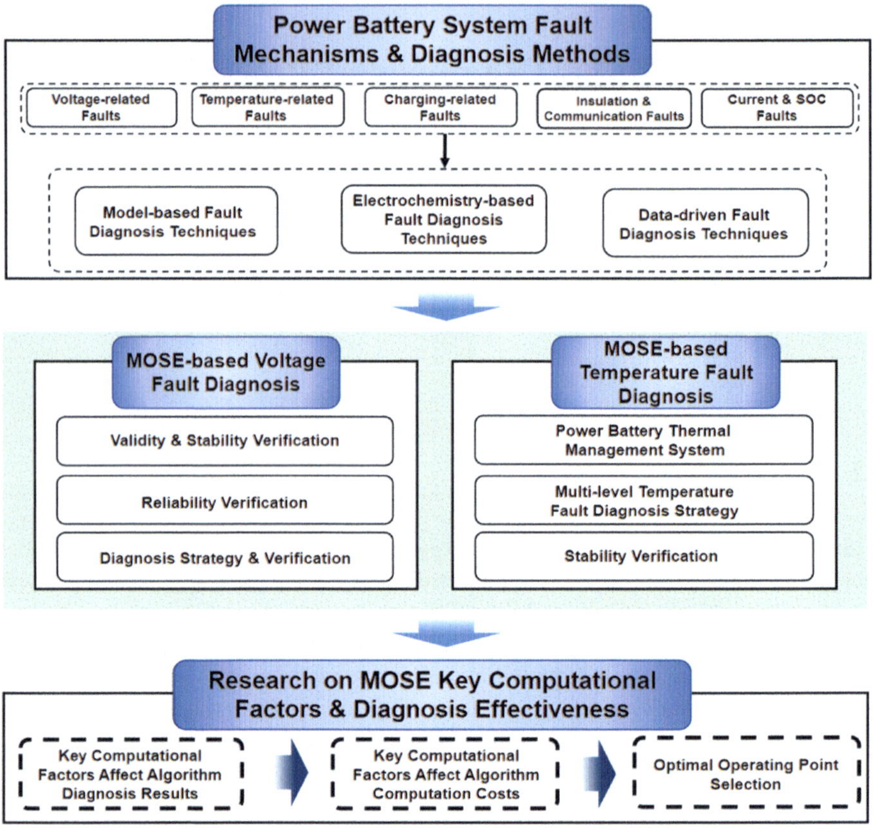

Fig. 5.1 Research on fault diagnosis of power battery system

5.2 Failure Mechanism and Diagnosis Method of Power Battery System

5.2.1 Analysis of Failure Mechanism of Power Battery System

Power battery system failures can be divided into several typical fault types such as voltage faults, temperature faults, charging faults, insulation and communication faults, SOC and current abnormalities. Below we will analyze these fault types and their respective causes.

1. Voltage Faults

   The main types of voltage faults in the power battery system include overvoltage, undervoltage, excessive voltage difference, and voltage jump. Overvoltage fault refers to the phenomenon where one or more battery cell voltages are significantly higher and exceed a certain safety threshold. The causes of such faults are generally due to collection errors, load and measurement unit (Load and Measurement Unit, LMU) balance function failure, low cell capacity or too fast charging voltage rise. Undervoltage fault refers to the phenomenon where one or more battery cell voltages are significantly lower and exceed a certain safety threshold. The possible causes of this type of fault are collection errors, LMU balance function failure, low capacity or too fast discharge voltage drop. Excessive voltage difference faults usually include dynamic and static excessive voltage differences, mainly manifested as the voltage of one or more cells reaching the full charge COV first during the charging of the power battery system; during the actual vehicle driving process, the throttling process makes the voltage of one or more cells drop faster than other cells; during the actual vehicle driving process, the brake process makes the voltage of one or more cells rise faster than other cells. The causes of excessive voltage difference faults are usually due to loose fastening connection nuts, contamination at the connection, high self-discharge rate of the cell, low cell capacity, and cell leakage. Voltage jump refers to the phenomenon of cell voltage jump during vehicle operation or charging, mainly due to collection errors, loose signal collection lines or LMU faults.

2. Temperature Faults

The main types of temperature faults in the power battery system include faults in the battery thermal management system (Battery Thermal Management System, BTMS), high temperature, low temperature, excessive temperature difference, and rapid temperature rise. BTMS faults are mainly divided into heating faults and cooling faults, manifested as the predetermined heating and cooling devices that cannot be normally opened when the battery box temperature reaches a certain high or low value. The causes are usually due to the failure of the heating/cooling relay or the battery group management module (Battery Management Unit, BMU), or the failure of the heating sheet and cooling fan. High temperature fault refers to the phenomenon where the temperature of one or more temperature points in the power battery system is high and reaches a certain safety alarm threshold during operation. This type of fault is often caused by temperature sensor faults, LMU faults, fan faults or overcharging. Low temperature refers to the phenomenon where the temperature of one or more temperature points in the power battery system is low and reaches the safety alarm threshold during operation. This type of fault is usually caused by temperature sensor faults or LMU faults. In addition, the phenomena and causes of the excessive temperature difference and rapid temperature rise faults are similar to high and low-temperature faults, and various faults are not independent of each other, but are interrelated and cause and effect. For example, rapid temperature rise faults generally occur with high temperature or excessive temperature difference faults, because local rapid temperature rise is very likely to cause high temperature and excessive temperature difference.

3. Charging Faults

   The main types of charging faults in the power battery system are divided into DC charging faults and AC charging faults. Based on the definition conditions of GB/T 27,930-2015,[1] if the power battery system cannot start charging, the charging gun jumps during charging, or the battery SOC cannot be successfully reset after charging, it can be considered that the battery has a charging fault. The causes of the fault include voltage and temperature class battery faults, abnormal CC1 ground resistance, BMU faults, and abnormal CC2 ground voltage.

4. Insulation and Communication Faults

   Insulation faults in the power battery system refer to abnormal insulation resistance. The causes of the fault include insulation misdiagnosis, water ingress in the battery box plug-in, excessive environmental humidity, battery leakage or controller insulation faults. Communication faults in the power battery system mainly manifest as communication faults of LUM, BMU and other components, and missing information of LMU or BMS. The causes of the fault are mostly abnormal LMU/BMU signal communication.

5. Current and SOC Anomalies

   Current anomalies in the power battery system usually manifest as Hall sensor and circuit faults, DC charger faults or BMS control logic failures. SOC anomalies in the power battery system include SOC jumps, inaccurate SOC estimation, no change in SOC or too fast/slow decline. The main causes of this type of fault are communication anomalies, inaccurate estimation methods, current anomalies and BMU faults.

Power battery system faults can be divided into two main types according to their location: single-cell battery faults and battery pack faults, as shown in the fault tree of the power battery system in Fig. 5.2 through fault analysis. Single-cell battery faults mainly refer to the side reactions produced by the electrochemical reactions between the electrodes and the electrolyte during the operation of the battery, such as lithium deposition, membrane puncture, electrolyte decomposition, etc., or battery faults caused by overcharging, over-discharging, overheating, short-circuiting and other misuse conditions. These can specifically manifest as a decrease in single-cell battery voltage, an increase in temperature, and a decrease in capacity, and may further trigger extreme dangers such as thermal runaway or even explosion of the battery.

Battery pack faults in the power battery system mainly occur in the battery packs of electric vehicles. Due to the limitations of existing technical conditions, the voltage of a single-cell battery will generally not exceed two digits under a defined volume,

[1] The current charging protocol, officially known as the "Communication Protocol between Non-Vehicle Conductive Chargers and Battery Management Systems for Electric Vehicles", was issued by the General Administration of Quality Supervision, Inspection and Quarantine of the People's Republic of China and the Standardisation Administration of China on 28th December 2015.

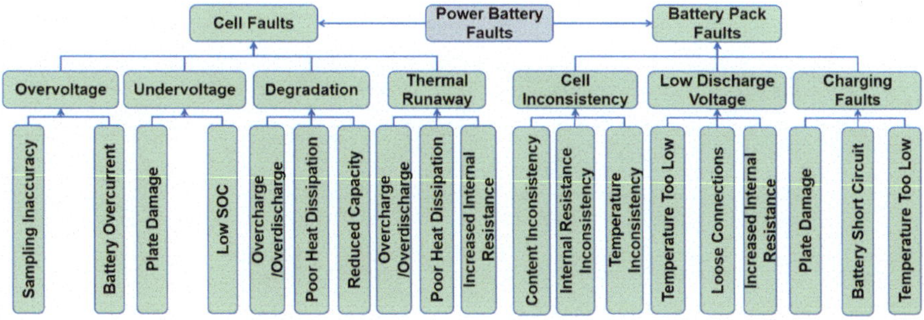

Fig. 5.2 Power battery system fault tree

and the energy of a single-cell battery cannot meet the power requirements of electric vehicles. Therefore, the energy source of electric vehicles mainly comes from battery packs (or battery packs) composed of many single-cell batteries in series or parallel. However, during the use of single-cell batteries in series and parallel, inconsistencies can cause some single cells to overcharge, over-discharge, and accelerate the aging of capacity. For example, during the charging of the power battery system, low-capacity single cells and high-capacity single cells are in overcharged and undercharged states respectively; during discharge, low-capacity single cells and high-capacity single cells are in over-discharged and under-discharged states respectively. In addition, power battery packs that have already shown inconsistencies will continue to increase their inconsistency levels with long-term work and use, resulting in a decline in overall performance and rapid decay of life.

5.2.2 Power Battery System Fault Diagnosis Method

1. **Fault Diagnosis Method**

   Fault diagnosis refers to the precise identification, alarm and post-processing of faults by combining the working principle characteristics of the diagnosis target and using technologies such as power electronics and computers. According to the views of the international authority on fault diagnosis, Professor Paul M. Frank of the University of Duisburg in Germany, fault diagnosis methods can generally be divided into methods based on analytical models, knowledge-based methods and signal processing-based methods. In recent years, with the continuous development of fault diagnosis technology, we can divide fault diagnosis methods into two major categories, namely quantitative analysis methods and qualitative analysis methods, as shown in Fig. 5.3.

Fault diagnosis based on qualitative analysis does not need to establish a mathematical model, and can roughly be divided into graph theory methods, qualitative simulation and expert system methods. Among them, fault diagnosis methods based

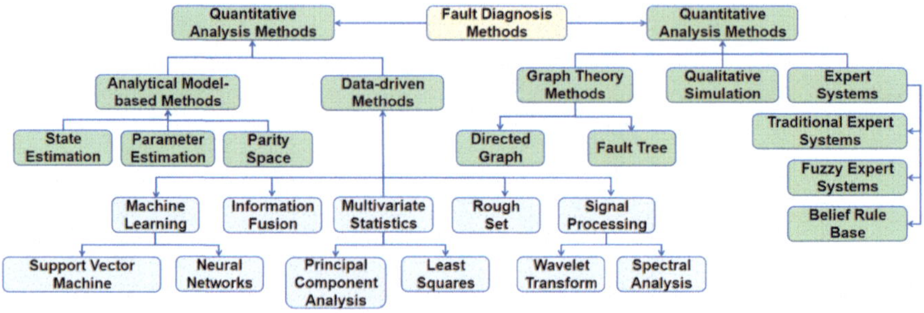

Fig. 5.3 Classification of fault diagnosis methods

on graph theory include symbolic directed graph methods and fault tree analysis methods. The advantage of this method is that it can obtain easily understandable diagnostic results, so its applicability is relatively wide, but the disadvantage is that it is difficult to accurately diagnose faults in large and complex systems. The fault diagnosis method based on the expert system is mainly based on the extensive engineering practice experience of industry experts, but the level of knowledge will have a great impact on the accuracy of this type of diagnosis method. Qualitative simulation technology is a grey system reasoning technology for dealing with incomplete information that has developed in recent years based on research in the field of artificial intelligence. This technology mainly studies how to determine the qualitative state of system variables using various qualitative constraints and qualitative state transmission rules, but its effectiveness depends on the amount of information provided by the system.

Fault diagnosis methods based on quantitative analysis include methods based on analytical models and data-driven methods. Among them, fault diagnosis methods based on analytical models can be divided into state estimation methods, parameter estimation methods and equivalent space methods according to the way residuals are generated. The main advantage of this type of method is that it can deeply understand the dynamic changes of the diagnosed system, which is convenient for real-time diagnosis and isolation of faults, but the disadvantage is that the diagnostic accuracy depends too much on the established mathematical model, and the modeling effect on non-linear models is not good. Data-driven fault diagnosis methods mainly include machine learning methods, rough set methods, multivariate statistical methods, information fusion methods and signal processing methods. Although this type of method has a wider range of applications than methods based on analytical models, it requires a large amount of computation and high requirements for the quantity and quality of historical data of the research object.

2. **Power Battery System Fault Diagnosis Method**

The fault diagnosis of the power battery system is usually based on the real-time collected current, voltage, temperature and other state information, using effective fault diagnosis methods to design a reasonable fault diagnosis system,

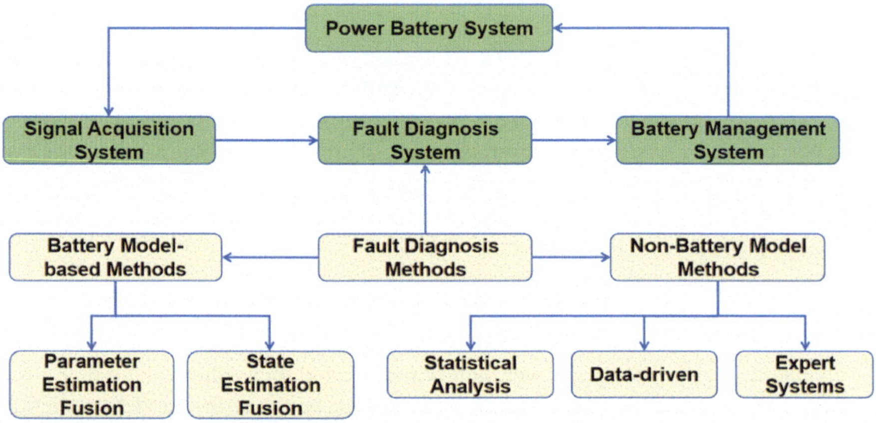

Fig. 5.4 Schematic diagram of the fault diagnosis process of the power battery system

and feeding back the fault diagnosis information to the BMS system, in order to take necessary protective measures or repair schemes, to ensure the safe, stable and reliable operation of the power battery system. The schematic diagram of the fault diagnosis process of the power battery system is shown in Fig. 5.4.

At present, the fault diagnosis methods of the power battery system can be divided into two major categories according to the application method: fault diagnosis methods based on battery models and non-battery model fault diagnosis methods, as shown in Fig. 5.4.

The fault diagnosis method of non-battery models does not require dynamic characteristic modeling of the power battery system, and can efficiently perform online iterative updates of battery parameters. However, compared to model-based fault diagnosis methods, this method relies too heavily on the quality and quantity of samples, and it is difficult to quantitatively describe the diagnosis results, so its universality and applicability are poor. Existing fault diagnosis methods of non-battery models can be roughly divided into three categories: based on statistical analysis, data-driven, and expert system-based methods.

Fault diagnosis methods based on statistical analysis can directly use statistical methods such as information entropy and normal distribution to analyze the collected battery current, voltage, and temperature data, and then set suitable abnormal coefficient thresholds to achieve online diagnosis of battery faults. However, this type of method's ability to recognize fault types is not good. Data-driven fault diagnosis methods can use the measurable data of the power battery system as input, with battery fault characteristics as output, and then construct data-driven models such as artificial neural networks and multi-model fusion strategies to achieve online precise diagnosis of power battery system faults. However, the accuracy of this type of method's fault diagnosis is highly dependent on the quality and quantity of training data, and it is necessary to use the power battery system big data platform to obtain sufficient training data. The power battery system fault diagnosis method based on expert systems consists of a real-time database, knowledge base, inference engine, and human–machine interface, among which the knowledge base is the core of the expert system; its adaptability needs to be regularly/irregularly updated and optimized online by reasonably using the actual operation data and maintenance records of the power battery system. However, the rules or models in the knowledge base can easily conflict with each other, and using expert systems for power battery system fault diagnosis requires the development of a decision-making method with strong self-learning ability and robustness.

3. **Model-Based Power Battery Fault Diagnosis Technology**

   The key to model-based power battery fault diagnosis is how to establish a high-precision, high-reliability power battery model, and then achieve fault diagnosis and early warning by comparing the difference between the model prediction value and the actual measured value. The battery models that are currently widely researched and applied mainly include electrochemical models, equivalent circuit models, and data-driven models. Among them, the design and calculation of the battery electrochemical model are relatively complex, so it is difficult to complete real-time online applications, and it is generally used for battery electrochemical mechanism analysis; the data-driven model is an empirical model, its performance is overly dependent on the quality of the research sample, and its universality is difficult to guarantee; the battery equivalent circuit model is a semi-empirical and semi-mechanism model, which has the advantages of low computational complexity and good predictive performance, and is also the most widely used fault diagnosis method based on battery models in existing battery management systems, usually integrating parameter estimation and state estimation methods to establish battery models.

The fusion parameter estimation method can establish normal battery models such as first-order equivalent circuit models and second-order equivalent circuit models, and the model parameters identified online can achieve power battery system fault diagnosis and also establish battery fault models, using the comparison of actual measured values and fault model prediction values to achieve fault diagnosis. The fusion state estimation method usually uses battery state parameters as the input and output of the model, combines the actual measured values of state parameters and filtering algorithms to obtain estimated values, and then calculates the residuals of the estimated values and their actual measured values to achieve fault diagnosis of the battery system. The key to this type of diagnostic method is how to design and develop high-precision, high-reliability, and high-stability state space models and state estimation methods.

4. **Power Battery Fault Diagnosis Technology Based on Electrochemical Principles**

   Fault diagnosis methods based on electrochemical principles monitor the internal parameters of the power battery, extract the feature vectors of the battery operation cycle as the basis for fault diagnosis, and combine the battery model to estimate the parameters and states of the power battery. Dey and others used an electrochemical model combined with thermodynamic principles to establish multiple sliding observers for different fault modes, and performed fault identification and diagnosis of power batteries through the electrochemical parameters extracted by the observers. Li and others used the P2D model of machine learning to identify electrochemical reaction parameters (diffusion coefficients, reaction rates, etc.) with external representations as inputs, and compared them with the calibration data of the three-electrode battery, which can diagnose certain battery safety risks and internal faults. Establishing an electrochemical impedance spectrum model and identifying the state of the battery using the parameters identified by this model is also one of the fault diagnosis methods based on electrochemical principles. However, these methods are all limited to the degree of mastery of the internal reactions of the battery. At present, the complexity of the internal reactions of the battery still limits the development of fault diagnosis technology based on electrochemical principles.

5. **Data-Driven Power Battery Fault Diagnosis Technology**

Data-driven fault diagnosis methods can mine and analyze the information in the data itself, and predict possible faults. The key lies in how to establish an accurate data model to simplify the diagnostic process, improve diagnostic accuracy, and reduce training difficulty. Data-driven methods avoid the complex process of establishing physical models and also avoid misdiagnosis and missed diagnoses caused by the failure of physical models. However, the diagnostic results of this method are difficult to interpret clearly from structural, physical, and chemical perspectives, and often require a large amount of historical data for training and verification to make the results accurate and reliable. Internal short circuit phenomena occurring in the battery can be diagnosed by identifying the voltage curve and correlation coefficient of the battery cell. Or a new data-driven fault diagnosis method based on a single battery voltage comparison. At the same time, methods based on machine learning are also one of the important ways of data-driven, but methods related to machine learning often require a large amount of data for training. A SOC prediction model of a lithium-ion battery system was established on the support vector machine, including a comparison with the support vector machine model and prediction simulation of the battery under fault conditions. In addition, voltage outliers in small-scale data sets are detected through angle variance tests to achieve fault diagnosis under small data sets. Regardless of the size of the data set, these data-driven methods are based on fault data for model training and then diagnosis.

5.3 Voltage Fault Diagnosis Based on MOSE

5.3.1 MOSE Algorithm

The concept of information entropy was first proposed by Laude Elwood Shannon in 1948. It is now widely used in thermodynamics, information science, and other fields to diagnose the degree of disorder in systems. It is also commonly referred to as Shannon entropy. Its typical calculation process is as follows:

$$H(X) = -\sum_{i=1}^{n} p(x_i)\log p(x_i) \tag{5.1}$$

where: x represents a random variable, corresponding to the set of all possible outputs X. $H(X)$ is the entropy value; $p(xi)$ represents the output probability function in the ith interval; n is the number of intervals.

The calculation of entropy is an iterative process, and the amount of calculation will increase as the data accumulates, which poses a challenge to the real-time diagnosis of power batteries online. Therefore, it is necessary to make appropriate modifications to the Shannon entropy algorithm to adapt to the needs of online fault diagnosis. This study has made modifications to the data reading method, calculation window length, and limit threshold selection, and the MSNE calculation process obtained after the modification is shown in Fig. 5.5.

5.3.2 Diagnostic Strategy and Effect Verification

1. **Validity and Stability Verification**

   By retrieving and exporting the historical operation data of vehicle A, its power battery system contains 98 single batteries, and the data collection time is from 10:50:16 to 11:10:16 on March 22, 2016, with a sampling interval of 1 s. When using the MSNE algorithm to process the collected data, it is found that the accuracy of MSNE is affected by the calculation window parameter K value, for example, the single cell voltage of vehicle A as shown in Fig. 5.5 is respectively $K = 10$ and $K = 300$ were used for the calculations. If the value of K is too small, the fluctuation characteristics of the battery voltage cannot be fully demonstrated, as shown in Fig. 5.6a; if the value of K is too large, it will result in too few iterations, causing abnormal voltage fluctuations to be effectively recognized, as shown in Fig. 5.6b. This study, through a large amount of trial and error and comparative verification, finally determined the calculation window size to be $K = 100$.

   Figure 5.7 shows the single-cell voltage of vehicle A and the corresponding Shannon entropy curve when $K = 100$. Figure 5.7a shows that some cells have noticeably low voltages, especially cell 62, which has a very noticeable low

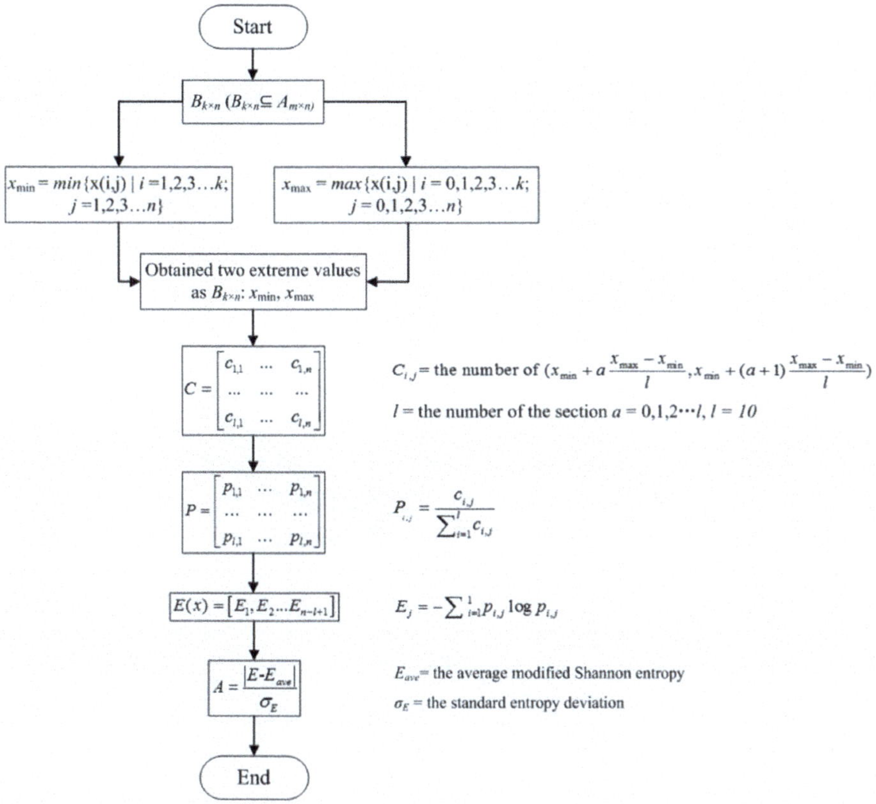

Fig. 5.5 MSNE calculation flowchart

voltage. Figure 5.7b shows that the entropy value of cell 62 is always higher than that of other cells during the process of over 800~900 iterations, and the corresponding voltage value and fluctuation amplitude of cell 62 in Fig. 5.7a are larger than those of other cells in this range. Therefore, it can be inferred that there may be abnormalities or faults in cell 62 in this range, but since all cell voltages are within the safe range, conventional threshold methods have difficulty diagnosing this kind of cell voltage abnormality.

The operating data of vehicle B, which triggered a single-cell overvoltage alarm (3.65~4.4 V, a secondary fault set by the platform), was exported from the platform. This vehicle is also a pure electric taxi operating in the urban area of Beijing, with its power battery pack consisting of 100 series-connected single cells. The data collection time was from 05:47:41 to 06:47:41 on 9 May 2016, with a data sampling interval of 1 s. Figures 5.8a and b show the voltage and Shannon entropy curves of vehicle B. As shown in Fig. 5.8a, the entropy curve shows that multiple cells, such as 50, 65, 86, 87 and 88, all showed varying degrees of abnormal voltage at the end of charging, manifested as the voltage

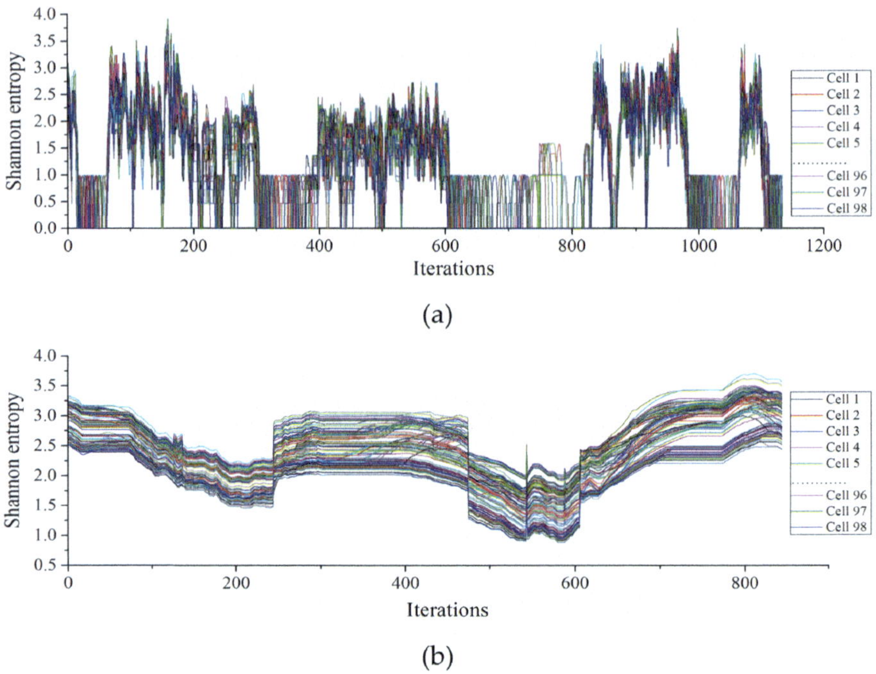

Fig. 5.6 Shannon entropy of vehicle A's single cell voltage with different K values: **a** $K = 10$; **b** $K = 300$

rising to the highest voltage earlier and faster at the end of charging. In addition, some cells also triggered an overvoltage alarm at the end of charging. Through the Shannon entropy curve in Fig. 5.8b, we can see that the Shannon entropy curve can effectively detect all abnormal cell voltages, and also verifies the stability of this method.

2. **Reliability Verification**

The operating data of vehicle C, which triggered a single-cell overvoltage alarm (exceeding 3.6 V, a secondary fault set by the platform), was exported from the platform. Its power battery system consists of 102 series-connected single cells, and the data collection time was from 17:01:58 to 17:59:27 on 8 May 2016, with a sampling interval of 10 s. Figure 5.9a and b show the voltage and Shannon entropy curves of vehicle C. Figure 5.6a shows that all cell voltages rise with the sudden increase in current at the end of charging, which is also the cause of the overvoltage alarm. It may be due to the gradual difference in the aging degree of each cell in the power battery system during actual operation, so the cell voltages in Fig. 5.9a have a certain inconsistency. However, the fluctuation trend of all cell voltages is almost the same, and no cell has abnormal voltage fluctuations different from other cells. The Shannon entropy curve shown in Fig. 5.9b shows

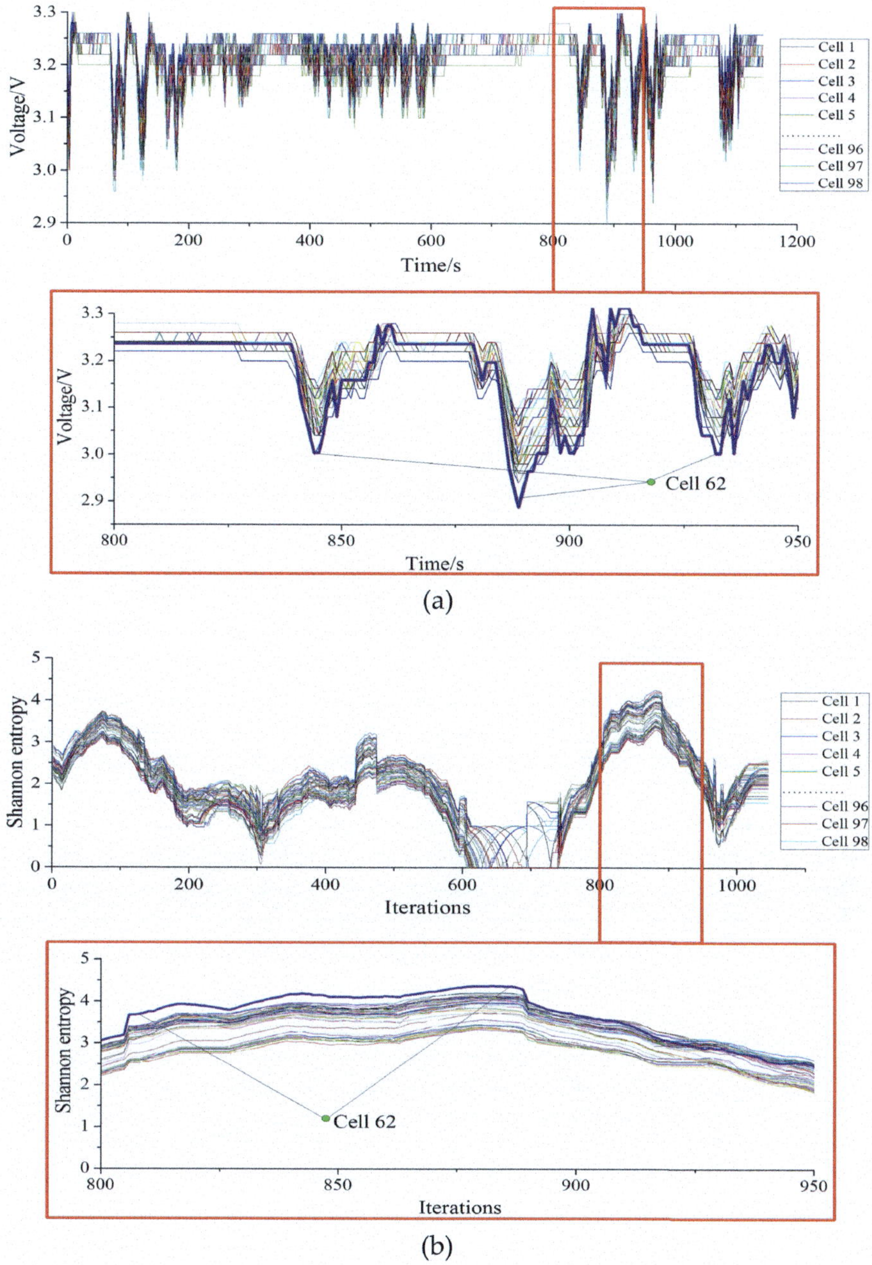

Fig. 5.7 Vehicle A's single cell voltage and Shannon entropy curve: **a** voltage curve; **b** Shannon entropy curve

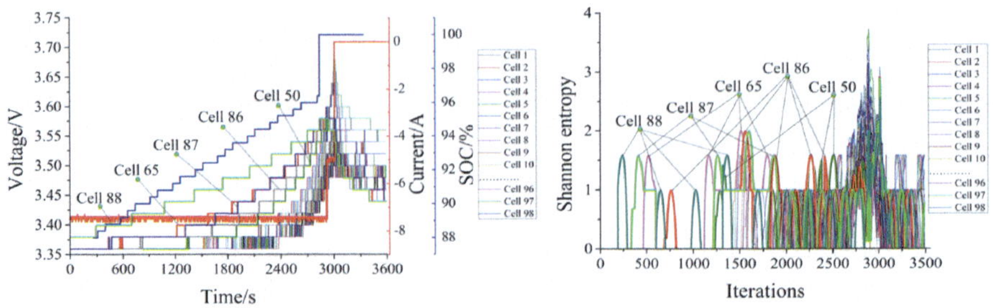

(a) (b)

Fig. 5.8 Vehicle B's single-cell voltage and Shannon entropy curve: **a** voltage curve; **b** Shannon entropy curve

that no cell's entropy curve has fluctuations significantly higher or lower than other cells. Therefore, this also verifies that the effectiveness of this method depends on the inconsistency and abnormal fluctuation of cell voltage, and has no direct relationship with the cell voltage value. The method proposed in this paper will not trigger false alarms in the absence of abnormal voltage.

3. **Diagnostic Strategy and Verification**

The operating data of vehicle D, which triggered a single-cell overvoltage alarm (exceeding 3.6 V, a secondary fault), was exported from the platform. Its power battery system consists of 100 single cells, and the data period was from 09:10:17 to 10:10:16 on 9 May 2016, with a data collection interval of 10 s. The overvoltage fault occurred at the moment of full charge. According to Fig. 5.10b, at the beginning, the Shannon entropy value of cell 71 is higher than other cells, but for a long time afterward, its Shannon entropy value is much lower than other battery cells, so cell 71 may have abnormal voltage. However, since the voltage of cell 71 is always within the normal voltage range, conventional voltage sensors cannot detect this abnormality.

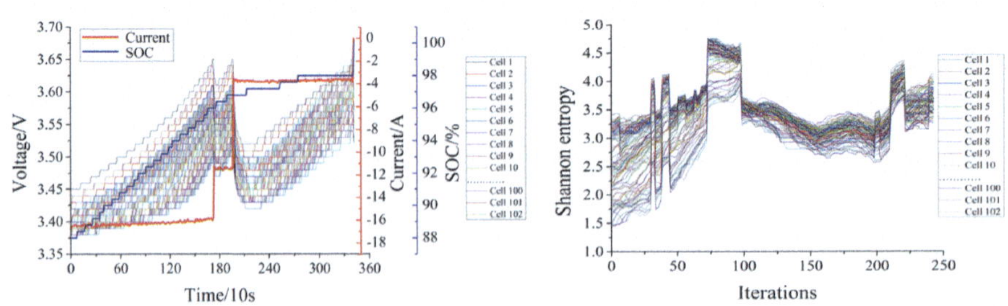

(a) (b)

Fig. 5.9 Vehicle C's single cell voltage and Shannon entropy curve: **a** voltage curve; **b** Shannon entropy curve

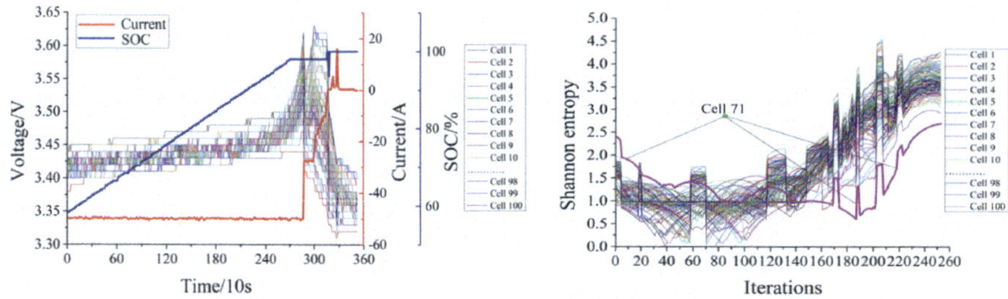

(a) (b)

Fig. 5.10 Vehicle D's single-cell voltage and Shannon entropy curve: **a** voltage curve; **b** Shannon entropy curve

Figure 5.11 shows the voltage and Shannon entropy curves one hour and one day before the aforementioned overvoltage alarm. Figure 5.11a is the voltage curve one hour before the overvoltage alarm (08:10:33–09:10:27 on 9 May 2016), and Fig. 5.11b is its MSNE curve. Vehicle D stopped running at 19:48:07 the day before the alarm, so we selected the voltage data one hour before the vehicle stopped running (18:48:17–19:48:17 on 8 May 2016), as shown in Fig. 5.11c. The Shannon entropy curve shown in Fig. 5.11d shows that cell 71 can still be detected. Therefore, we can effectively diagnose the location of the cell that is about to experience a voltage anomaly one hour or one day in advance.

The inconsistency of the power battery system can usually be measured by the standard deviation of the individual cell voltage, so the standard deviation of the Shannon entropy of the individual cell voltage can also reflect the inconsistency of the individual cells. Unlike the individual cell voltage, the Shannon entropy value of the individual cell voltage is related to the fluctuation state of the voltage, and it cannot be guaranteed to always be within a fixed size range. Moreover, the voltage fluctuation of the faulty cell may be the most intense or the most gentle at certain times, so it is difficult to detect minor voltage faults simply by the standard deviation of the voltage or its Shannon entropy. In addition, the complex and variable voltage fluctuation state is not conducive to objectively evaluating the abnormal level based on entropy value, and it also poses a challenge to the real-time diagnosis and early warning of power battery faults based on actual vehicle operation. Comparing the standard deviation of the Shannon entropy values of the individual cell voltages of the aforementioned Vehicles C and D one hour before, as shown in Fig. 5.12, it can be found that the two vehicles have a large difference in the fluctuation range of the entropy value standard deviation due to different battery states. Therefore, it is not advisable to directly use Shannon entropy or its standard deviation to evaluate the abnormal level of individual cell voltage in real-time. We need to explore a highly accurate and universally applicable method for evaluating abnormal voltage.

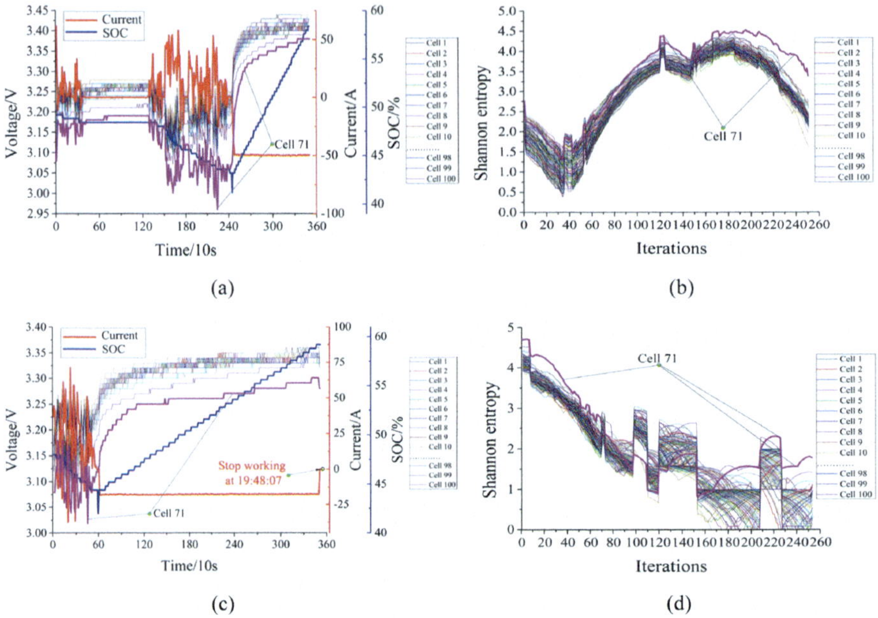

Fig. 5.11 Voltage and Shannon entropy curves of individual cells one hour and one day before the overvoltage alarm of Vehicle D: **a** individual cell voltage one hour before; **b** Shannon entropy curve one hour before; **c** individual cell voltage one day before; **d** Shannon entropy curve one day before

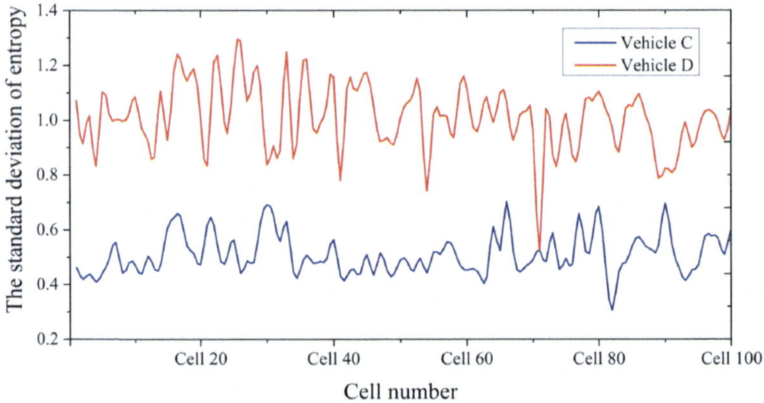

Fig. 5.12 Standard deviation of entropy values of all individual cell voltages of Vehicles C and D

Early studies on the analysis of power battery voltage characteristics have proven that the individual cell voltage of power batteries follows a normal distribution. Therefore, statistical tools based on standard deviation and normal distribution, such as the Z score, can be used to set the abnormal coefficient of individual cell voltage. The Z score (Z-score) is also known as the standard score, and is expressed by the formula:

$$Z = (x - \mu)/\sigma \tag{5.2}$$

where x is a specific score, μ is the mean, σ is the standard deviation, and Z represents the distance between the original score x and the population mean u. The Z score can answer the question: "How many standard deviations is a given score from the mean?" Scores above the mean will get a positive standard score, and scores below the mean will get a negative standard score. The voltage abnormality coefficient based on the Z score can be set as:

$$A = (E - E_{ave})/\sigma_E \tag{5.3}$$

where E is the improved Shannon entropy value corresponding to a certain monomer voltage, E_{ave} is the average Shannon entropy value of all monomer voltages, σ_E is the standard deviation of all monomer Shannon entropy values. Normally, the battery voltage fluctuates within a fixed range under safe conditions, so the corresponding abnormality coefficient also fluctuates within a fixed range under safe conditions. If there is an abnormal fluctuation in the monomer voltage and it exceeds a certain threshold, then regardless of the size of the voltage, its abnormal change can be diagnosed through the abnormality coefficient.

In summary, the advantages of this method are: (1) it overcomes the shortcomings of traditional threshold diagnosis indicators being single and having short alarm times, and can effectively diagnose hidden abnormal monomers within the standard safety threshold range; (2) it concentrates all monomer voltage fluctuation characteristics within a specific abnormality coefficient range, and effectively diagnoses voltage faults of different levels by setting abnormality coefficient thresholds.

In order to achieve accurate voltage fault diagnosis, a large amount of real vehicle data was collected from the big data platform, and the proposed abnormality coefficient determination method was analyzed and verified by trial and error, finally obtaining the multi-level diagnosis strategy for real vehicle power battery system voltage faults as shown in Fig. 5.13.

According to the set diagnosis, the voltage abnormality coefficient can be divided into three levels, as shown in Fig. 5.13. Among them:

Level 3: When all monomer abnormality coefficients meet $|A| \leq 3.5$, the power battery voltage of the detected vehicle is normal and in a safe state.

Level 2: When any monomer's abnormality coefficient reaches $3.5<|A|<4$, the power battery of the detected vehicle is in a second-level warning state, with

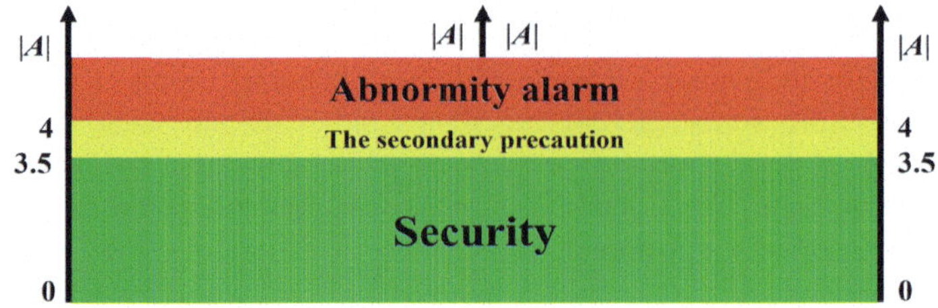

Fig. 5.13 Multi-level diagnosis strategy for power battery system voltage faults

abnormal voltage but temporarily no risk of voltage faults. In this case, it can continue to be observed for a few hours, as it may be a false alarm caused by a sudden change in vehicle operating status. When the voltage abnormality coefficient drops to $|A| \leq 3.5$, the second-level warning can be eliminated.

Level 1: When any monomer's abnormality coefficient reaches $|A| \geq 4$, the power battery of the detected vehicle issues a first-level abnormal alarm. This battery group detects abnormal voltage, and if measures are not taken in time, more serious voltage faults or even thermal runaways may occur.

As shown in Fig. 5.14, the abnormality coefficient curve of vehicle D at different times, where the abnormality coefficient of the 71st monomer reached $|A| \geq 4$ at the time of the alarm, one hour before the alarm, and one day before, all belong to the third level warning according to the diagnosis strategy. The 71st monomer had already shown voltage abnormalities at least one day before the overvoltage alarm, as shown in Fig. 5.14c. Therefore, based on the proposed multi-level diagnosis strategy, monomer batteries with abnormal voltage can be diagnosed early.

As shown in Fig. 5.15, the voltage abnormality coefficient curve of vehicle D at different times after removing the 71st monomer voltage, it can be seen from Fig. 5.15a that in addition to the diagnosed 71st monomer, the voltages of monomers 24, 26, 82, etc. also have slight abnormalities, and there may also be some abnormal monomer voltages that are not visible to the naked eye; from the abnormality coefficients of the three periods in Fig. 5.15b–d it can be seen that even if the Shannon entropy is calculated without the 71st monomer voltage, the abnormality coefficients of all other monomer voltages always satisfy $|A| \leq 4$, none exceed the first-level threshold, only monomers 13, 17, 19, 21, 32, 33, 54, 82, etc. exceed the second-level threshold. Therefore, this result also proves the reliability of the method proposed in this paper, and its diagnosis result depends on whether there is abnormal voltage fluctuation in all monomer voltages.

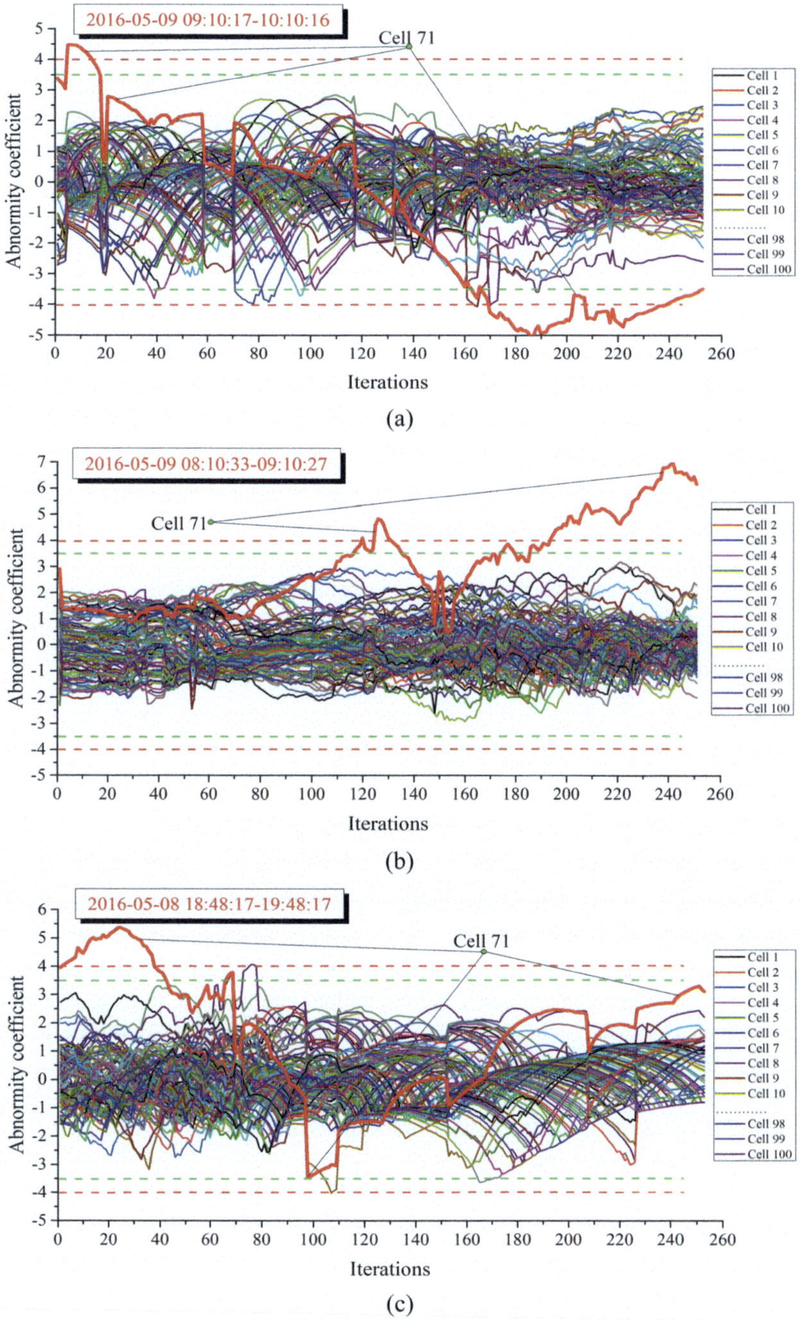

Fig. 5.14 Voltage abnormality coefficient curve of vehicle D at different times: **a** One hour when the overvoltage fault alarm occurs; **b** One hour before the fault alarm; **c** One day before the fault alarm

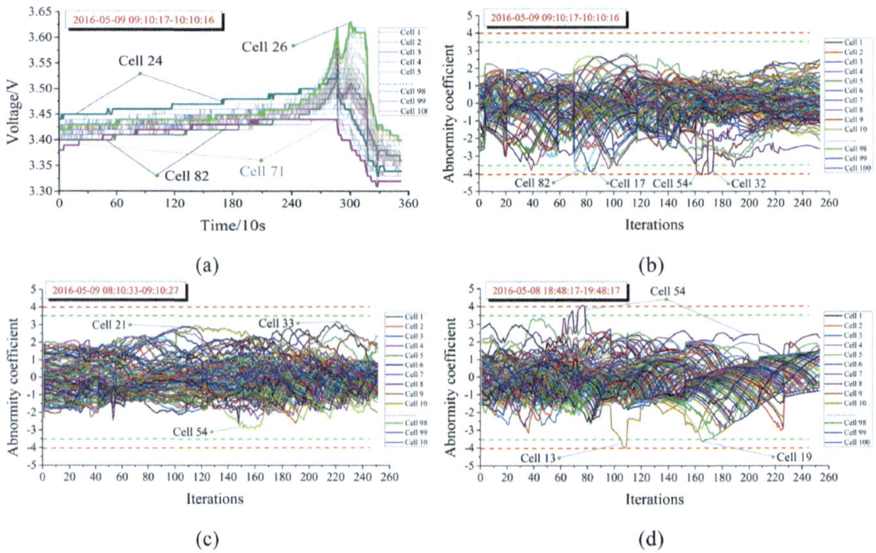

Fig. 5.15 Voltage abnormality coefficient curve of vehicle D at different times after removing the 71st monomer voltage: **a** Voltage at the time of the fault; **b** Abnormality coefficient at the time of the fault alarm; **c** Abnormality coefficient one hour before the fault alarm; **d** Abnormality coefficient one day before the fault alarm

5.4 Temperature Fault Diagnosis Based on MOSE

5.4.1 Power Battery Thermal Management System

The temperature control of the power battery system is crucial to the driving performance and safety of electric vehicles. If the battery heating rate exceeds the heat dissipation rate and causes the temperature to exceed the safe range, it may trigger a series of extreme phenomena such as battery leakage, smoke, jetting, combustion and even explosion, etc., which are thermal runaway conditions. Normally, when the power battery system undergoes thermal runaway, it is directly manifested as an increase in the reaction rate of the internal battery materials. In addition, the aging rate and failure probability of the battery are also greatly affected by abnormal battery temperatures, such as too high temperature, too large temperature difference, and too fast temperature rise, etc. Therefore, to ensure the safe operation of the power battery system, it is necessary to adopt a reasonable battery thermal management (Battery Thermal Management, BTM) strategy in electric vehicles and introduce accurate and efficient fault diagnosis algorithms.

A well-designed BTMS has the function of balancing the temperature of the battery pack in the power battery system, and can keep it within the range required for work. Normally, for a power battery system with only a few battery modules, a BTM system using air as a heat transfer medium is simpler in structure than a BTM system using liquid as a heat transfer medium, and the cooling/heating effect is also more effective; however, the opposite is true when a power battery system has more battery modules. The schematic diagrams of different types of BTMS using air and liquid are shown in Fig. 5.15. Figure 5.12a shows three types of air-based BTMS, including passive air cooling, passive air cooling/heating, and active air cooling/heating, which can be achieved by natural air convection or forced air convection. Compared with air, liquid has a higher thermal conductivity and specific heat capacity, so liquid-based BTMS is considered a more effective solution for the thermal safety management of battery systems, as shown in Fig. 5.16b. Liquid BTMS can also be divided into passive (i.e., relying solely on the surrounding environment) or active (i.e., built-in power supply for heating or cooling at over-cooling or overheating temperatures). By determining the thermal management control strategy through the electronic control unit, BTMS can use air for heating/cooling ventilation, or use liquid as a cooling/heating insulation layer. In addition, phase change materials are also another option for cooling/heating the battery system as a cold storage device.

Existing research on battery thermal safety management mainly focuses on the optimization of BTMS strategies, and research on temperature fault warnings is also limited to monitoring battery temperature changes under laboratory conditions through temperature sensors. However, for the early detection and warning of temperature faults in the power battery system during actual vehicle operation, there is still a lack of systematic theory or research methods, and most of the existing battery management technologies are based on conventional threshold methods for temperature detection, which cannot identify those with smaller amplitudes or hidden

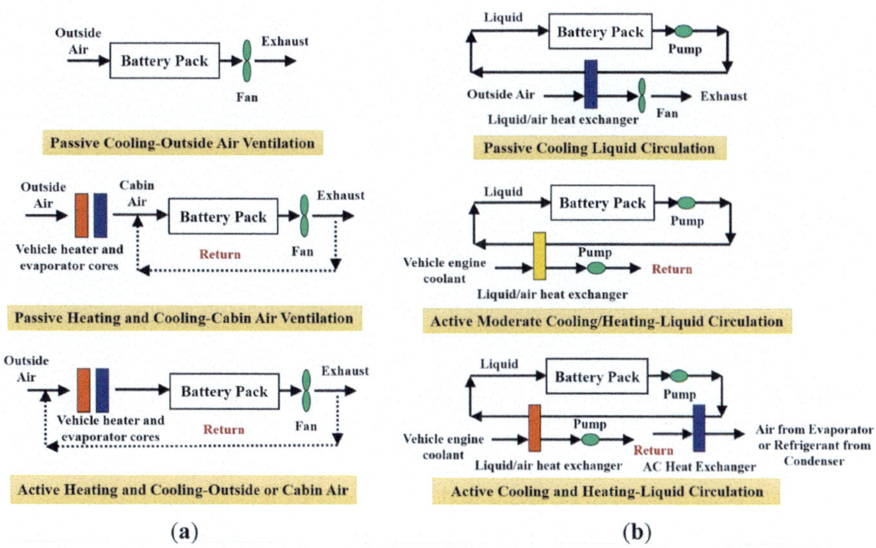

Fig. 5.16 Schematic diagram of air and liquid BTMS: **a** three types of air BTMS; **b** three types of liquid BTMS

abnormalities or faults within the normal temperature range, so the warning effect for thermal runaway of power batteries is not good. In addition, because the temperature probes in the battery pack are often arranged on the outside of the module, under the thermal management system, they are greatly affected by environmental temperature, thermal management opening, etc., and cannot reflect the battery's own temperature state, so there are still some problems with offline fault diagnosis using temperature data. The faults diagnosed by this type of method cannot rule out misdiagnoses caused by battery problems, thermal management system problems, or environmental abnormalities.

5.4.2 Diagnostic Strategy and Effect Verification

1. Diagnostic Strategy

   Referring to the evaluation strategy for voltage abnormalities mentioned above, by using the Z score to set the abnormality coefficient of the probe temperature of the power battery system, statistical analysis of big data from actual vehicle operation shows that the probe temperature of the power battery system does not follow a normal distribution, so it is not reasonable to directly use the Z score to set the threshold of the abnormality coefficient. For this reason, we use Boxplot to display its abnormality coefficient, the principle of which is shown in Fig. 5.17. As is well known, the methods based on normal distribution, 3 σ rule or Z score are premised on the assumption that the data follows a normal distribution, and the effectiveness of using these methods to judge outliers in non-normally distributed data is minimal. Boxplot does not need to impose any restrictions on the data, and its judgment of outliers is based on quartiles (Q1 and Q3 in Fig. 5.16) and interquartile range (distance between Q1 and Q3). Boxplot can allow up to 25% of the data to be distributed arbitrarily without causing noticeable disturbance to the position of the quartiles, so the quartiles have a certain resistance, making the identification results of Boxplot for outliers more objective.

   In order to achieve an accurate diagnosis of temperature faults in the power battery system, this study collected a large amount of actual vehicle operation data, and set multi-level temperature abnormality coefficients through trial and error, and proposed a multi-level diagnostic strategy for temperature faults in the power battery system. The temperature abnormality coefficient is divided into the following three levels:

   Level 3: When all the Boxplot median lines of the probe temperature anomaly coefficients are less than or equal to 1, the power battery probe temperature of the vehicle being tested is normal and in a safe state.

   Level 2: When the Boxplot median line of the anomaly coefficient of any temperature probe reaches more than 1 and less than 1.2, the power battery of the vehicle being tested is in a secondary warning state, with temperature anomalies

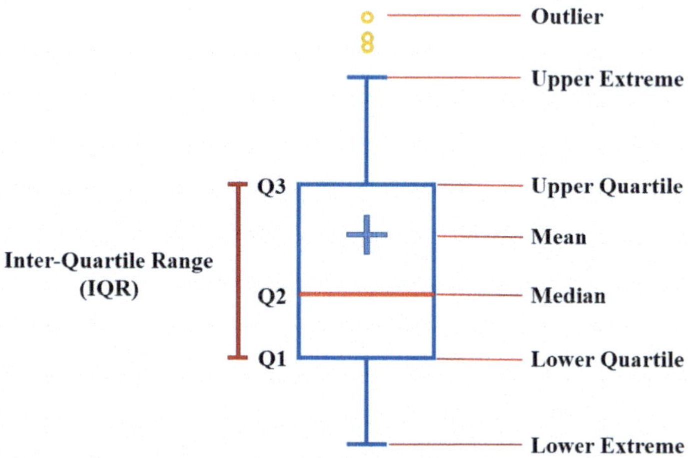

Fig. 5.17 Schematic diagram of Boxplot working principle

but temporarily no risk of temperature-related faults. In this case, it can continue to be observed for a few hours, as it may be a false alarm caused by a sudden change in the vehicle's operating state. When the Boxplot median line of the anomaly coefficient drops below 1, the secondary warning can be eliminated.

Level 1: When the Boxplot median line of the temperature anomaly coefficient of any temperature probe is greater than or equal to 1.2, the power battery of the vehicle being tested will issue a primary anomaly alarm. This battery group detects abnormal probe temperature changes, and if measures are not taken in time, more serious temperature faults or even thermal runaways may occur.

2. Over-Temperature Fault Diagnosis

Retrieve and export the historical data of vehicle E on 6th March 2017. This vehicle is a pure electric sanitation vehicle driving in the urban area of Beijing, with a maximum speed of 45 km/h, and the battery type is a lithium iron phosphate battery with 120 cells in series. The working hours of the vehicle on that day were 07:05:29–14:44:20 (with some non-working periods in between), and at 11:07:20 in the morning, it experienced a high-temperature fault alarm of $T >$ 45 °C (a level three alarm recognized by the platform). The power battery pack has 16 temperature probes installed at different positions, with a data collection interval of 20 s. The temperature and SOC curves of vehicle E are shown in Fig. 5.18. From the figure, it can be seen that the temperature of probes 1 and 9 has a significantly different fluctuation trend compared to other probes, and probe 1 triggered a high-temperature fault alarm during the vehicle's operation. This is likely due to a fault in the battery cell or module at the location of probe 1, but as 45 °C is still within the safe temperature range, it is impossible to detect the faulty temperature probe before the alarm is triggered using conventional threshold methods.

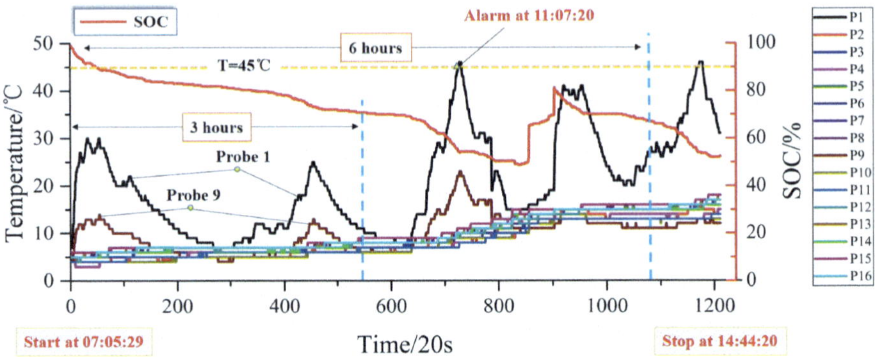

Fig. 5.18 Temperature and SOC of vehicle E

As can be seen from the above, the choice of the length of the calculation window K has a significant effect on the entropy value. If K is too small, the original features of temperature fluctuation will not be fully displayed; on the contrary, if K is too large, the iterations will become too few to identify abnormal temperature fluctuations. Through trial and error and analysis of a large amount of real vehicle data, we still choose $K = 100$ as the calculation window length.

After the vehicle starts for a while, the working state of the power battery system will gradually stabilize, and the temperature of all temperature probes will also be in a gradually stable state. At this time, it is difficult to detect and isolate abnormal temperatures and their probe locations in a short time. Based on the anomaly coefficient mentioned in the previous section, we select the monitoring data from the initial start to a certain driving time of the vehicle each day for anomaly coefficient evaluation. According to the trial and error method, we initially determined 3 h as the shortest time window for temperature anomaly coefficient evaluation. Figures 5.19a and b respectively show the anomaly coefficient curves of vehicle E for the initial 3 h and 6 h on 6th March 2017. The two pictures in Fig. 5.19 show that the anomaly coefficients of probes 1 and 9, especially probe 1, are significantly larger than those of other probes, which also proves that the proposed method can accurately identify the location of the probe with abnormal temperature.

The Boxplots of the anomaly coefficients of vehicle E for the initial 3 h and 6 h are shown in Figure 5.20a and b respectively. The results show that probes 1 and 9 can be easily detected, and the anomaly coefficient of probe 1 is much larger than that of probe 9 and other temperature anomaly coefficients. According to the set temperature anomaly coefficient threshold, it is found that the Boxplot median lines of probes 1 and 9 exceed the primary anomaly threshold of 1.2 in both the initial 3 h and 6 h so that this method can diagnose the abnormal temperature probe at least 4 h before the high-temperature alarm.

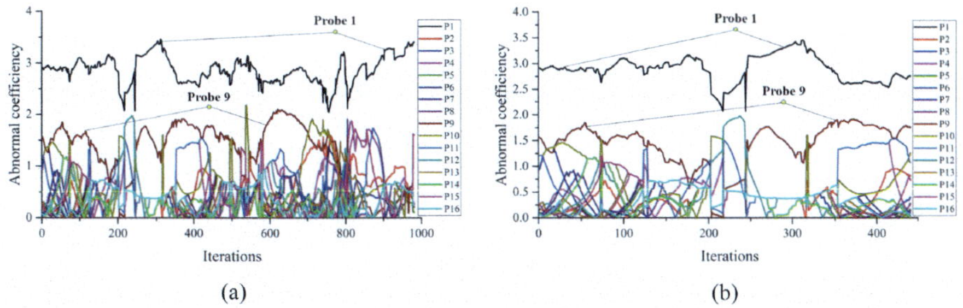

Fig. 5.19 Anomaly coefficients of vehicle E: **a** initial 3 h; **b** initial 6 h

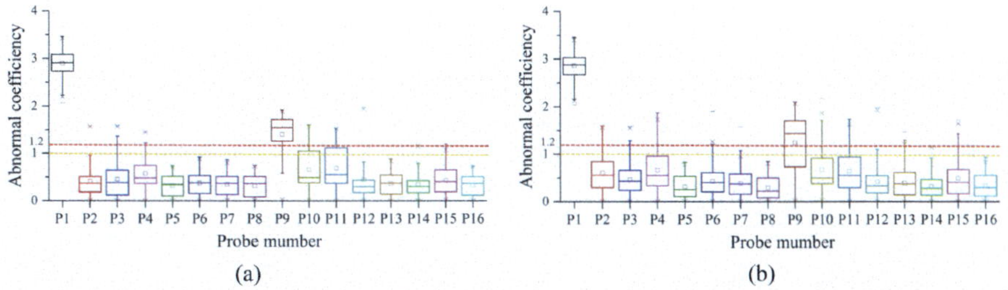

Fig. 5.20 Boxplot of temperature anomaly coefficients of vehicle E: **a** initial 3 h; **b** initial 6 h

3. Excessive Temperature Difference Fault

Retrieve and export the historical data of vehicle F on 2nd November 2016. The working hours of the vehicle on that day were 07:55:57–23:59:54, and a TD (Temperature difference) > A 5 °C temperature difference is too large, triggering a fault alarm (a secondary alarm recognized by the platform). The power battery pack also has 16 temperature probes installed at different locations, with data collected every 10 s. The daily temperature curve is shown in Fig. 5.21. As can be seen from the figure, after approximately 7 h of operation on the day, probe 11 gradually deviates from the collective probe temperature, showing a significantly lower temperature than the other probes. This is the direct trigger for the excessive temperature difference fault alarm. However, at this time, the temperature of all probes is still within the safe temperature range of 0~30 °C. The conventional temperature threshold method cannot diagnose the abnormal probe 11 before the alarm occurs.

Figure 5.21 shows the abnormal coefficient and Boxplot of vehicle F on 2nd November 2016 for the initial 3 h, 6 h, and 9 h. As can be seen from Fig. 5.22a and b, there are many temperature probes with obvious abnormal coefficients within the initial 3 h, and the extreme values of the abnormal coefficients are large. If the battery voltage abnormal coefficient evaluation strategy is followed, probes 1, 2, 9, 10, and 16 may all have abnormalities, which is clearly inconsistent with the original voltage curve. However, this problem can be avoided through the Boxplot

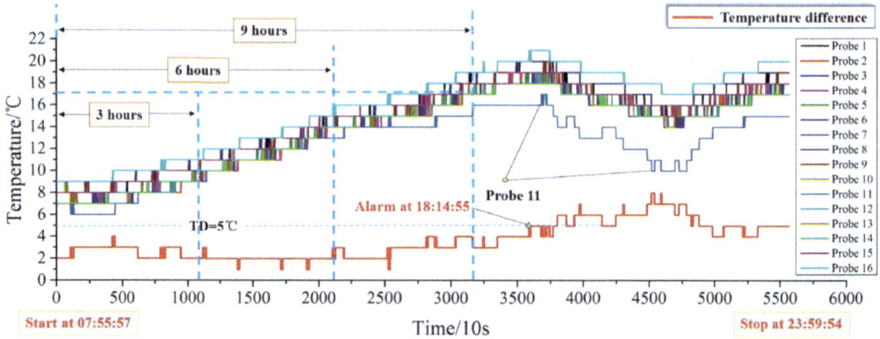

Fig. 5.21 Vehicle F probe temperature on 2nd November 2016

in Fig. 5.22b, where it is found that the median of the Boxplot of the abnormal coefficients of all probe temperatures has not exceeded the secondary threshold. Figure 5.22b shows that the median of the Boxplot of the abnormal coefficients of all probe temperatures in the initial 6 h has not exceeded the primary threshold, but the median of probe 11 has reached the primary threshold, indicating that probe 11 has already shown abnormal fluctuations within the initial 6 h. This abnormality is currently invisible in the voltage curve in Fig. 5.21. At the same time, this also reflects the rationality and effectiveness of using Boxplot for early diagnosis of temperature faults.

Figure 5.22e and f show that the temperature of probe 11 has shown obvious abnormalities within the initial 9 h, and the median of the Boxplot of the abnormal coefficient has significantly exceeded the primary threshold. Therefore, the occurrence of excessive temperature difference faults can be effectively avoided through the diagnosis of temperature abnormalities, thereby preventing the possible occurrence of more serious temperature faults.

4. Stability Verification

In order to verify the stability of the proposed method for diagnosing temperature faults in power battery systems, battery data for vehicle F on 1st November 2016 was further exported from the platform. The vehicle's working hours on the day were from 10:51:05 to 23:36:38, and a TD > 5 °C excessive temperature difference fault alarm occurred at 17:12:15 after more than 9 h of driving. The probe temperature curve of vehicle F on 1st November 2016 is shown in Fig. 5.23, and its fluctuation characteristics are very similar to the temperature fluctuation characteristics of the vehicle one day later on 2nd November.

The abnormal coefficient and Boxplot of vehicle F on the day for the initial 3 h are shown in Fig. 5.24. Figure 5.24a shows that the temperature of probe 1 has shown abnormal fluctuations, but these fluctuations are hidden in the normal temperature curve in the initial 3 h shown in Fig. 5.23, and are always within the safe temperature range, so it is difficult to detect them in advance. Figure 5.24b shows that the median of the Boxplot of probe 11 has reached

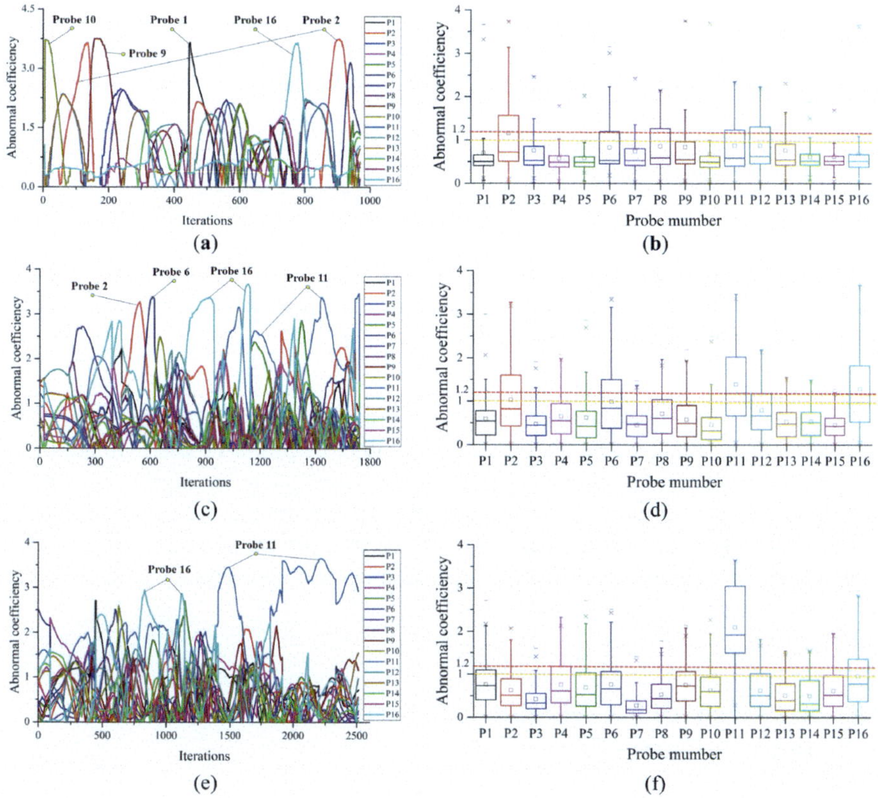

Fig. 5.22 Vehicle F abnormal coefficient curve and Boxplot at different starting times: **a** initial 3 h abnormal coefficient; **b** initial 3 h abnormal coefficient Boxplot; **c** initial 6 h abnormal coefficient; **d** initial 6 h abnormal coefficient Boxplot; **e** initial 9 h abnormal coefficient; **f** initial 9 h abnormal coefficient Boxplot

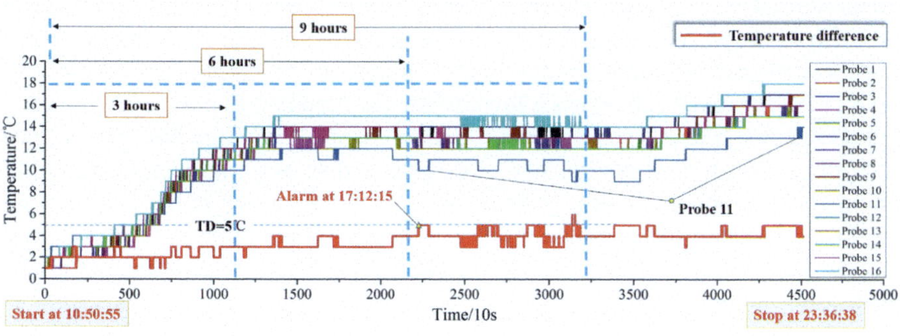

Fig. 5.23 Vehicle F probe temperature on 1st November 2016

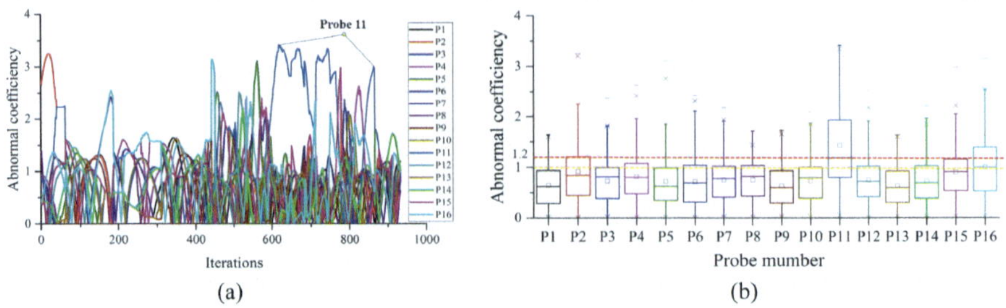

Fig. 5.24 Vehicle F probe temperature abnormal coefficient and box plot on 1st November 2016: **a** abnormal coefficient; **b** box plot

the primary threshold, indicating that our proposed temperature fault diagnosis method can diagnose the abnormal temperature of probe 11 early. In addition, during the process of optimizing the abnormal coefficient threshold, we verified a large amount of data from the big data platform, including historical data of faults such as excessive temperature difference, rapid temperature rise, and high temperature. The results all show that the proposed method has good accuracy and stability.

5.5 Research on MOSE Key Calculation Factors and Diagnostic Effects

As shown in Fig. 5.25, MOSE maintains a good fault diagnosis effect when facing abnormal cells. The key calculation factor of MOSE is the calculation window. The abnormal coefficient curve of the abnormal cell is significantly higher than other cells under most calculation windows, and the distance between the abnormal curve and the normal curve gradually increases with the increase of the calculation window value. However, when the calculation window grows to a certain value, the growth of the curve will gradually slow down. Through fitting analysis, the abnormal coefficient curve of the abnormal cell shows a logarithmic function distribution evolution rule.

This rule is based on a small number of samples, and its validity has been verified under the verification of several sets of data. This discovery has significant implications for the application of algorithms like Shannon entropy in the field of fault diagnosis. However, this rule is after all based on a small number of samples, and to make this method applicable to more types of batteries, it requires more data verification and rule exploration.

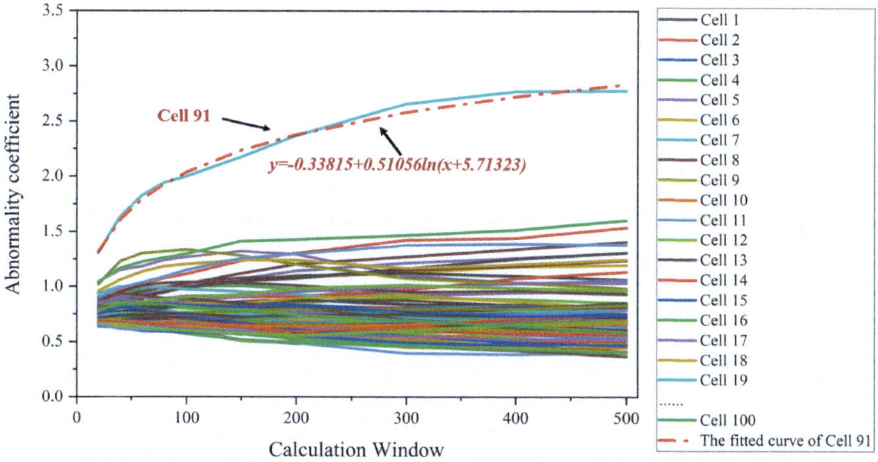

Fig. 5.25 Relationship between MOSE fault diagnosis effect and calculation window size

5.6 Summary

This chapter provides a detailed analysis of various types of faults in the power battery system, such as voltage faults and temperature faults, and their triggering causes, clarifies the fault mechanism of the power battery system, and conducts a qualitative and quantitative analysis of existing fault diagnosis methods.

Based on the MOSE algorithm, a fault diagnosis model for the power battery system has been established, overcoming the shortcomings of the traditional threshold method diagnosis index being single, the alarm time being short, etc., and achieving early and accurate diagnosis of hidden faults within the conventional safety threshold range. The results show that voltage faults and temperature faults one day before the conventional battery fault alarm can be accurately diagnosed.

For the characteristics of voltage faults and temperature faults in the power battery system, the Z-score and Boxplot methods based on statistics are used to set the abnormal coefficient threshold and multi-level diagnosis strategy, respectively, to achieve qualitative classification of real vehicle power battery fault characteristics and quantitative evaluation of fault levels. The effectiveness, stability, and reliability of the method have been verified by big data from real vehicle operation.

Chapter 6
Research on Risk Warning of Power Battery System

6.1 Overview

The accurate early warning of safety risks in power battery systems is of great significance for preventing serious failures and optimising battery maintenance cycles. However, the "mechanical–electrical-thermal" safety risk features of power battery systems are complex and variable, with multiple factors interdependent. The extraction and location of early safety risk signals for real vehicle power battery systems are challenging, and the robust application effect of safety risk warnings under complex operating environments is unsatisfactory. This chapter will extract the multi-time scale and time–frequency features of early risk signals of power battery systems based on MMSE and NDWD algorithms, study the multi-level risk warning strategy of real vehicle power battery systems, and verify the effectiveness, stability and robustness through experiments and real vehicle data (Fig. 6.1).

6.2 Safety Risk Analysis of Power Battery System

6.2.1 Failure Mode of Power Battery System

The frequent occurrence of fires and even explosions of power battery systems under vehicle-mounted conditions in recent years indicates that the losses caused by battery failure and faults are huge and heavy, so it is urgent to carry out in-depth research on FMFMA of power battery systems. Table 6.1 summarises the common failure modes and fault mechanisms of lithium-ion power batteries from a structural perspective.

Conducting research on power battery system failure modes is of great significance for improving battery life, vehicle safety, and reducing the cost of electric vehicle use. Power battery system failure modes can be divided into three types: cell failure, BMS failure, and Pack system integration failure. Cell failure modes mainly

J. Hong, *Big Data Analysis and Application of Power Battery Operation Safety*, Key Technologies on New Energy Vehicles, https://doi.org/10.1007/978-981-96-4399-8_6

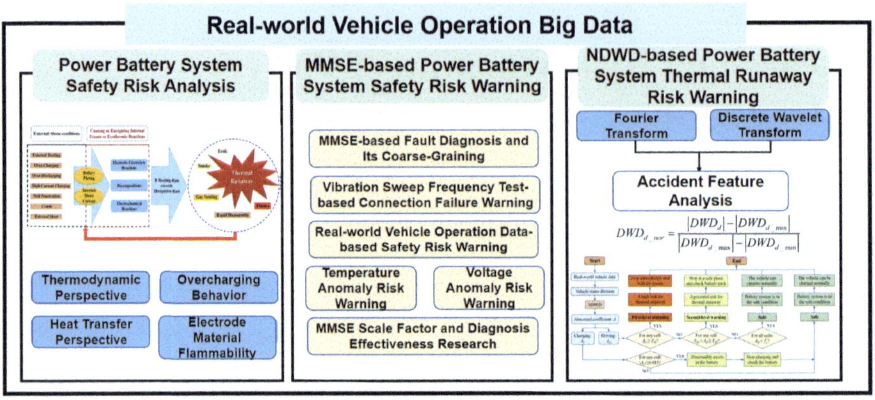

Fig. 6.1 Research on Risk Warning of Power Battery System

include positive and negative short circuits, single cell leakage, negative electrode lithium precipitation, cell swelling, capacity decay, low-temperature capacity reduction, poor capacity consistency, and excessive self-discharge. Once a cell failure occurs, it usually leads to serious consequences, even endangering personal safety. BMS failure modes generally result in severe consequences such as voltage detection failure, temperature detection failure, current and insulation detection failure, and electromagnetic compatibility issues. Pack system integration failures mainly include busbar failure, connector failure, fuse failure, and contactor sticking.

In response to various power battery system failure modes, major battery manufacturers and related researchers need to understand battery performance more fully and continuously improve production processes, starting from the design and manufacturing stage to ensure the safety of the power battery system as much as possible. In addition, the most crucial barrier to ensuring battery safety is the correct use, try to avoid extreme misuse behaviours such as thermal misuse, electrical misuse, and mechanical misuse of the power battery system, which can effectively avoid various safety risks during the use of the power battery system and truly enhance the safety and reliability of electric vehicles.

6.2.2 Power Battery System Safety Risks

The main manifestations of safety issues in electric vehicle power battery systems are leakage, smoke, combustion, and even explosions, as shown in Fig. 6.2. The root cause of these problems is the triggering of thermal runaway inside the battery. Some external factors, such as overcharging, fire sources, squeezing, puncturing, short circuits, etc., can also cause safety issues in the battery. The safety risks of power battery systems can usually be analysed from several different perspectives.

Table 6.1 Common failure modes and fault mechanisms of power battery system

Battery component	Failure mode	Cause of failure	Common symptoms	Failure mechanism	Accident probability	Severity level
Negative electrode (active material)	SEI film thickening	Chemical reduction deposition	Increased charge transfer impedance, reduced capacity and power	Chemical side reactions between lithium ions, negative electrode active material, and electrolyte solvent	High	Low
	Active particle fragmentation	Mechanical stress	Reduced capacity and power	Excessive stress from lithium-ion deintercalation	Medium	Low
	Low electrode porosity	Physical decay	Increased ion diffusion impedance, reduced capacity and power	Changes in internal electrode space	Medium	Low
	Lithium dendrite formation on the negative electrode surface	Chemical reduction deposition	Dendrite penetration of the separator causing an internal short circuit in the battery	Low-temperature charging or high current charging	Low	High
Negative electrode (current collector)	Formation of copper particles or copper dendrites	Chemical corrosion and reduction deposition	Increased internal resistance of the battery, reduced power, decreased current density	Over-discharge of the battery	Low	High
Positive electrode (active material)	SEI film thickening	Chemical corrosion and reduction deposition	Increased charge transfer impedance, reduced capacity and power	Chemical side reactions between lithium ions, positive electrode active material, and electrolyte solvent	High	Low
	Active particle fragmentation	Mechanical stress	Reduced capacity and power	Excessive stress from lithium-ion deintercalation	Medium	Low

(continued)

Table 6.1 (continued)

Battery component	Failure mode	Cause of failure	Common symptoms	Failure mechanism	Accident probability	Severity level
	Decreased electrode porosity	Physical decay	Increased ion diffusion impedance, reduced capacity and power	Changes in internal electrode space	Middle	Low
	Gas production, battery bulging	Chemical thermal decomposition	Capacity reduction	Battery overcharge or internal short circuit	Low	High
Positive electrode (collector)	Aluminium collector corrosion	Chemical corrosion	Battery internal resistance increases, power decreases, current density decreases	Battery overcharge	Low	Middle
Separator	Separator perforation	Mechanical damage	Continuous heat production, battery bulging, rapid voltage drop	Internal dendrite formation, battery subjected to external pressure or puncture, etc	Low	High
	Separator micropore closure	Thermal melting	Unstable charging and discharging	Battery internal temperature too high	Low	High
Lithium ions	Reduction of lithium ions, thickening of SEI film	Electrolyte reduction, lithium metal generation	Capacity reduction	Chemical side reactions between lithium ions, active materials, and electrolyte solvents	High	Low
Electrolyte	Reduction of lithium concentration in electrolyte	Chemical corrosion and reduction deposition	Increase in battery internal resistance, may lead to thermal runaway	Chemical side reactions between lithium ions, active materials, and electrolyte solvents	High	Low
Organic solvent	Battery gas bulging	Chemical thermal decomposition	Increase in battery internal resistance, may lead to thermal runaway	Overcharge or excessive environmental temperature	Low	High

(continued)

Table 6.1 (continued)

Battery component	Failure mode	Cause of failure	Common symptoms	Failure mechanism	Accident probability	Severity level
	Thickening of SEI film	Chemical corrosion and reduction deposition	Increase in charge transfer impedance, reduction in capacity and power	Chemical side reactions between lithium ions, active materials, and electrolyte solvents	High	Middle
Welding end	External lead corrosion	Chemical corrosion	Continuous heat production, battery bulging, rapid voltage drop	Accidental short circuit of positive and negative leads	Low	High
	Lead breakage	Mechanical fatigue	Disconnection of external equipment	Battery open circuit	Low	Middle
Battery casing	Positive and negative short circuit	Mechanical stress	Continuous heating, battery bulging, rapid voltage drop	External load or stress	Low	High

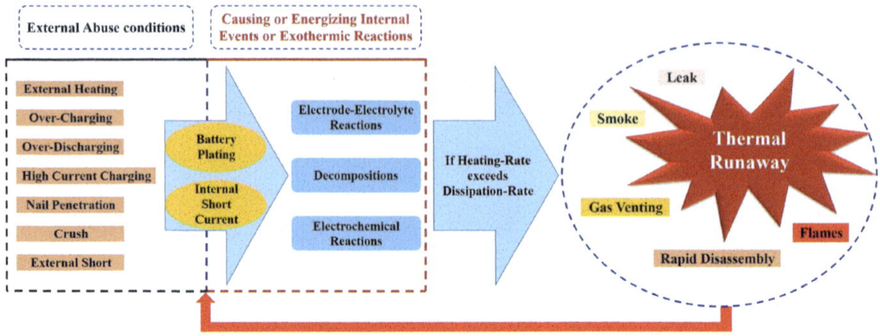

Fig. 6.2 Power battery system safety risks

1. Thermodynamic Perspective

Existing research confirms that, in addition to the battery's negative electrode, the surface of the positive electrode material is also covered with a thin SEI film, which has a significant impact on all aspects of the performance of lithium-ion power batteries. It needs to be emphasised that, from the perspective of the Fermi energy level,[1] the lithium-ion power battery system is thermodynamically unstable, and its stable operation depends on the SEI film on the surface of the positive and negative electrodes isolating further reactions between the positive and negative electrodes and the electrolyte. Therefore, the safety risk of the power battery system is directly related to the integrity of the SEI film on the surface of the positive and negative electrodes.

2. Heat Transfer Perspective

The "mechanical–electrical–thermal" misuse behaviour of power batteries is very likely to trigger various dangerous side reactions and generate excessive heat. When the battery temperature rises to 130 °C, the negative electrode SEI film decomposes; when the internal local temperature rises above 200 °C, the SEI film on the surface of the battery positive electrode decomposes and oxygen is released; when the battery temperature reaches above 240 °C, the lithium-embedded negative electrode and the binder will have a violent exothermic reaction. The decomposition of the SEI film on the surface of the negative electrode will cause a violent exothermic reaction between the lithium-embedded negative electrode and the electrolyte, which will cause the battery temperature to rise and then trigger the thermal runaway of the battery. It can be seen that precise control of battery temperature is crucial to the safety of the power battery system.

[1] The Fermi level refers to the highest energy level filled with electrons in the solid energy band at absolute zero temperature. For solid samples, since the vacuum energy level is related to the surface condition and is easy to change, this energy level is used as the reference energy level, and the electron binding energy refers to the energy difference between the energy level where the electron is located and the Fermi energy level.

3. Electrode Material Flammability

The organic solvent of the power battery has the characteristics of flammability and low flash point, so various misuse behaviours are likely to cause the battery to burn or even explode. According to incomplete statistical analysis, the probability of power batteries burning is much higher than the probability of explosion, but the explosion of the battery is usually accompanied by burning. When the power battery bulges and cracks, the moisture and oxygen in the air and the lithium-embedded negative electrode are also very likely to have a violent chemical reaction, and the release of a large amount of heat at this time will also cause the battery to burn. In addition, negative electrode materials, positive electrode carbon materials, and separators also have a certain degree of flammability.

4. Overcharging Behaviour

For power battery systems, overcharging can lead to serious safety risks in many aspects, such as damage to the crystal structure of the positive electrode material, lithium plating on the negative electrode, and intensified reaction of the positive electrolyte, leading to the risk of thermal runaway. Therefore, preventing overcharging is extremely important for the safety of power battery systems. Precise control of charging voltage is currently the main protective measure to prevent overcharging, mainly achieved through the design of positive electrode capacity limitation. Because the completion degree of the charging process of the graphite negative electrode is difficult to detect, the monitoring of the negative electrode voltage is relatively difficult. However, limiting the capacity of the positive electrode can ensure that the negative electrode has sufficient extra capacity and effectively prevent the occurrence of lithium plating on the negative electrode.

6.3 Risk Warning for Power Battery Systems Based on MMSE

Under normal circumstances, the BMS is mainly responsible for the energy management and state estimation of the power battery system. The purpose of risk warnings for power battery systems is to timely diagnose abnormalities and notify the driver to avoid safety accidents. Research on risk warnings for battery modules and battery packs, which are closer to real vehicle power battery systems, started late. Many battery abnormal features are not obvious in their early stages, and many of the risk warning studies that have been carried out lack the ability to identify abnormal locations based on regular time scales in the early stages of risk. To make up for these shortcomings, this section will propose a risk warning method for power battery systems based on the MMSE algorithm. This method can real-time and quickly detect multiple early hidden abnormal signals, which are usually undetectable by conventional diagnostic methods.

6.3.1 MMSE Algorithm

The traditional method of assessing the regularity of a time series is usually achieved by evaluating the frequency of repeated patterns. However, these methods cannot draw a direct correspondence between regularity and complexity, and conventional entropy-based measures will increase with the increase of randomness. Since the maximum value will be assigned to unrelated random and unpredictable signals (white noise), traditional entropy-based algorithms may lead to misdiagnosis when applied to time series in the field of physiological medicine. In addition, different types of faults in power battery systems will show different complexities on different time scales. When the power battery system is operating normally, the vehicle's own vibration, coupled with the influence of external factors such as weather, road conditions, and driving behaviour, all kinds of battery state signals obtained are highly complex and random on different time scales. Multi-scale entropy (MSE) can represent the complexity of signals from a macro perspective and reveal their detailed features at different scales, making it one of the ideal choices for analysing the complex characteristics of various fault signals in power battery systems.

The calculation of MSE is based on sample entropy, which measures the complexity of a time series by measuring the probability of new patterns in the signal. That is, the larger the value of the sample entropy, the more complex the sequence, and vice versa. The original data is coarsened, and then the sample entropy values on each scale are grouped into a sequence to generate the multi-scale entropy of the time series. If two sequences are of the same scale, but the entropy of one sequence is higher than the other, then the former is more complex than the latter. The steps for calculating MSE are as follows:

1. Coarse-graining process

   Let the original data be $X_i = x_1, x_2, \ldots, x_N$, where N is the length of the data. The original time series is divided into non-overlapping windows based on the time scale, here called the scale factor τ, and then the coarse-grained series is established as follows:

$$y(j) = \frac{1}{\tau} \sum_{i=(j-1)\tau+1}^{j\tau} x_i \quad (\mathrm{j} = 1,2, \cdots, N/\tau) \tag{6.1}$$

In fact, $y(1)$ is the original time series, X_i divided into τ coarse-grained sequences of length N/τ, namely $\{y(j)\}$ (where τ is a positive integer not exceeding N/τ).

Coarse-graining of data means taking the average of different numbers of consecutive data points to create signals of different scales/resolutions. Generally, coarse-graining processing is divided into two forms: one is non-overlapping, jumping τ data each time, then taking the average of τ data to generate new data. The second is overlapping, which jumps $1 - \tau$ data each time, then takes the average of τ data. The above coarse-graining process is non-overlapping. When

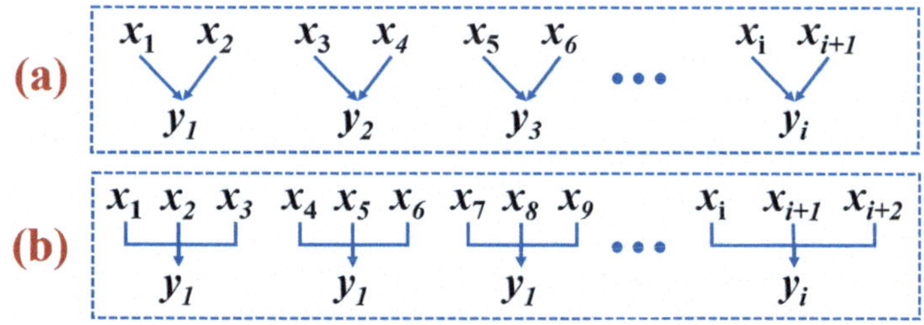

Fig. 6.3 Schematic diagram of coarse-graining: **a** $\tau = 2$; **b** $\tau = 3$

the scale factor $\tau = 1$, the result of coarse-graining data is the original time series; when $\tau = 2$, the coarse-grained time series is calculated by calculating the average of two consecutive time points, forming the coarse-grained result as shown in Fig. 6.3a; when $\tau = 3$, the coarse-grained time series is composed of the average of three consecutive time points, as shown in Fig. 6.3b, and so on.

2. Calculate the sample entropy of the coarse-grained sequence

Under the same similarity tolerance, calculate the sample entropy of the coarse-grained sequence and convert it into a function of the scale factor τ. For the coarse-grained sequence $\{y(1), y(2), \ldots, y(M)\}$, where M is the sequence length, let $M = j$, the calculation of sample entropy is as follows:

1. Given the pattern dimension is m, a m-dimensional vector is composed from the original sequence:

$$Y(i) = [y(i), y(i+1), \ldots, y(i+m-1)] \quad (i = 1, 2, \ldots, M+m-1) \tag{6.2}$$

2. Define the distance between $Y(i)$ and $Y(j)$:

$$d(i,j) = \max_{k=1}^{k=m-1} |y(i+k-y(j+k))| \tag{6.3}$$

3. Given the similarity coefficient threshold r, count the number of $d(i,j)<r$ (this is the number of template matches) and the ratio of this number to the total number $N-m+1$ ratio, denoted as $P_i^m(r)$:

$$P_i^m(r) = \frac{[d(i,j) < r]}{N-m+1} \quad (1 \le j \le N-m, j \ne i) \tag{6.4}$$

Then, calculate the average of all i:

$$P^m(r,N) = \frac{1}{N-m+1} \sum_{i=1}^{N-m+1} P_i^m(r) \tag{6.5}$$

4. For $m+1$, repeat steps ①~③, denoted as $P_i^{m+1}(r)$

$$P^m(r) = \frac{1}{N-m+1}\sum_{i=1}^{N-m+1} P_i^m(r) \tag{6.6}$$

Then, the sample entropy of the sequence can be obtained as:

$$SaEn(m, r) = \lim_{N\to\infty}[-\ln\frac{p^{m+1}(r)}{p^m(r)}] \tag{6.7}$$

When N takes a finite value, the estimated value of the sample entropy at sequence length N can be obtained:

$$SaEn(m, r, N) = -\ln\frac{p^{m+1}(r)}{p^m(r)} \tag{6.8}$$

Using the above formula, the sample entropy of the subsequence at each scale can be calculated, and then a function with scale τ as the independent variable and sample entropy $SaEn$ as the dependent variable can be established, as shown in formula (6.5)–(6.9), to obtain the multi-scale entropy-scale factor curve, and the complexity characteristics of the time series of battery temperature data can be analysed from this.

$$MSE(\tau) = SaEn\left(y^{(\tau)}, m, r\right) \tag{6.9}$$

Obviously, MSE is related to the scale factor τ, the embedding dimension m and the similarity coefficient r. In this paper, referring to relevant literature, based on experience, we take $m = 2$, $r = 0.2\sigma$ (where σ is the standard deviation of the original sequence).

The length of the coarse-grained sequence in formula (6.1) depends on the length of the time series, so the larger the scale factor, the shorter the length of the coarse-grained sequence. As the length of the coarse-grained sequence decreases, the deviation of the entropy value will gradually increase, and the estimation error will also increase. In order to avoid information omission, for the composite MSE, the MMSE method was proposed using the following composite coarse-graining.

1. The Coarse-Grained Sequence is Defined as Follows:

$$y_{k,j}^{(\tau)} = \frac{1}{\tau}\sum_{i=(j-1)\tau+k}^{j\tau+k-1} x_i \tag{6.10}$$

2. For each scale factor τ, calculate the sample entropy of each coarse sequence y_k^τ ($k = 1, 2, \ldots, \tau$), then obtain the MMSE under the scale factor τ by averaging the k entropy values:

$$MMSE(X, \tau, m, r) = \frac{1}{\tau}\sum_{k=1}^{\tau} SaEn\left(y_k^{(\tau)}, m, r\right) \tag{6.11}$$

When calculating the coarse-grained sequence of the standard MSE, the coarse-graining process will shorten the length of the time series as the scale increases. When the time series is too short, it may lead to incorrect entropy estimates or undefined entropy values. The MMSE proposed in this paper integrates the information of all coarse-grained sequences, overcoming the shortcomings of the standard multi-scale entropy that only considers a single coarse-grained sequence, and avoids entropy fluctuations caused by the sliding average process. Therefore, the processing results of MMSE are more consistent than the standard MSE. In addition, MMSE can be applied to unrestricted large-scale data, such as the large data of real vehicle power batteries studied in this project. These unique advantages provide a solid foundation for carrying out precise and efficient early warnings for different types of battery safety risks. Next, we will use experimental data and real vehicle operation data to analyze the multi-level risk warning effect of MMSE for connection failure risks, temperature and voltage class fault risks.

6.3.2 Connection Failure Warning Based on Vibration Sweep Experiment

1. Experimental Platform and Data Preprocessing

   To collect battery connection failure experimental data, a battery vibration experiment test platform as shown in Fig. 6.4 was set up, including the Digtron battery test system (BTS-600), vibration test bench, signal collector, multiple voltage sensors and a computer for processing data. The vibration test bench provides a vibration environment for the battery system, and the signal collector is responsible for collecting multi-channel voltage and current data. During the experiment, the battery was charged/discharged at a constant current of up to 45A. The experimental battery is a square battery with a nominal capacity of 50Ah. Five single cells are connected in series to form a battery module, where cells 1 to 4 are effectively measured, and the nut of the connecting piece of cell 1 is slightly loose before the experiment begins.

Due to the vibration of the vibration table, the data collection process will produce a large amount of vibration noise. For this reason, we used the Discrete Cosine Transform (DCT) method to pre-process the collected battery voltage data. When the charge and discharge experiment began under vibration conditions, we collected battery voltage data for 30 charge and discharge cycles. The battery voltage and current before and after filtering are shown in Fig. 6.5, and the single-cell voltage after filtering for six different charge and discharge cycles is shown in Fig. 6.6. It can be seen from the two figures that the voltage after DCT filtering is smoother

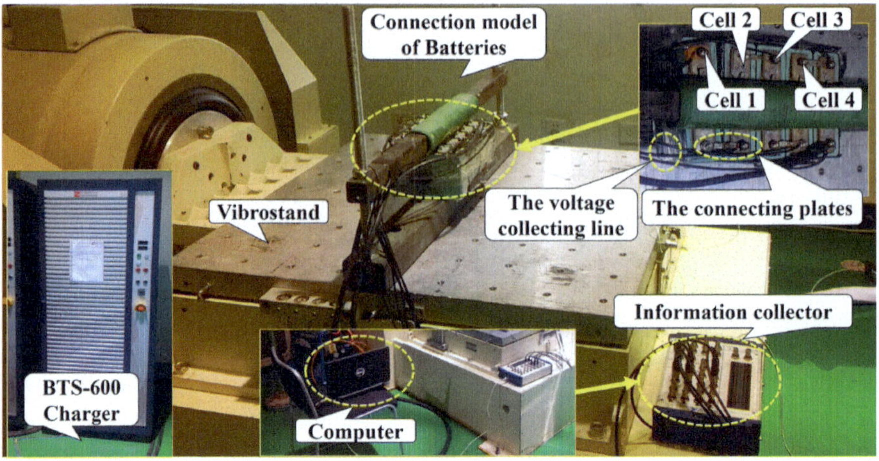

Fig. 6.4 Battery connection failure experimental platform

than before filtering, and effectively filters out deviations caused by system noise and measurement noise while retaining the true value of the voltage. The metal connections between the single cells are in good condition during the initial cycle stage, so the consistency of the single-cell voltage is good. However, as the charge and discharge proceed, it can be seen that the nut of cell 1 is obviously loose, the false connection of the connecting piece of cell 1 becomes more and more serious, and its voltage curve shows a significant large-scale sudden change fluctuation, and some electric sparks appear at the connecting piece of cell 1 in the 20th cycle, indicating that the false connection is becoming more and more serious, and the connection of cell 1 has failed.

2. Sensitivity Evaluation

   Figure 6.7 shows the MMSE curves for different charge and discharge cycles, where the maximum scale factor is $\tau = 20$. The MMSE of the four single cells gradually increases as the scale factor increases, and there is no particularly obvious difference between the four MMSE curves between 1 and 5 cycles. When the experiment proceeded to the 10th cycle, the MMSE curve of cell 1 showed abnormal fluctuations, and the value of MMSE gradually became higher than the other three cells, and it was measured that the local temperature of its connecting piece rose from the original 30–72 °C. After 15 cycles, the voltage fluctuated violently, and this trend of MMSE became more obvious. At the same time, we can also see that as the constant current charge and discharge proceed, the inconsistency difference between the MMSE curves becomes more prominent.

In this study, the Z-score method is a good choice for evaluating the sensitivity of MMSE, but the battery performance dataset studied contains multiple scale features, and the sensitivity of different scale factors is also different under different battery

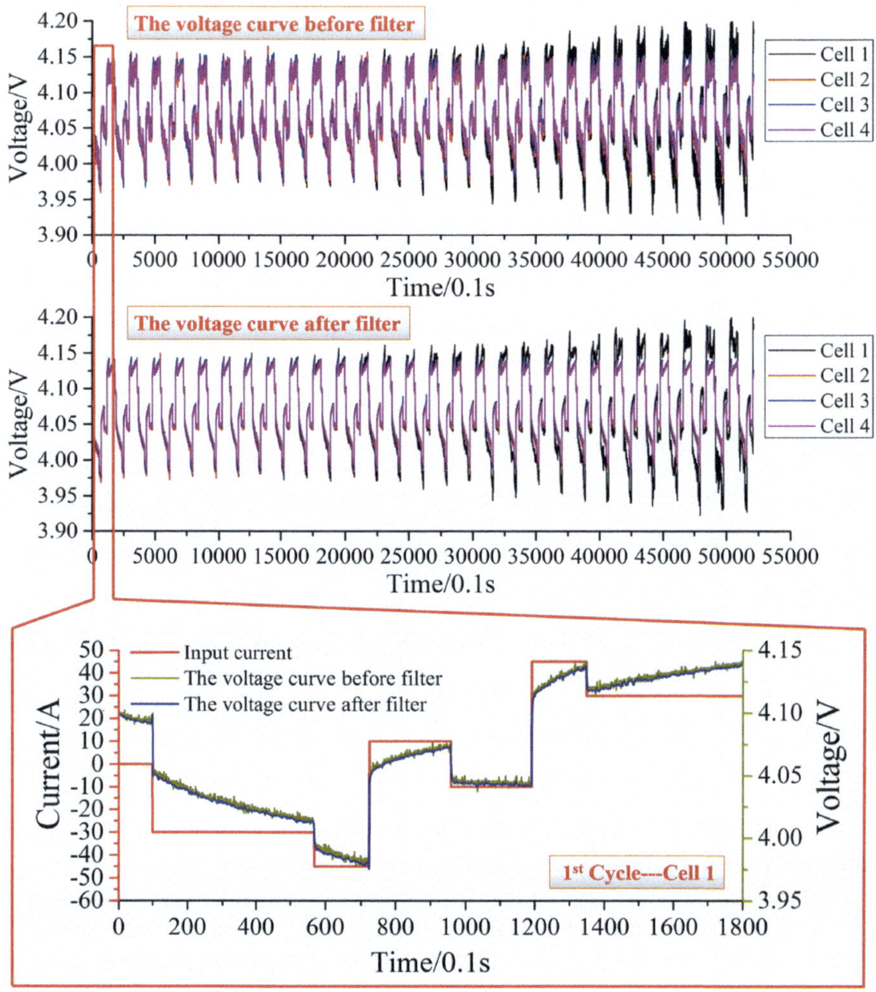

Fig. 6.5 All single-cell voltages and charge and discharge current steps before and after filtering

states. In order to improve the consistency, stability and accuracy of sensitivity evaluation, and to avoid the curse of dimensionality,[2] based on the *Z*-score method, a method for evaluating MMSE sensitivity was proposed.

[2] The Curse of Dimensionality typically refers to a phenomenon where the computational load increases exponentially with the increase in dimensions when dealing with vector computation problems. The Curse of Dimensionality was first proposed by Richard E. Bellman when considering optimization problems, used to describe various problem scenarios encountered due to the exponential increase in volume when the dimension of (mathematical) space increases, making it difficult to analyze and organize data in high-dimensional spaces (usually hundreds or thousands of dimensions).

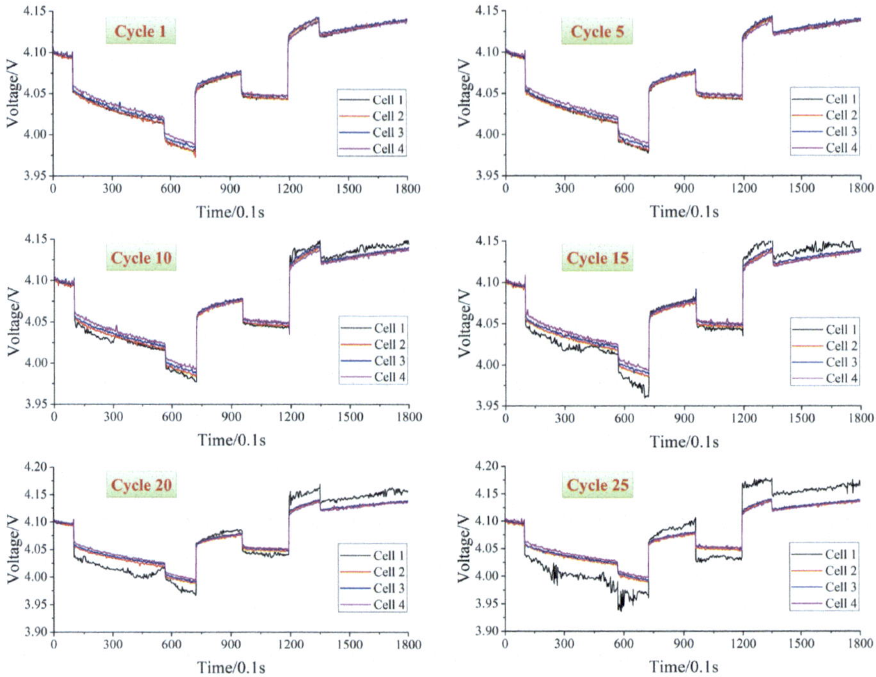

Fig. 6.6 Single cell voltage after filtering for different charge and discharge cycles

In order to improve the evaluation efficiency, all sensitivities are converted into positive values. Then, to enhance the sensitivity of anomaly evaluation, a preliminary screening is performed on the standard deviation of MMSE for different calculation targets (such as battery voltage, probe temperature, etc.) on all scales (for example, the 20 scales in this article). The MMSE of the five scales with the smallest standard deviation is filtered, and then the average value of the remaining other scale MMSE is calculated, finally obtaining the sensitivity factor, as shown below:

$$Sen_j = \sum_{i=1}^{\widehat{N}} \left| \frac{MMSE_i - MEAN}{\sigma} \right| \tag{6.12}$$

$$SenF = \sum_{j=1}^{\widehat{S}} \frac{SenF_j}{\widehat{S}} \tag{6.13}$$

where $\widehat{N}$ is the number of research targets, such as the number of battery cells. *MEAN* is the average *MMSE* under a certain scale, σ is the standard deviation of all MMSEs, $\widehat{S}$ is the number of scale factors after filtering, i is the research target number, j is the

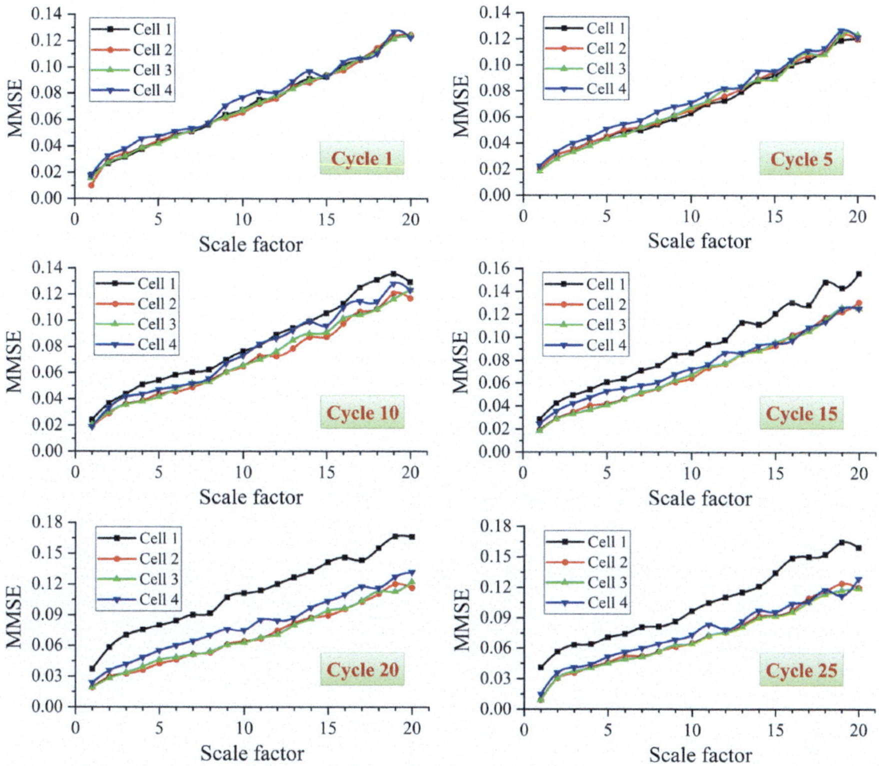

Fig. 6.7 MMSE curves for different charge and discharge cycles

scale factor number, Sen_j is the sensitivity of the j th scale, $SenF$ is the sensitivity factor.

Figure 6.8 shows the sensitivity of four single cells in different charge and discharge cycles. We can see that the sensitivity factors of all single cells are limited within a specific range in different charge and discharge cycles. When the false connection is not serious, the abnormal voltage is not obvious, and the sensitivity difference between different single cells is not large, as shown in the 5th cycle; from the 10th cycle, the sensitivity of the No. 1 single cell gradually becomes prominent and significantly higher than other single cells, and it is more evident than the above voltage curve and MMSE curve, which also reflects the effectiveness of using sensitivity to detect abnormal single cells and the feasibility of evaluating abnormal levels.

Figure 6.9 shows the sensitivity factors of each single cell in four different charge and discharge cycles. From this, we can see that the false connection in the No. 1 single cell gradually becomes serious after 10 cycles. In addition, the sensitivity factor can improve the evaluation efficiency of the remaining effective scales by filtering the inefficient scales, so as to diagnose the position of the single cell with

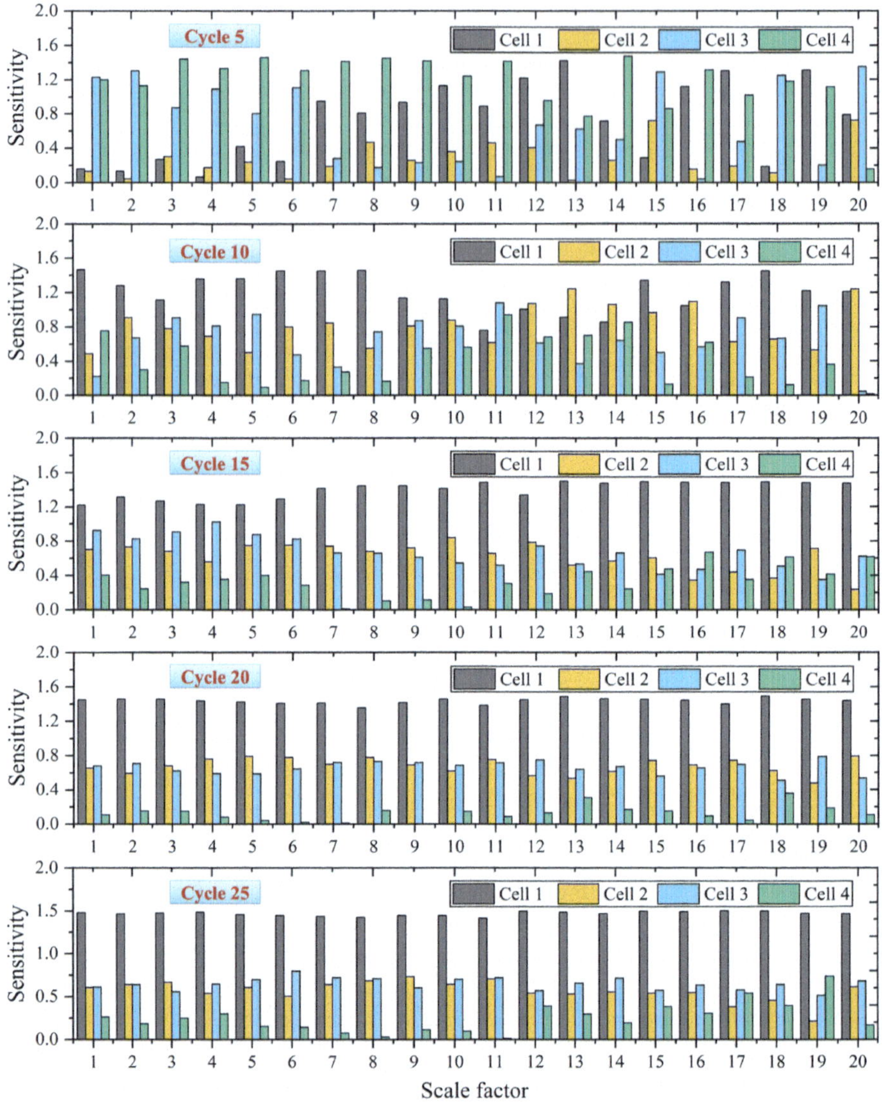

Fig. 6.8 Sensitivity of different charge and discharge cycles

false connection more timely and accurately. Since the fluctuation range of the power battery voltage is usually within a certain range, the obtained safety voltage sensitivity factor will also be stable within a certain safe range. By setting a threshold for the sensitivity factor to evaluate the voltage's abnormal risk level, it can judge whether the battery voltage is a safety risk, such as setting the threshold to 1.1 in Fig. 6.9.

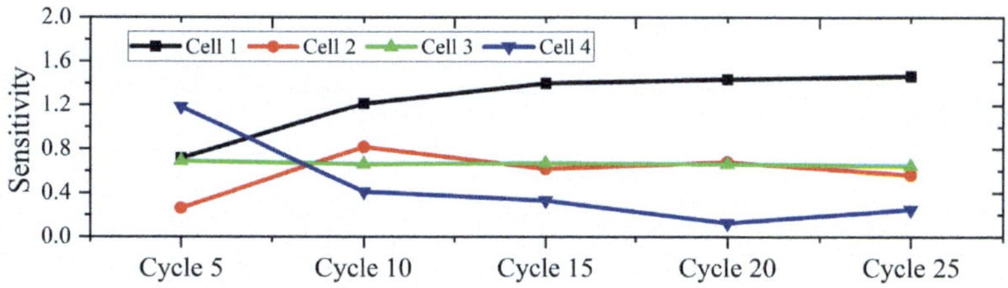

Fig. 6.9 Sensitivity factors of single cell voltage in different charge and discharge cycles

6.3.3 Safety Risk Warning Based on Actual Vehicle Operation Data

1. Voltage Abnormal Risk Warning

The operation data of vehicle G on August 9, 2016, was collected from the platform. The power battery pack of the vehicle is composed of 100 single cells, and there are 16 temperature probes inside and outside the battery box. For convenience of calculation, we selected two driving/charging cycles of vehicle G on August 9, 2016; at the end of the second driving/charging cycle, a secondary alarm of single cell voltage greater than 3.6 V occurred in the 91st single cell. The voltage and SOC curves of the power battery system of vehicle G are shown in Fig. 6.10.

Taking the data of the first driving/charging cycle without triggering an alarm as the calculation object, the MMSE and sensitivity curves of all single cells are shown in Fig. 6.11a and b, respectively, both of which indicate that the 91st single cell with abnormal voltage can be accurately detected through the first driving/charging cycle.

The sensitivity factor of single-cell voltage is shown in Fig. 6.12. It can be found that the sensitivity factor of the 91st single cell is significantly higher than that of other single cells, indicating that the MMSE method can diagnose the abnormal single cell early before the overvoltage alarm. By using a trial and error method to verify a large amount of actual operation data, the risk threshold

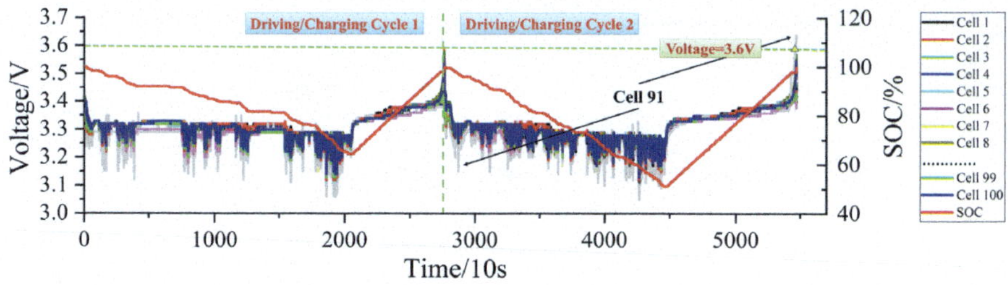

Fig. 6.10 Voltage and SOC curves of the power battery system of vehicle G

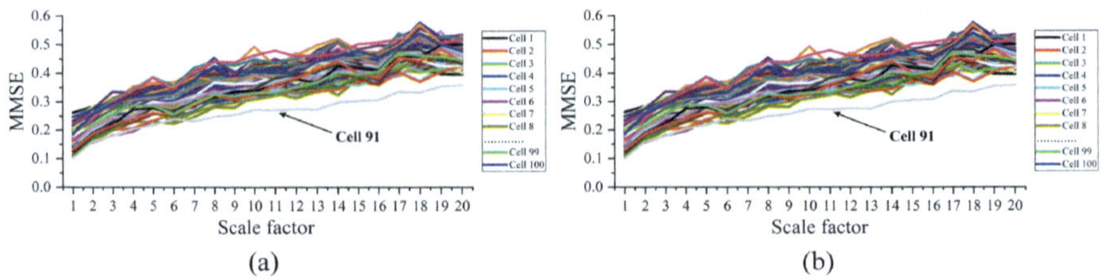

Fig. 6.11 MMSE and sensitivity curves of the power battery system of vehicle G: **a** MMSE; **b** sensitivity

of single-cell voltage sensitivity factor is set to $SenF = 3$, as shown in Fig. 6.12. This means that when the sensitivity factor of any single cell meets $SenF \geq 3$, it indicates that the single cell voltage has safety risks and needs to be checked or replaced in time. Continued operation may cause battery safety failure; when the sensitivity factors of all battery voltages meet $SenF < 3$, it indicates that the power battery system is in a safe state.

The operation data of vehicle H on February 21, 2016, was exported from the platform. The vehicle model and power battery parameter configuration are the same as those of vehicle G. For the convenience of calculation, we selected two driving/charging cycles of this day; during the second driving/charging cycle, two single-cell batteries (No. 84 battery and No. 100 battery) issued a third-level under-voltage warning of a voltage lower than 3.0 V. The voltage and SOC curves of the power battery system of vehicle H are shown in Fig. 6.13.

Using the data from the first driving/charging cycle without triggering an alarm as the calculation object, the MMSE and sensitivity curves of all single cells are shown in Fig. 6.14a and b, respectively. Both figures indicate that the abnormalities of cell 84 and cell 100 can be accurately detected. The sensitivity factors of all cells are shown in Fig. 6.15. It can be seen that the sensitivity factors of cell 84 and cell 100 are significantly higher than those of other cells, and both exceed the safety risk threshold $SenF = 3$. These results show that even if multiple cell voltages are abnormal, the proposed MMSE method can accurately diagnose all abnormal cells before the fault alarm is triggered.

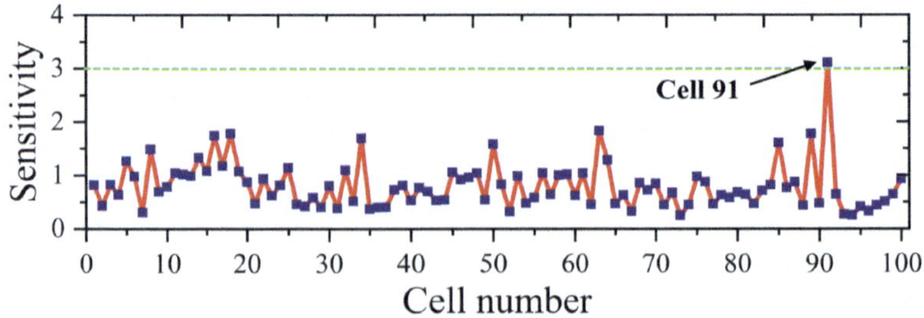

Fig. 6.12 Sensitivity factor of the power battery system of vehicle G

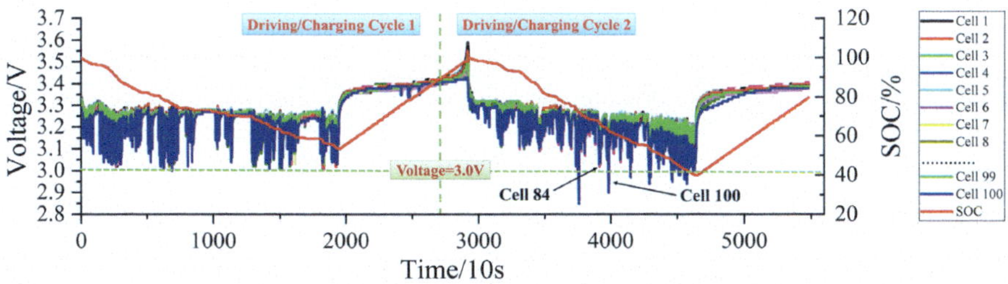

Fig. 6.13 Voltage and SOC curves of the power battery system of vehicle H

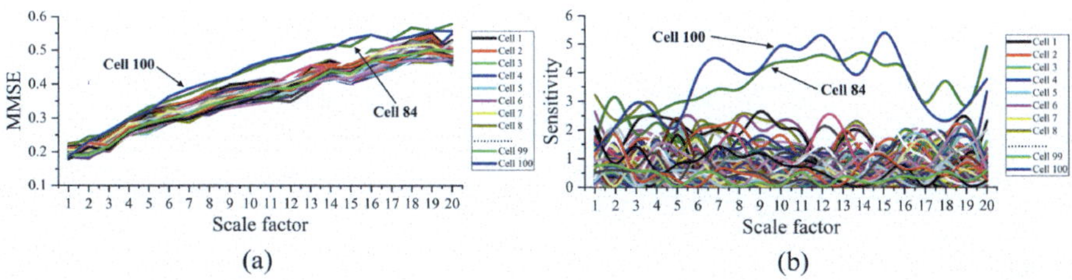

Fig. 6.14 MMSE and sensitivity curves of the power battery system of vehicle H: **a** MMSE; **b** sensitivity

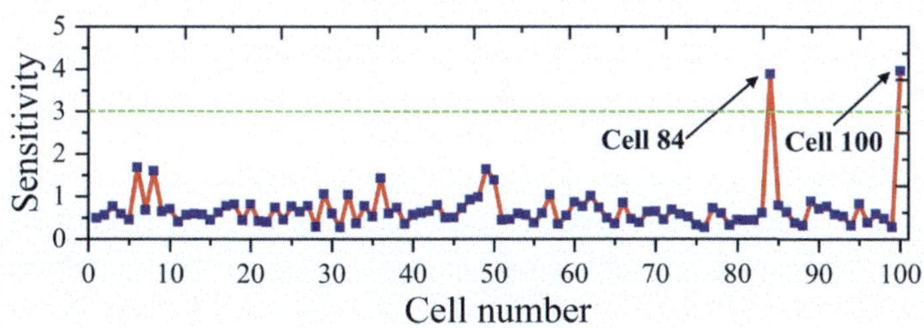

Fig. 6.15 Sensitivity factor of the power battery system of vehicle H

To verify the robustness of the MMSE method after losing some feature information, the G car data was processed with sparse variable sampling intervals (sampling interval, SI) of 30 s, 60 s, and 180 s, respectively, resulting in three sets of sparse sampling interval data. The voltage and SOC curves are shown in Fig. 6.16. It can be seen from the figure that as the sampling interval increases, the basic fluctuation characteristics of the voltage do not change significantly, but the density of effective information is greatly reduced, and some feature information also disappears.

Subsequently, using the data from the first driving/charging cycle as the calculation object, the MMSE and sensitivity factors of different sampling intervals were calculated, as shown in Figs. 6.17 and 6.18. Figure 6.17 shows that as the sampling interval increases, the MMSE curve becomes more chaotic; especially

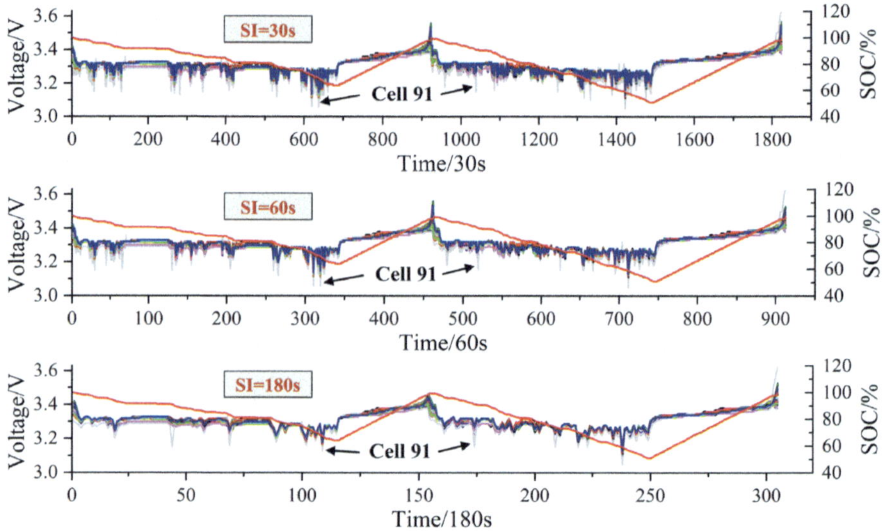

Fig. 6.16 Voltage and SOC curves of the G car power battery system with three extended sampling intervals

when SI = 180 s, due to the loss of too much information, it is difficult to detect the abnormal cells that cause overvoltage with the naked eye. Figure 6.18 shows that when SI = 30 s and SI = 60 s, we can still detect the abnormal cell 91 in advance through the safety risk threshold of the sensitivity factor, but when SI = 180 s, due to the loss of too much effective information, the sensitivity factor of cell 91 does not exceed the safety risk threshold, even though it is higher than the sensitivity factors of other cells. In summary, the MMSE method proposed in this paper can allow for an appropriate extension of the sampling interval, reduce the storage burden, and also improve the efficiency of risk warning. Therefore, the MMSE method can adapt to the sparse sampling requirements of the electric vehicle big data monitoring platform, providing technical support for the safety risk warning research of sparse big data of real vehicle power battery systems with large time spans and low sampling frequencies.

2. Temperature Abnormality Risk Warning

The operating data of vehicle I of the same model as the G car was exported from the platform from June 1 to June 3, 2016. There are 16 temperature probes distributed inside/outside its battery box. The probe temperature, total voltage, SOC, and maximum temperature difference curves of the I car power battery system are shown in Fig. 6.19. This car had a second-level alarm (temperature difference greater than or equal to 5°C) on June 2. From the fluctuation of the probe temperature, it can be seen that probe 16 and probe 12 consistently occupy the positions of the highest and lowest temperatures, and the maximum temperature difference alarm is caused by these two temperature probes. However, the

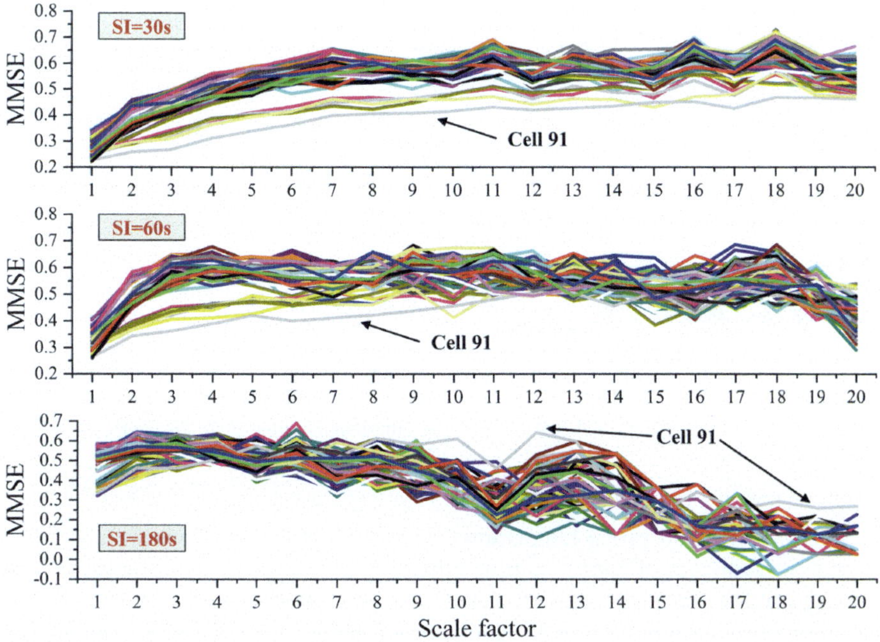

Fig. 6.17 MMSE curves of the G car power battery system at different sampling intervals

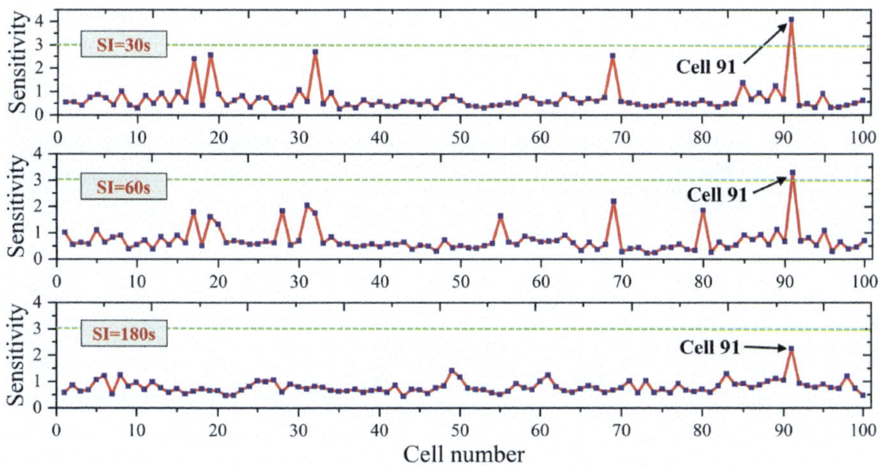

Fig. 6.18 Sensitivity factors of the G car power battery system at different sampling intervals

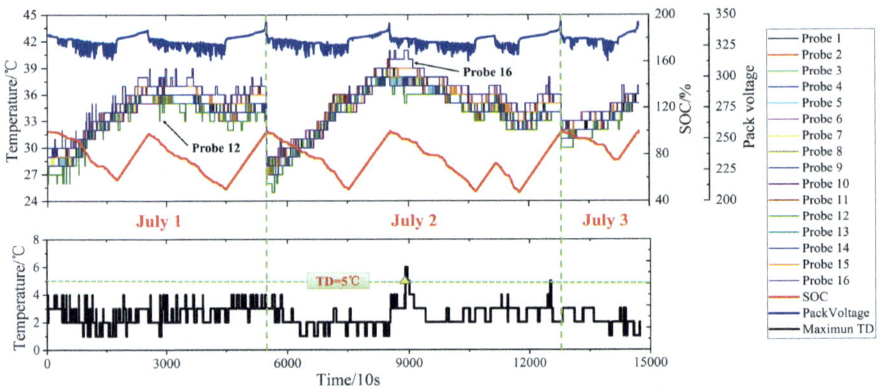

Fig. 6.19 Probe temperature, total voltage, SOC, and maximum temperature difference of the I car power battery system

consistency of all probe temperatures is always good, and they are all within the safe temperature range, so it is difficult to give early risk warnings for such temperature abnormalities with conventional methods.

Due to the relatively small fluctuation of the probe temperature, the data of the first day without triggering a warning was selected as the calculation object. The MMSE and sensitivity of all probe temperatures are shown in Fig. 6.20a and b. The MMSE and sensitivity curves can very clearly show the abnormal temperature characteristics of probe 12 and probe 16. Figure 6.21 shows the sensitivity factors of the probe temperature of the I car power battery system. From it, we can see that the sensitivity factors of probe 12 and probe 16 are much higher than those of other temperature probes, and both exceed the safety risk threshold. Therefore, the proposed MMSE method can detect the two abnormal probes before the fault of excessive temperature difference occurs.

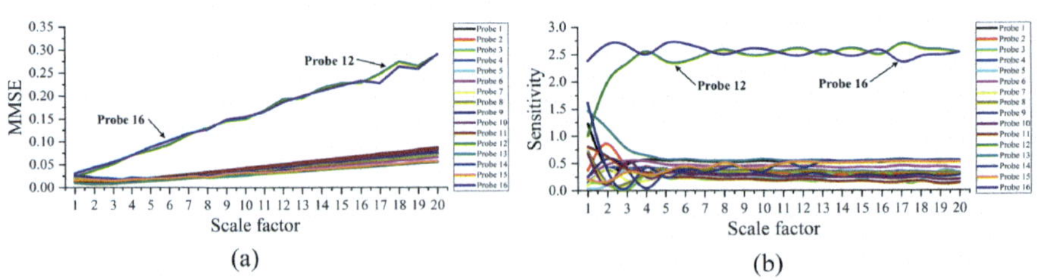

Fig. 6.20 MMSE and sensitivity curves of the probe temperature of the I car power battery system: **a** MMSE; **b** Sensitivity

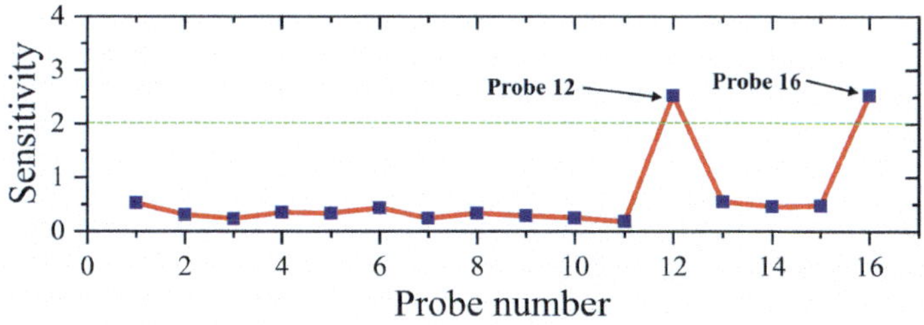

Fig. 6.21 Sensitivity factors of the probe temperature of the I car power battery system

6.3.4 *Research on MMSE Key Calculation Factors and Diagnostic Effects*

As shown in Fig. 6.22, the key calculation factor of MMSE is the scale factor. The choice of scale factor has a decisive impact on the diagnostic effect. When the scale factor is small, the diagnostic effect improves as the scale factor increases, but as the scale factor continues to increase, the diagnostic effect will gradually deteriorate. Through fitting analysis, it was found that the diagnostic effect of MMSE and the scale factor show an approximately normal distribution relationship. This paper uses the Gaussamp function for fitting.

This rule is based on a small number of samples, and the validity of the rule has been verified under the verification of several sets of data. This discovery has significant implications for the application of multi-scale entropy and other entropy algorithms in the field of fault diagnosis. However, this rule is after all based on a

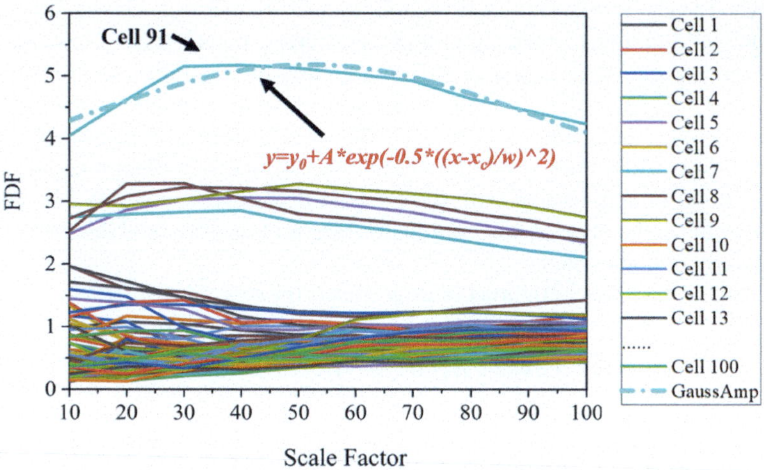

Fig. 6.22 Relationship between MMSE diagnostic effect and calculation window size

small number of samples. If we want to make this method applicable to more types of batteries, we need more data for verification and rule exploration.

6.4 Risk Warning of Thermal Runaway in Power Battery Systems Based on NDWD

6.4.1 *Time–Frequency Conversion Algorithm*

Based on the research of thresholds, algorithms, models, etc. in the time domain space, this study uses the frequency domain features of power batteries to carry out frequency domain feature analysis and safety risk warning research of power battery systems. The commonly used time–frequency conversion methods include Fourier transform, Laplace transform, *Z* transform, wavelet transform, etc., among which Fourier transform and wavelet transform are the two most commonly used transform algorithms.

1. Fourier Transform

   Frequency generally refers to the rate of change of something. If a certain mathematical or physical variable changes very quickly, we can consider its frequency to be high, otherwise it is low, and if the variable remains unchanged, the frequency is 0 or it can be considered as no frequency. Through the time–frequency conversion of the Fourier transform, some information that is difficult to be found in the time domain can be easily seen from the frequency domain, so it is often used in the diagnosis of medical conditions.

   The Fourier transform is a reversible transform, that is, it allows the original signal and the transformed signal to be converted to each other. However, only one type of information is available at any time in this transformation, and the frequency domain after the Fourier transform does not contain the original time information, so the time domain after the inverse transform cannot obtain the time information before the transformation. In fact, this phenomenon can be understood in conjunction with the formula of Fourier transform. The Fourier transform can be divided into continuous Fourier transform and discrete Fourier transform, among which:

   The formula for continuous Fourier transform is:

$$F(w) = F\left[f(t)\right] = \int_{-\infty}^{\infty} f(t)e^{-iwt}dt \tag{6.14}$$

$$f(t) = F^{-1}[F(w)] = \frac{1}{2\pi}\int_{-\infty}^{\infty} F(w)e^{iwt}dw \tag{6.15}$$

   The formula for the discrete Fourier transform is:

$$X[k] = \sum_{n=0}^{N-1} x_n e^{-i2\pi kn/N} \tag{6.16}$$

$$x_n = \frac{1}{N} \sum_{n=0}^{N-1} X[k] e^{i2\pi kn/N} \tag{6.17}$$

where, $F(w)$ is the image function of $f(t)$, $f(t)$ is the image function of $F(w)$, $X[k]$ represents the transformed data, x_n is the sampled analog signal, and N is the number of sampling points. We can see that because the integral in the Fourier transform is from negative infinity to positive infinity, any changes in the frequency component at any time will affect the global result of the integral. The same is true for the discrete Fourier transform, which is why the Fourier transform is generally not suitable for analyzing non-stationary signals. Various parameters of the power battery system, especially the fault signals, are definitely some non-stationary signals, so the Fourier transform has no advantage in the frequency domain analysis of power batteries. For example, in the 20th cycle of the power battery connection failure experiment analyzed above, the result of the discrete Fourier transform is shown in Fig. 6.23. From the figure, it can be seen that whether in the low-frequency, medium frequency or high-frequency stage, after the time–frequency conversion using the Fourier transform, the abnormal signal of the 1st single cell with a loose connection is actually weakened, so the diagnostic effect of the abnormal signal based on the discrete Fourier transform is not ideal.

2. Discrete Wavelet Transform and Decomposition

   Wavelet analysis was proposed by J. Morlet of France in 1974 based on Fourier analysis and is a rapidly developing new field in applied mathematics and engineering disciplines. As the name suggests, a wavelet is a small waveform, which has both oscillatory and attenuating characteristics. The expansion of any signal

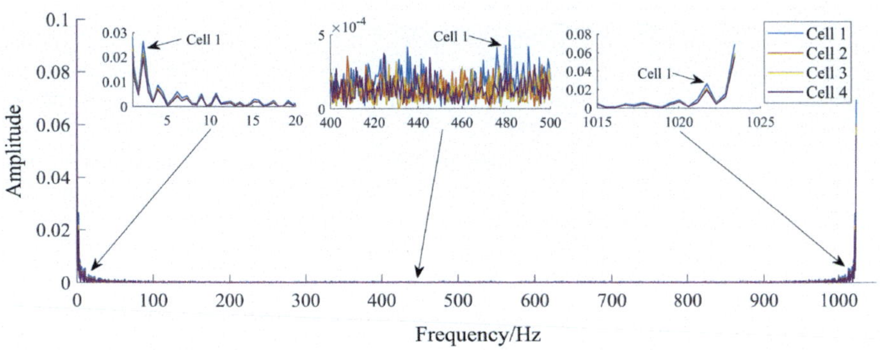

Fig. 6.23 Discrete Fourier transform result of Cycle 20

or function on a continuous wavelet basis is called a continuous wavelet transform. The wavelet transform, by adopting gradually refined sampling steps for different frequency components, can focus on any detail of the signal, so it has good time–frequency localization characteristics and can handle weak or abrupt signals well, making it particularly suitable for analyzing non-stationary distorted waveforms.

Any signal or function $f(t)$ defined in function space or signal space $L^2(R)$, in the expansion of the continuous wavelet basis $\varphi_{(\alpha,\tau)}(t)$, $WT_f(\alpha,\tau)$ is called $f(t)$ Continuous wavelet transform.

$$WT_f(\alpha,\tau) = \frac{1}{\sqrt{|\alpha|}}\varphi\left(\frac{t-\tau}{\alpha}\right)\int_R f(t)\varphi_{(\alpha,\tau)}(t)dt \tag{6.18}$$

where:

$$\varphi_{(\alpha,\tau)}(t) = \frac{1}{\sqrt{|\alpha|}}\varphi\left(\frac{t-\tau}{\alpha}\right) \tag{6.19}$$

and satisfies:

$$C_\varphi = \int_R \frac{\varphi(v)}{|v|}dv < +\infty \tag{6.20}$$

The scale parameter α and the translation parameter τ determine the time–frequency characteristics of the wavelet transform. Because the coefficients of the continuous wavelet transform are coupled and repeated, it is necessary to further discretize them in practical applications, resulting in the discretized wavelet transform and discrete α and τ parameters. The steps for discretization are as follows:

(1) Power level discretization of the scale parameter α

$$\alpha = \alpha_0^m \tag{6.21}$$

(2) Discretisation of the translation parameter τ

$$\tau = k\alpha_0^m\tau_0 \tag{6.22}$$

where, $\alpha_0 > 0$, $\tau_0 > 0$, m and k are all integers. After discretizing the scale and translation parameters, a discrete wavelet basis can be obtained. $\varphi_{\alpha_0^m,k\tau_0}$ and the discrete wavelet transform $WT_f\left(\alpha_0^m, k\tau_0\right)$, as follows:

$$\varphi_{\alpha_0^m,k\tau_0} = \alpha_0^{-\frac{m}{2}}\varphi\left(\alpha_0^{-m}t - k\tau_0\right) \tag{6.23}$$

$$WT_f\left(\alpha_0^m, k\tau_0\right) = \alpha_0^{-\frac{m}{2}}\int_R f(t)\varphi\left(\alpha_0^{-m}t - k\tau_0\right)dt \tag{6.24}$$

The basis of the Fourier transform is the cosine function, while the basis of the wavelet is various shaped wavelets, that is, it regards the entire waveform as the superposition of multiple wavelets of different positions and widths. The scale parameter α controls the small stretch, and the translation parameter τ can control the translation of the wavelet.

Through the above analysis, the Fourier transform is not suitable for handling non-stationary signals and can only obtain a spectrum, but most waveforms in reality are non-stationary waveforms. The frequency domain information only provides the components of the waveform and does not provide the order of the components. Therefore, pure Fourier analysis has a poor resolution for non-stationary signals whose frequency changes with time.

Wavelet analysis can obtain a time–frequency spectrum, including signal decomposition (i.e., wavelet decomposition) and reconstruction (i.e., wavelet transform) two stages, which makes wavelet analysis very suitable for analyzing non-stationary time series such as the power battery signals in this paper. Wavelet decomposition is to find the best mapping from the actual signal space to the wavelet function space in order to obtain the best recovery of the original signal. The low-frequency signal and high-frequency signal obtained after decomposition can perfectly restore the original signal through reconstruction, so it is widely used in denoising various signals such as images. In addition, after wavelet decomposition, high-frequency and low-frequency components are obtained. The low-frequency component obtained from the low-pass filter can be called the "approximate component", and the high-frequency component obtained from the high-pass filter can be called the "detail component"; the positions of the two components correspond to the original signal, so the signal after wavelet decomposition can perfectly reconstruct and locate the details in the original signal, and can also quantize the coefficients of each layer of wavelet decomposition by selecting an appropriate threshold.

Taking the 20th cycle of the virtual connection fault in the power battery as an example, the result after discrete wavelet decomposition is shown in Fig. 6.24. The low-frequency signal shown in Fig. 6.24a basically retains the characteristics of the original signal as shown in Figure (Cycle 20 above), and the high-frequency signal shown in Fig. 6.24b can effectively display the abnormality and sudden change of the signal of the No. 1 cell. Therefore, it is of important potential engineering application value to monitor and warn the safety status of the actual vehicle power battery system in real-time based on the discrete wavelet decomposition method.

6.4.2 Accident Feature Description

Taking a purely electric vehicle that caught fire in the second half of 2019 (based on corporate privacy confidentiality, the following accident time and other information have been offset and desensitized) as the research object, the car caught fire shortly after driving after completing fast charging at the charging pile. The car is equipped with a ternary lithium battery, in which 91 individual batteries are stacked

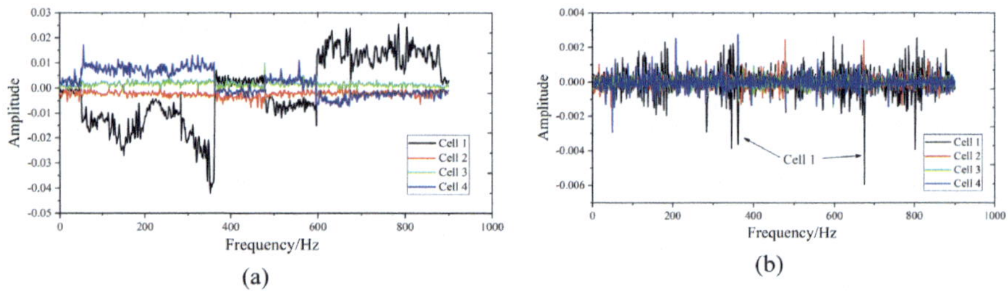

Fig. 6.24 Wavelet decomposition results of Cycle 20: **a** Low-frequency information; **b** High-frequency information

and connected in series and parallel to form a battery pack. Figure 6.25 is a schematic diagram of the distribution and damage of the individual cells in the power battery system under study after the accident. A represents the rear of the battery pack, B represents the front of the battery pack, and C and D represent the right and left sides of the battery pack, respectively. It is not difficult to see from the figure that the right side and the rear of the battery pack are basically intact after the accident, but there are perforations in the front, and the other parts on the left are basically intact but severely damaged at about 1/3.

Analyze the data from the 15 days prior to the thermal runaway, the voltage curve is shown in Fig. 6.26. On the day of the accident, the car was only charged once, and the thermal runaway occurred a few minutes after starting to drive. The voltages of cell 9, cell 7, and cell 13 first dropped rapidly, followed by a gradual sharp drop in the voltages of cell 23, cell 19, and cell 5. This also reflects the starting position and direction of the thermal runaway, which corresponds to the distribution of the individual cells in the above Fig. 6.24 (cells 7 and 13 are both distributed near cell 9). In addition, in the 15 days before the thermal runaway, the vehicle underwent about 20 different depth driving/charging cycles, and the voltage fluctuations during the charging and driving stages were significantly different. Furthermore, for a period

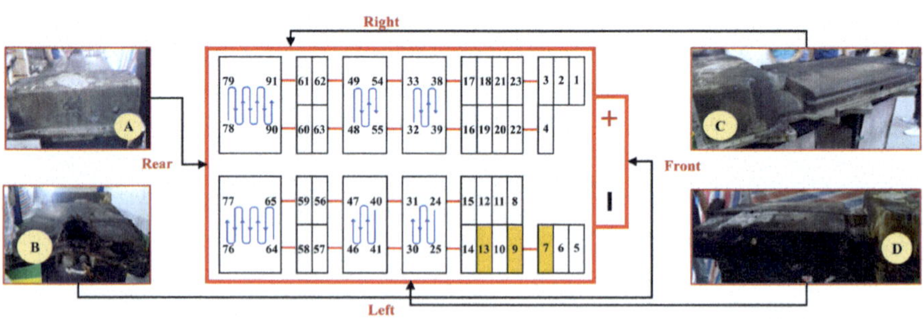

Fig. 6.25 Schematic diagram of the distribution and damage of the individual cells in the power battery system under study

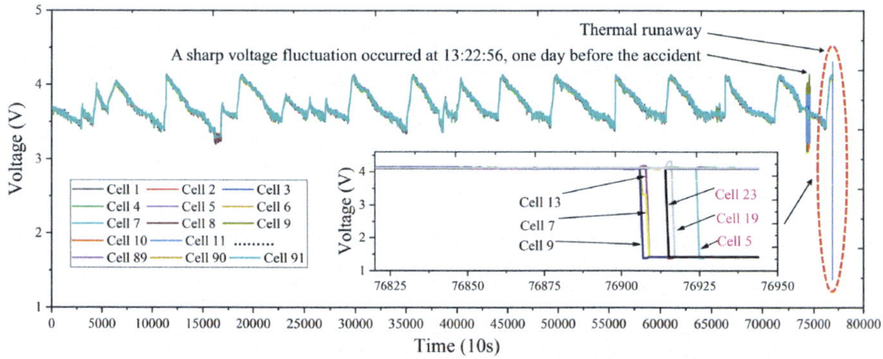

Fig. 6.26 Voltage curve 15 days before thermal runaway of the vehicle under study

after 13:22:56 on the morning of the day before the accident, the vehicle experienced a short period of severe voltage fluctuation while driving. This short-term severe fluctuation may be due to various reasons such as the sudden deterioration of the health status of the battery system under study before approaching thermal runaway, encountering some sudden situation, data collection error, or overly rough data preprocessing. However, even if there are some errors in the collected data, it will only affect the authenticity of the original data and will not have any impact on the effectiveness of the proposed algorithm and the generality of the application object.

The battery voltage characteristics of the charging state and driving state during the actual vehicle operation process are very different, so the actual vehicle operation data is divided into driving data and charging data, and some specific date charging and driving data are selected for analysis. Figure 6.27 shows the one-hour single-cell voltage curve under the vehicle charging state on the day of the accident, 6 days ago, and 15 days ago, and Fig. 6.28 shows the one-hour single-cell voltage curve under the vehicle driving state 1 day before, 3 days before, 7 days before, and 15 days before the accident (there is only a small amount of charging data and no driving data on the day of the accident and 6 days ago). Figure 6.27 shows that the voltage fluctuations of all cells under the charging state on the day of the accident, 6 days ago, and 15 days ago are small, and the consistency is very good. Figure 6.28 shows that the single-cell voltage curve under the driving state 1 day before, 3 days before, 7 days before, and 15 days before the accident has very obvious fluctuations, and 1 day before, 3 days before, and 7 days before the accident, especially 1 day before the imminent thermal runaway, cells 7, 9, and 13 have more obvious abnormal fluctuations compared to other cells, which is related to the complex and variable vehicle driving state. However, all cell voltage fluctuations are still within the safe threshold in the conventional time domain, so it is difficult to diagnose these abnormal cells with conventional time domain methods.

Figures 6.29 and 6.30 are the discrete high-frequency wavelet curves of the three charging intervals and four driving intervals analyzed above. The discrete wavelet curves of all cells under the charging state on the day of the accident, 6 days ago,

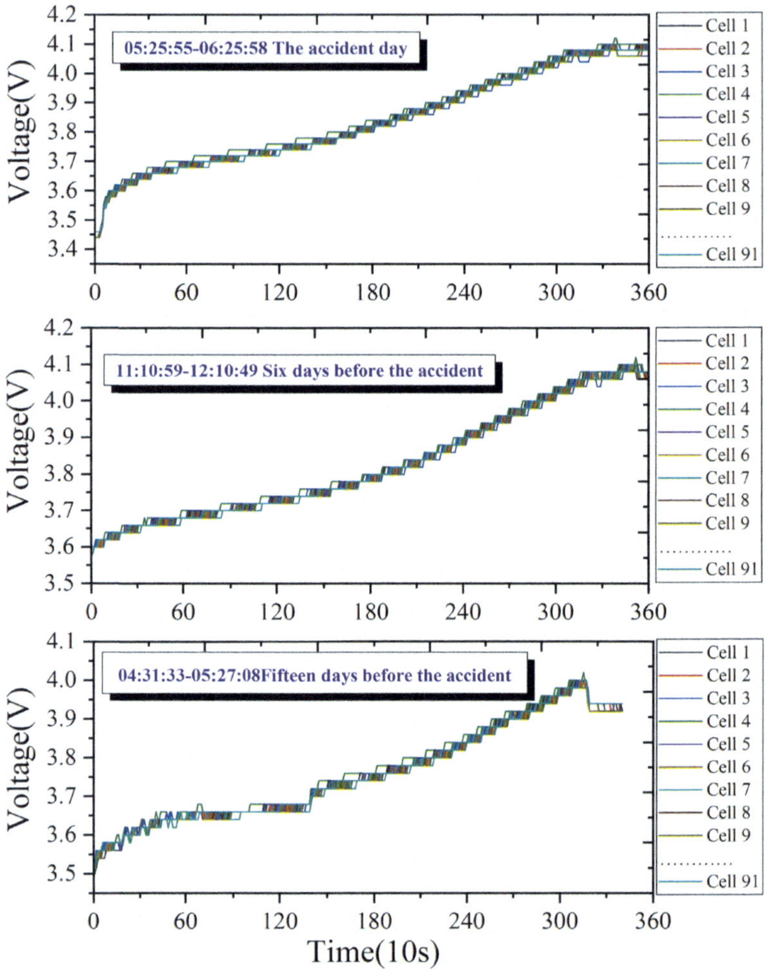

Fig. 6.27 Voltage under charging state three days before thermal runaway

and 15 days ago have small fluctuations and good consistency, which corresponds to the voltage curve in Fig. 6.27. Due to the dynamic variability of the vehicle driving state, the discrete wavelet curves under the driving state 1 day before, 3 days before, 7 days before, and 15 days before the accident all have severe fluctuations, and the discrete wavelet values of each cell battery vary in size due to different degrees of abnormality. 3 days before and 7 days before the accident, cells 7 and 9 both have abnormal fluctuations; 1 day before the accident, cells 7, 9, and 13 all have relatively obvious abnormal fluctuations, which are consistent with the voltage curve in the above Fig. 6.28. Therefore, the DWD method is stable and effective for early risk prediction and abnormal cell positioning of thermal runaway of power battery systems.

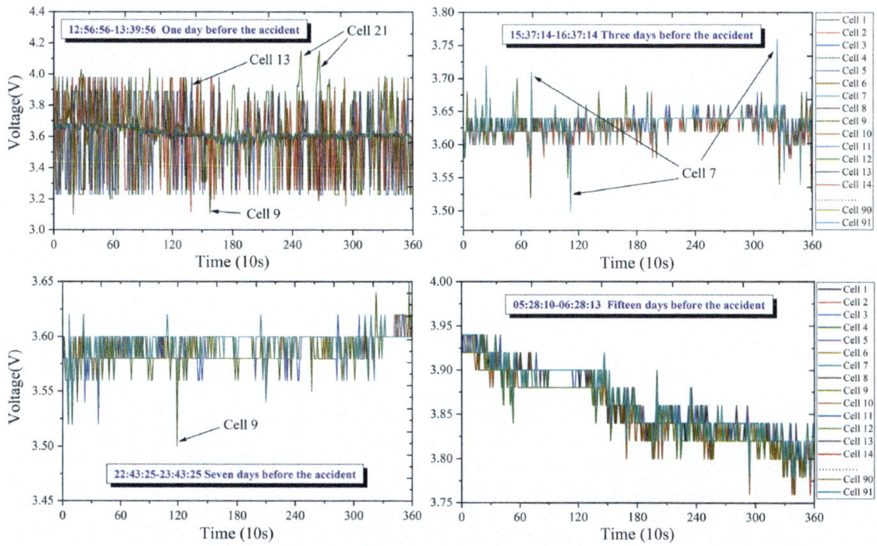

Fig. 6.28 Voltage under driving state four days before thermal runaway

6.4.3 *Analysis of Warning Results*

Through the above two discrete high-frequency wavelet graphs, it can be found that the power battery system has a large difference in voltage and wavelet characteristics during the charging and driving stages due to different vehicle operating conditions. Based on a large amount of actual vehicle operation data from the big data platform, combined with the normalization method, this paper has developed a multi-level risk warning strategy based on NDWD through multiple groups of actual vehicle operation data verification and trial and error, and has set corresponding abnormal risk coefficient thresholds for the two vehicle states of driving and charging, as shown in Fig. 6.31. Since the discrete wavelet values under different battery types or different battery states are not fixedly distributed within a certain size range, this paper normalizes the discrete wavelet values under the vehicle driving state, and obtains a first-level risk warning threshold of 0.8 and a second-level risk warning threshold of 0.6 through a large amount of actual vehicle data verification. Since the vehicle condition is always in a relatively stable state that is completely different from the driving state during the charging state, an abnormal risk coefficient threshold of ± 0.015 is obtained through a large amount of actual vehicle data verification. The rules for setting abnormal risk coefficients for the two states are as follows:

The abnormal risk coefficient for the driving state is:

$$DWD_{d\_nor} = \frac{|DWD_d| - \left|DWD_{d\_\min}\right|\,\Big|}{\left|DWD_{d\_\max}\right|\,\Big|\left|DWD_{d\_\min}\right|\,\Big|\,\Big|} \tag{6.25}$$

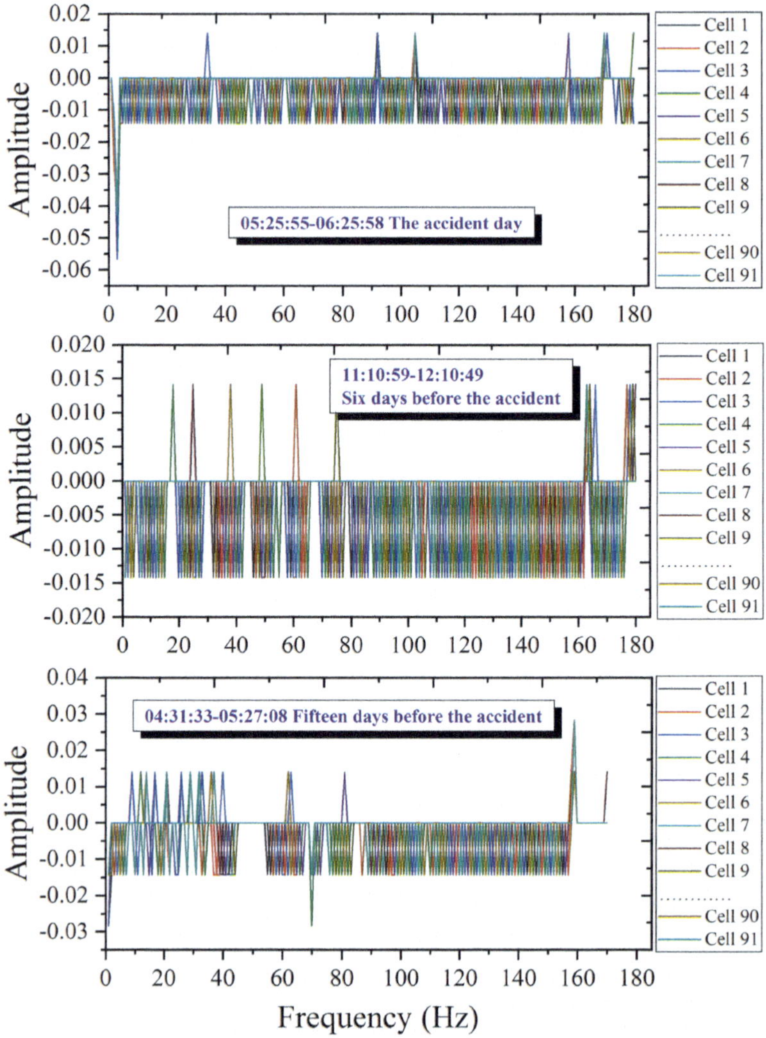

Fig. 6.29 Discrete high-frequency wavelets under charging state three days before thermal runaway

$$A_D = DWD_{d\_nor} - DWD_{d\_ave} \tag{6.26}$$

where, $DWD_{d\_\min}$ For the minimum value of the discrete wavelet of the monomer voltage; $DWD_{d\_\max}$ For the maximum value of the discrete wavelet of the monomer voltage; DWD_{d_nor} For the discrete wavelet value of the monomer voltage after normalization DWD_{d_ave} For the average value of the discrete wavelet of the monomer voltage; A_D For the abnormal risk coefficient under the vehicle driving state.

The abnormal risk coefficient under the driving state is:

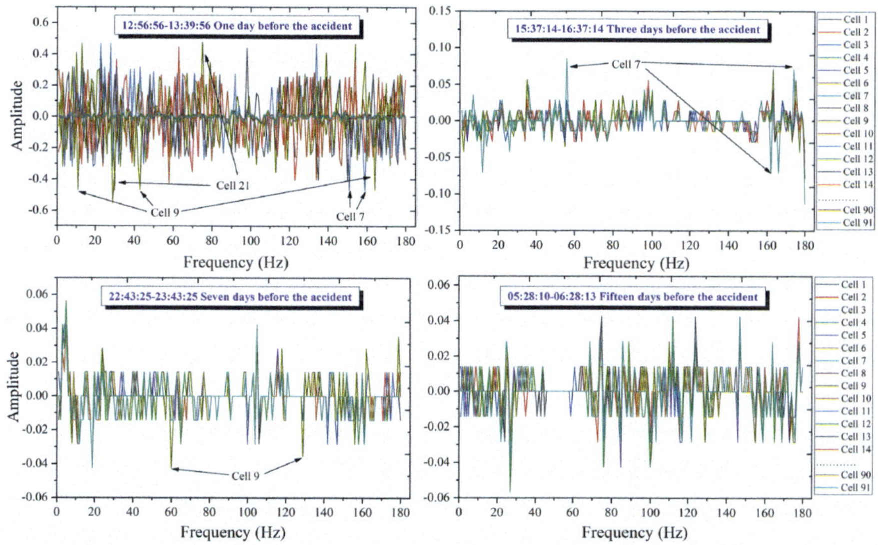

Fig. 6.30 Discrete high-frequency wavelets under driving state four days before thermal runaway

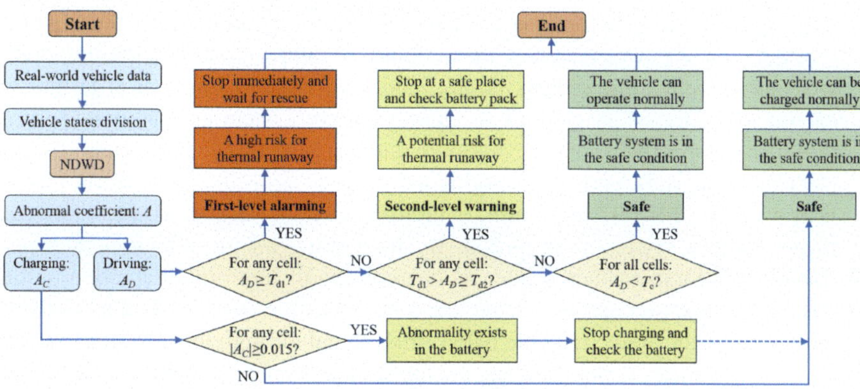

Fig. 6.31 Multi-level risk warning strategy based on NDWD

$$A_C = DWD_c - DWD_{c\_ave} \tag{6.27}$$

where, DWD_c For the discrete wavelet value of the monomer voltage; DWD_{c_ave} For the average value of the discrete wavelet of the monomer voltage; A_C For the abnormal risk coefficient under the vehicle driving state.

The vehicle under study was validated, and the abnormal charging coefficient curves for the day of the accident, 6 days before, and 15 days before were obtained, as shown in Fig. 6.32; the abnormal driving coefficient curves for 1 day before, 3 days before, 7 days before, and 15 days before the accident, as shown in Fig. 6.33. The three-stage charging abnormal risk coefficient always lies between −0.015~0.015, which is consistent with the charging voltage data characteristics. From the four-stage driving abnormal risk coefficient, it can be seen that all cells were normal 15 days before the accident. 3 Days and 7 days before the accident, the abnormal risk coefficients of cell 7 and cell 9, respectively, exceeded the second-level risk warning threshold. 1 day before the accident, cells 7, 9, and 13 all reached or exceeded the first-level alarm threshold, which is very close to the occurrence of thermal runaway. Therefore, the multi-level risk warning strategy based on NDWD can diagnose the abnormal cell 9 seven days before the accident through the second-level risk warning, and through reasonable troubleshooting, repair or replacement measures, it can avoid the risk of thermal runaway in advance and ensure the safety of the driver and passengers. In summary, the NDWD method can not only achieve online assessment and warning of the risk of thermal runaway of the real vehicle power battery system but also achieve reverse tracking of the cause and location of the accident through accident investigation and historical data analysis.

6.5 Summary

(1) Starting from the structural perspective, the common failure modes and failure mechanisms of power batteries are summarised, and a three-level failure mode analysis of cell failure, BMS failure, and Pack system integration failure is carried out. The safety risks of the power battery system are elaborated from multiple perspectives such as thermodynamics, heat transfer, flammability of electrode materials, and overcharging behavior.

(2) Based on the MMSE algorithm, a multi-level risk warning strategy for the power battery system is proposed. By integrating all the effective information of the coarse-grained sequence, overcoming the entropy fluctuation in the sliding average process, and achieving a macroscopic complexity analysis and multi-time scale analysis of the battery risk signal through low-efficiency scale filtering. Multi-level risk warning coefficients and safety thresholds are set for different types of safety risks. Through the verification of battery connection faults based on experiments, voltage class and temperature class fault data based on real vehicles, it is proved that this method can carry out online multi-level evaluation, early identification and precise warning of various safety risks of "mechanical–electrical–thermal" power battery systems. In addition, through the analysis of large sampling intervals such as 30 s, 60 s, 180 s, it is proved that the MMSE method can adapt to the characteristics of large sampling time span and unstable sampling frequency of the real vehicle big data monitoring platform, providing technical support for the safety risk warning research based on the sparse sampling characteristics of the real vehicle power battery system.

(3) Based on the NDWD algorithm and real vehicle thermal runaway data, a multi-level risk warning strategy for the real vehicle power battery system is proposed. NDWD overcomes the problem of missing time information in

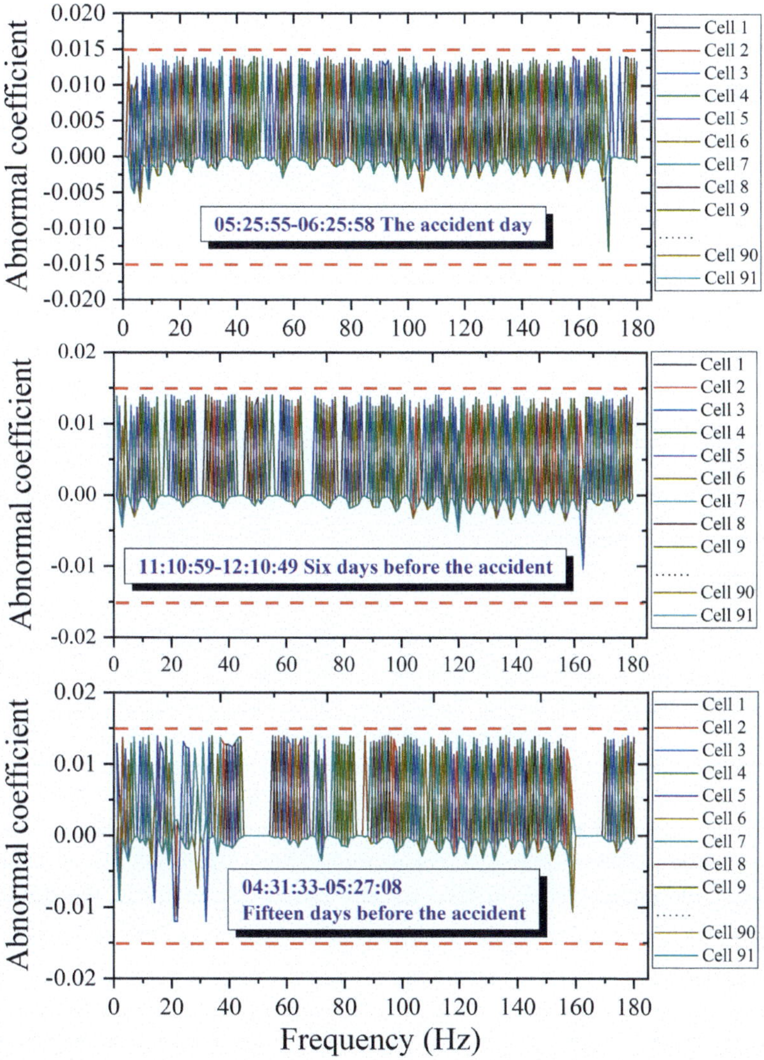

Fig. 6.32 Three-day charging abnormal risk coefficient

the frequency domain of conventional time–frequency conversion methods. Its time–frequency localization characteristics are good at dealing with weak or sudden information, suitable for analyzing non-steady-state distortion waveform problems such as real vehicle power battery signals. The low-frequency signal and high-frequency signal obtained after decomposition can perfectly restore the original signal through reconstruction, and its high-frequency detail component can achieve online identification and precise positioning of early risk signals of the power battery system. Combined with the normalization method,

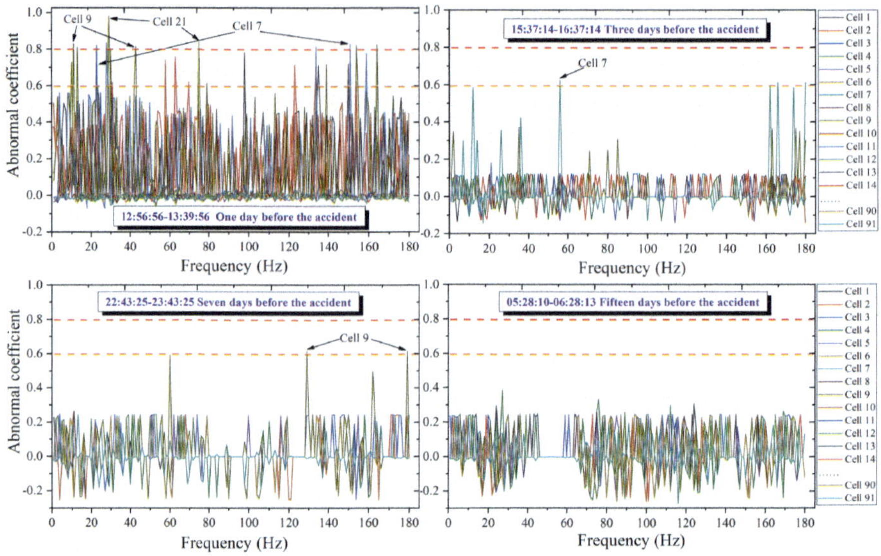

Fig. 6.33 Four-day driving abnormal risk coefficient

abnormal risk coefficient thresholds are set for driving and charging states respectively. Through the verification of real vehicle data of a thermal runaway self-ignition accident, abnormal cells can be diagnosed seven days before the accident through the second-level risk warning. Therefore, the NDWD method can achieve reverse tracking of the cause and location of thermal runaway accidents through accident investigation and historical data analysis. Through reasonable troubleshooting, repair or replacement measures, it can avoid the risk of thermal runaway in advance and ensure the safety of the driver and passengers.

Chapter 7
Research on Safety Control Strategy of Power Battery

7.1 Overview of This Chapter

The safety control of the power battery system is of great significance to the safe and reliable operation of electric vehicles. In order to achieve efficient safety control of the power battery system of the actual vehicle, various battery fault diagnosis and risk warning models need to be efficiently integrated with the Battery Management System (BMS). This chapter will systematically summarize and analyze the common safety issues of the power battery system and their solutions, study the multi-model fusion theory and method of multi-fault early warning of the power battery system, propose a safety control strategy for the full lifecycle operation of the power battery system of the actual vehicle, and the overall analysis framework is as follows (Fig. 7.1).

7.2 Analysis of Safety Issues of Power Battery

7.2.1 Safety Issues of Power Battery and Solutions

During the operation of the actual vehicle, the power battery system is affected by operating conditions and external environment, and there are various potential safety risks. These safety issues have high randomness, complex inducements, poor repeatability, and few measurable parameters. This article proposes some solutions to these problems, as shown in Fig. 7.2.

1. Cell Safety For cell safety issues, standardization of cells and battery modules can greatly reduce repetitive manual work and actively enhance cell safety. Here, standardization refers to the stipulation of uniform technical standards in product quality, variety specifications, and parts universality to adapt to

J. Hong, *Big Data Analysis and Application of Power Battery Operation Safety*, Key Technologies on New Energy Vehicles, https://doi.org/10.1007/978-981-96-4399-8_7

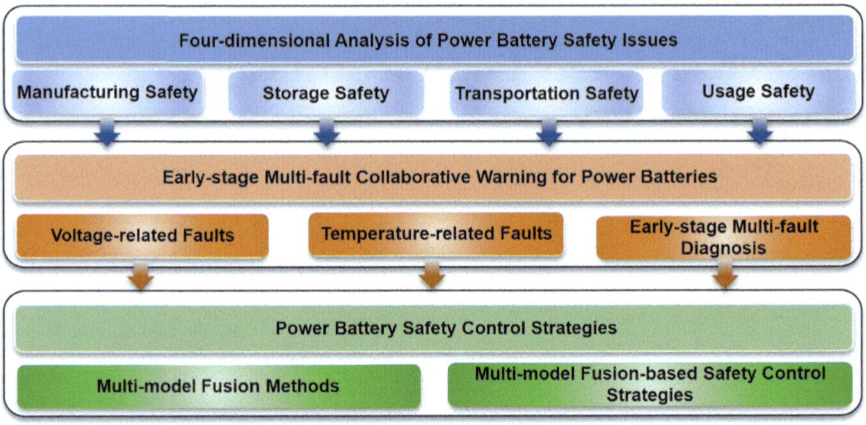

Fig. 7.1 Overall analysis framework of power safety control strategy

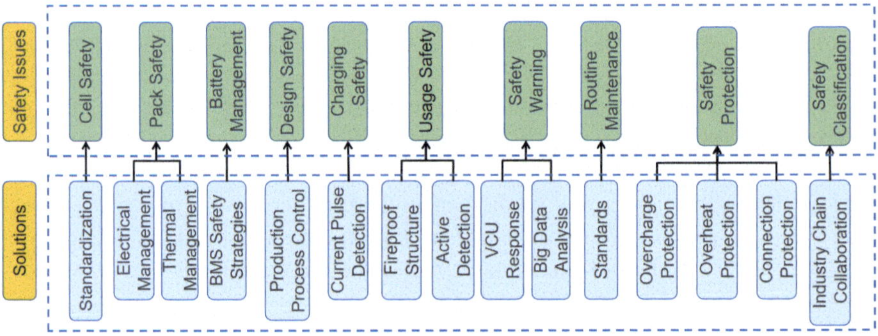

Fig. 7.2 Common safety issues and solutions of power battery system

the needs of scientific development and organized production. Standardization can improve cell safety in many ways, such as production design standardization can expose and deal with cell design safety issues, improving the standardization level of production equipment can accelerate the product iteration speed of equipment manufacturing enterprises, and also, the experience of excellent equipment and parts suppliers can be promoted in the industry to comprehensively improve the safety technology level of the battery industry.

2. Group Safety For the group safety issues of the power battery system, the focus is on improving the group safety of the power battery system from the dimensions of reasonable electrical management, thermal management design, and improving battery welding technology. The BMS, through small current balance control, can reduce temperature and voltage differences among cells, thereby requiring accurate detection and real-time warnings for all cells.

3. Battery Management For battery management issues, the focus is on designing reasonable BMS safety control strategies. In order to improve the battery safety management plan, it is necessary to formulate corresponding safety goals for the power battery system, and verify its effectiveness and rationality through multi-level verification of actual vehicle operation data. In addition, it is also necessary to integrate the functions of thermal runaway risk warning and timely fire extinguishing on the basis of efficient control of BMS to ensure the safety of the power battery system.
4. Design Safety For the design safety issues of the power battery system, the focus is on carrying out scientific battery design and manufacturing. Different types of batteries have their own advantages and disadvantages in terms of materials and structure, and different materials and structures have different impacts on battery safety. We need to make the design of the power battery system scientific and manufacturing precise, so as to make the power battery system as safe as possible when it leaves the factory.
5. Charging Safety For charging safety issues, one of the most effective methods currently is to perform current pulse detection on the power battery system at the beginning of charging. This method can detect the inconsistency problem of the power battery system as early as possible and issue a safety warning in advance. In addition, a quick large current detection can be performed at the end of charging when 80% ~ 90% of the charging is completed, and then the difference in DC internal resistance change can also detect the single cell battery with poor consistency.
6. Usage Safety For the usage safety issues of the power battery system, the most important thing is to be able to perform timely safety control when a single cell experiences thermal runaway, ensuring the personal safety of the driver and passengers. In terms of delaying thermal runaway, fireproof materials can be used or the battery box structure can be treated with fireproof and heat insulation to delay the thermal runaway expansion process from a single cell to the entire battery pack; in terms of increasing the escape time for the driver and passengers, it is necessary to add easy-to-open safety escape devices. Additionally, active monitoring of heat, smoke, and chemical emissions enables early detection of a single cell's thermal runaway. This timely alert allows the driver and passengers to evacuate safely.
7. Safety Warning For the safety warning issues of the power battery system, a quick response to the problems of the power battery system can be made through the Vehicular Communication Unit (VCU). For example, the driver can be allowed to pull over or take some necessary safety protection measures, and the BMS information collected by the VCU can be uploaded to the cloud for big data analysis in time, and then through the autonomous learning of the VCU intelligent learning algorithm, the reliability of the power battery system safety warning can be improved.
8. Daily Maintenance For the daily maintenance of the power battery system, it is very urgent and necessary to improve the relevant daily maintenance standards and related legal provisions. At present, the national standards related to the

regular inspection cycle (three months, half a year or a year) and inspection items (battery box sealing, connection reliability, etc.) of the power battery system are not perfect. Therefore, the state or the automotive industry should gradually formulate and continuously improve the mandatory inspection standards for the power battery system, and the vehicle manufacturers should also cooperate in formulating mandatory inspection items or daily maintenance manuals.

9. Safety Protection For the safety protection of the power battery system, the focus is on protection against overcharging, overheating, and connection safety risks. In the face of battery overcharging, protection can be provided through the intervention of BMS, charger, and vehicle VCU; for battery overheating, reducing the cell density and connection impedance can fundamentally reduce the risk of overheating, and strengthening cooling measures can reduce the risk of thermal runaway during charging; for battery charging connection failure, daily inspection and maintenance can be used to determine whether there is a problem with the contact of the plug and socket, and then regular replacement can effectively solve the charging connection problem.
10. Safety Grading For the safety level division of the power battery system, it is necessary for vehicle manufacturers and parts companies to accelerate the implementation of industrial chain cooperation and form a specialized division of labor cooperation system. Starting from the production part approval process (PPAP), advanced product quality planning (APQP), process failure mode and effects analysis (PFMEA) and other links, according to the relevant requirements of the automotive industry, clarify their respective responsibilities, and achieve the safety status of the power battery system throughout its life cycle can be monitored and traced.

7.2.2 Four-Dimensional Analysis of Power Battery Safety Management Scheme

(1) **First Dimension: Safety Management of Power Battery Manufacturing**

There are many product specifications and models of power batteries, and the production processes are different from each other, but generally speaking, the production process is as follows:

(1) Slicing: The main process includes slurry making, coating, rolling, cutting, and baking;
(2) Assembly: winding, casing, baking, injection, sealing;
(3) Formation: cleaning, inspection, X-ray detection, aging, capacity sorting;
(4) Packaging: inspection, coding, packaging, warehousing;

The risks in the production and processing of power batteries mainly have two aspects, one is the fire and explosion of a single battery itself, the main reason is the short circuit between the battery electrodes, the electrolyte generates gas and burns and explodes in a single battery; the other aspect is in the production process, after a single battery thermal runaway and fire and explosion, the heat is transferred to the surrounding batteries, causing the heated battery electrolyte to leak, forming explosive steam in a certain space, catching fire or catching fire and not reaching the explosion limit range and exploding, which can cause serious accidents. Therefore, to prevent power battery accidents, attention should be paid to the analysis of battery raw materials, battery structural design, and production process. The choice of materials and the design of the battery structure are complex and have a great impact on battery safety. For example, the size of the positive electrode sheet, negative electrode sheet, and separator in the power battery has a great impact on the safety of the battery. When designing the battery, the dimensions should be arranged as follows: the separator should be larger than the negative electrode sheet, which in turn should be larger than the positive electrode sheet. This is because if the quality of the separator is not good, it may shrink when heated, or the size is smaller than the electrode sheet, at this time, the positive and negative electrodes contact and cause an internal short circuit, thereby causing battery safety problems.

Not every process in the entire power battery production process needs special attention. For convenience, the accident risks of power batteries are divided into ordinary risks and special risks. Ordinary risks are risks that generally exist in enterprises, such as electrical safety, mechanical safety, fire safety, and safety of dangerous chemical production and storage; special risks only consider safety issues related to lithium-ion battery production. From the overall analysis of the process, fires and explosions of power batteries generally occur after charging, but the causes of accidents are closely related to the quality of the previous process. For example, the control of the moisture content of the slurry and electrolyte, and the casing operation of the electrode and separator. The following lists the various risks that may occur during the production process.

(1) Positive Electrode Sheet Drying The primary components of the positive electrode sheet are lithium salt and the solvent N methylpyrrolidone (NMP). After being made into a slurry, these materials are coated onto an aluminum foil, then calendered and dried in a drying oven. The drying process primarily involves vaporizing the NMP, which is subsequently condensed back into liquid for recovery. The main risk lies in the calendering and drying stages. NMP has a boiling point of approximately 93°C, while the drying oven operates at temperatures between 110 °C and 120 °C. If the heating system and exhaust equipment fail to operate in proper sequence, or if the ventilation ducts are not adequately maintained, a fire hazard may occur.
(2) Coating After the electrolyte (electrolyte salt LiPF6) is made into a slurry, it is coated on an aluminum sheet, then rolled and dried in a drying oven. The drying process mainly involves drying the NMP into vapor, which is then condensed into a liquid for recovery. The main risk lies in the rolling and drying process. The boiling point of NMP is around 93°C, and the drying temperature in the oven is 110~120°C. If the heating temperature and the exhaust equipment fail

to start in sequence, or if the ventilation duct is not properly maintained, a fire accident may occur.

(3) Injection The main components of the electrolyte (electrolyte salt $LiPF_6$) solvent are alkyl carbonates, such as dimethyl carbonate (DMC), diethyl carbonate (DEC), and ethyl methyl carbonate (EMC), which are all low-boiling liquids with fire hazards. The injection site is relatively enclosed, and the storage of such hazardous chemicals needs to consider preventing fire, explosion, poisoning, and suffocation accidents.

(4) Formation and Capacity Sorting In the formation workshop, minor leaks in the batteries, major leaks, and individual battery combustion may occur. In the formation cabinet, batteries are charged at different voltages, some are operated in series, and measures need to be taken to deal with smoking or burning batteries in time. A single battery may smoke, but the harm caused by timely treatment is limited. During the capacity sorting process, the batteries are generally not in series, but there may also be situations where a single battery smokes and triggers a sprinkler. For the issue of installing sprinkler water after the battery smokes during the formation and capacity sorting process, a better solution needs to be found. In addition, accidents such as employees inhaling excessive concentrations of gas, causing adverse symptoms, electrical power lines sparking or burning, and sudden power outages in the workshop may occur.

(5) **Aging** Aging generally refers to the placement of the battery after the first charge and formation after the assembly and injection are completed. It can be normal temperature aging or high temperature aging, both of which are to stabilize the properties and composition of the SEI film formed after the first charge. The normal aging temperature is 25°C, the high temperature aging varies from factory to factory, some are 38°C, and some are 45°C, the time is between 48 ~72 h. The aging workshop places a large number of charged batteries at high temperatures for a long time, the battery separator is prone to aging and failure, and it is also easy to cause large-scale fires or explosions due to insufficient safety distance. In the battery production process, the aging process has the highest risk. Currently, the structures of aging rooms in different enterprises are different, and the safety measures taken are also different. This is the key to preventing fire and explosion accidents.

(6) **Packaging** The battery is already in a charged state before packaging, and is often stacked. Under such circumstances, defective battery products may self-ignite, or may cause larger-scale fires. On the other hand, the packaging workshop will store a large amount of packaging materials, and the packaging materials themselves are flammable materials, which can easily cause battery fires.

(7) **Raw Material Warehouse** The main risks in the raw material warehouse include the electrolyte and electrolyte packaging in the liquid raw material warehouse, which may leak or rupture during handling. The electrolyte leaks during production due to pipe detachment or cracking; waste liquid dumping in the cleaning room; all containers are not tightly capped or poorly sealed, etc. Therefore, these materials should be managed as general hazardous chemicals.

(2) **Second Dimension: Safety Management of Power Battery Storage**

Currently, there are many types of power batteries. In the daily storage of lithium batteries, it is necessary to base on relevant professional knowledge, and also need to be careful and cautious. As a metal that is particularly prone to chemical reactions, lithium batteries are prone to swelling and combustion, so if they are not handled according to regulations during storage, serious accidents such as combustion and explosion may occur. Clearly and accurately qualifying the characteristics of the stored lithium batteries can effectively formulate warehousing or distribution operation plans that are in line with product characteristics for warehousing and extended distribution services. Further provide theoretical and technical support to avoid accidents such as explosions and fires caused by improper storage of lithium battery goods. The following points should be noted during the storage of power batteries.

(1) The battery should be stored in a ventilated, dry, and cool environment.
(2) The stacking height of the battery box should not exceed 3 layers.
(3) Do not display or store the battery in places where it is directly exposed to sunlight or rain.
(4) The battery should be stored in its original packaging. If the packaging is opened, it needs to be repackaged and stored according to requirements.
(5) The battery storage charge state is 40% ~ 60%, the environmental temperature requirement is −20°C ~45°C (the recommended best is 0 ~ 35°C), and it should avoid contact with corrosive substances.
(6) When not in use for a long time, it is recommended to do a standard charge and discharge maintenance every about 3 months.
(7) During storage, disconnect the Manual Service Disconnect (MSD) and store it safely; ensure that the battery system's positive and negative terminals, as well as all ports, are properly protected to prevent electric shock.
(8) Avoid bumps, falls, and flooding during the storage of the battery system.
(9) There must be at least two sets of fire extinguishers in the battery storage area, two iron shovels, and the stored sand should not be less than 1 cubic meter, and the sand should be kept dry.

If the stored power battery causes combustion due to improper storage, changes inside the battery, or improper storage environment (high temperature, open flame, etc.), the following emergency measures should be taken.

(1) During the smoking stage, when the battery smokes but does not catch fire, it should be covered with sand and left for more than 1 h until it is completely extinguished.
(2) During the initial combustion stage, if the entire box of batteries catches fire, a combined method should be used to extinguish the fire. That is, first use a fire extinguisher to extinguish the open flame, then use sand to cover and extinguish the battery fire, and leave it for more than 2 h until it is completely extinguished.
(3) During the combustion stage, the combustion stage generally refers to multiple boxes of batteries catching fire. At this time, direct rescue is not advisable. The following measures should be taken:

A: Organize personnel to extinguish the fire.

B: Pay attention to personal safety and evacuate unrelated personnel.

C: Under windless conditions, on-site personnel will move the unburned batteries to a safe area, and at the same time, organize forklifts to move the unburned batteries to a safe area. Leave a fire gap of 6 m wide, and extinguish the fire or call the police as appropriate.

(3) **The Third Dimension: Safety Management of Power Battery Transportation**

Most of the current research on the safety of power batteries considers the safety under use, and the safety under transportation environment conditions is not considered enough. At present, all power battery transportation in China is carried out by road, and there are two types of exports, air and sea. Among them, air transportation has the best timeliness, but small capacity and high price. Sea transportation is relatively cheap, with larger capacity, but a long transportation cycle. Comparatively speaking, the timeliness of rail transportation is more than three times that of sea transportation, and it also has the advantages of large capacity and moderate cost. It is the best transportation method for long-distance and large-volume export of lithium batteries. However, fire incidents involving power batteries still occur during shipment due to non-standard operational practices. Transportation regulatory departments in various countries are particularly cautious about the transportation of power batteries, and many logistics companies explicitly state that they do not carry power batteries. The safe transportation of large quantities and high-energy power batteries has become a global focus issue, directly affecting the trade circulation of our country's power batteries, drones, and new energy vehicles.

The "Recommendations on the Transport of Dangerous Goods—Model Regulations" (21st edition) formulated by the United Nations is the benchmark for global dangerous goods transportation safety standards and norms, and many governments formulate dangerous goods transportation rules based on this. Since 2006, the "Model Regulations" has introduced lithium-ion batteries and gradually developed into the current lithium battery transportation regulation system. The "Model Regulations" classifies lithium battery transportation into lithium-ion batteries and lithium metal batteries. There are currently the following problems in lithium battery transportation:

(1) The inherent safety has not been resolved The chemical composition of the power battery determines its flammability. Although the safety standards and measures for power batteries are constantly advancing, the various factors that cause thermal runaway of power batteries are still difficult to completely avoid. Conflicts still exist between the demands of high energy density battery systems and established safety requirements. Different transportation modes have varying safety standards and risk tolerances. The causes of thermal runaway of power batteries are complex, and the existing countermeasures are not perfect. Unpredictable risks pose a great safety threat to air transportation.
(2) The transportation test system is not perfect The UN38.3 series test is the basic standard to ensure the safety of lithium battery transportation. This standard is

based on consumer-type small batteries, and does not propose targeted requirements for high-energy, large-mass, and multi-specification power batteries. The existing test clauses are incomplete and the technical indicators are lacking. Although it is being revised gradually, it still lags behind the development of power batteries.

(3) Lack of transportation safety evaluation capability. There is a limited number of independent labs capable of testing power batteries. Most third-party laboratories for lithium battery transportation testing do not have the testing capabilities for power batteries. According to regulations, power batteries with a larger energy density can seek not to perform UN38.3 series tests, which also leads to the lack of testing capabilities in various laboratories. The safety evaluation of power battery transportation should be based on valid test data. The current situation is that it is difficult to find a suitable laboratory for testing and obtain response data. Without effective data, appropriate test methods and standards cannot be established, and evaluation standards cannot be established. In addition, with the development of related industries, a large number of experimental, damaged, defective, recalled, and recycled power batteries need to be transported. These power batteries are not well-suited for UN38.3 series tests before transportation, therefore, a reasonable safety evaluation is particularly important.

(4) Insufficient risk control measures Appropriate packaging can effectively control the hazards of dangerous goods within the packaging, and achieve safe transportation. Due to the high energy density and large volume of power batteries, at present, there are no suitable packaging materials or forms to control the danger within the packaging, that is, the power battery is completely burned within the packaging and does not affect other items. Internationally, there has been small-scale use of designs such as fireproof covers or fireproof bags in the cabin, and similar products in China are still in the experimental stage. These packaging methods have not solved the problem of large amounts of toxic gases and dust produced by the combustion of power batteries. The packaging requirements are largely generic guidelines and serve only as provisional safety assurances. These requirements do not have specific technical indicators, lack experimental verification, and lack certification from authoritative institutions. This is also one of the main reasons why many modes of transportation have problems with the safety of power battery transportation.

(5) Lack of effective flame retardant and fire extinguishing methods The effectiveness of the flame retardant materials inside the power battery and the packaging materials outside still needs further verification. Studies have shown that 5 min after the thermal runaway of the bus power battery, the fire will spread to the car; at the same time, it is considered that the time for all personnel to evacuate is about 1 min. It is recommended that 5 min be used as the fire resistance time of the bus power battery. If under road transport conditions, 5 min is enough for road transport emergency evacuation, but it is far from enough for emergency under sea, railway and air transport. The fire extinguishing method of the power battery has not been determined, whether it is a large amount of water or gas fire

extinguishing agent, it is difficult to obtain enough in a short time during transportation. The high temperature, dense smoke and toxic gases produced by the combustion of power battery fuel will quickly cause casualties and equipment damage.

China's power battery industry is booming, and the large-scale transportation of power batteries is inevitable. Because power batteries have significant characteristics such as high specific energy, high consistency, and high voltage, the safety of transportation is different from that of consumer lithium-ion batteries. At present, there are still problems in the transportation of power batteries, such as the inherent safety has not been solved, the combustion mechanism and hazards are unclear, the transportation test system is not perfect, the transportation safety evaluation ability is lacking, the risk control measures are insufficient, and the effective flame retardant and fire extinguishing methods are lacking. In response to these problems, it requires the cooperation of government departments and enterprises, the close cooperation of manufacturing and transportation industries, and breakthroughs in multi-faceted research. Reasonably solve the safety issues of power batteries to meet transportation needs.

(4) **Fourth Dimension: Safety Management of Power Battery Use**

With the strong support of the Chinese government, China's new energy vehicle industry has made rapid progress, but safety accidents also occur from time to time. The power battery is one of the core components of new energy vehicles, and its safety and stability are crucial to the safe and reliable operation of electric vehicles. The safety risks faced by the power battery system are mainly related to the characteristics of its internal components and the external use and operating environment, which constitute the main challenges of product safety design. The risks that may occur during the use of power batteries include:

(1) Chemical Safety If the cell undergoes thermal runaway, it may cause electrolyte leakage, fire and combustion, but its destructive power is far from comparable to that of explosives. When explosives explode, energy is released in a very short time, so the power is huge, while the thermal runaway of the cell, the release of its energy is a gradual process, plus the battery pack of electric vehicles is composed of many cells in series and parallel, usually only 1 or a few cells fail, there is enough warning and disposal time.

For the cell, how to ensure chemical and thermal stability under various operating conditions and usage scenarios, and ensure that no safety risks are generated, this is a problem that must be considered and solved. The situations to consider include:

A: Normal working conditions within the rated range;

B: Long-distance transportation and long-term storage;

C: Extreme situations, such as overcharging, over-discharging, squeezing, puncturing, and burning of the cell;

In all situations, reasonable design goals must be determined for the safety of the cell, and they must be incorporated into the development process of the cell. For other components of the power battery system, chemical safety also involves chemical corrosion caused by electrolyte or coolant leakage (which may cause internal short circuits), salt spray corrosion, flame retardancy, and harmful gas emissions.

(2) Electrical Safety For the electronic and electrical system inside the power battery pack, electrical safety is the first consideration, and all safety risks related to "electricity" must be considered in safety issues.
(3) Functional Safety Functional safety refers to the battery management system, ensuring that the battery management system will not cause the failure of the safety system under any random failure, system failure, or common cause failure, thereby causing casualties, environmental damage, and equipment property loss; The safety protection function of the BMS should ensure normal operation whether under normal conditions or in the presence of specific faults. In order to ensure that the BMS reaches a certain level of functional safety, it must be based on electrical/electronic/programmable electronics, combined with other technologies in the system, fully consider the system's application environment, effectively identify hazards that affect the function of safety, thereby formulating reasonable safety goals, decompose the safety goals layer by layer, get safety requirements, and implement/distribute to each component in the system.

The content related to functional safety, the automotive industry's international standard for functional safety "ISO26262", has been implemented and applied by major foreign car companies. The domestic national standard for functional safety has issued the relevant "Road Vehicle Functional Safety". It can be seen that the composition of the power battery system is still quite complicated, including chemical objects such as battery cells, complex electronic and electrical systems and thermal management systems, and traditional mechanical parts. The professional types involved are very many, coupled with harsh operating environments, the safety risks faced are also very wide. The following points should be noted when using power batteries:

(1) When using and cleaning new energy vehicles, protect the power battery pack from impact and water ingress.
(2) When the meter shows low power, charge the power battery in time to avoid affecting the use of the power battery due to insufficient power, or even shorten the life of the power battery.
(3) When charging the power battery, the ambient temperature should be between 10°C ~30°C, and the battery should be well ventilated. Lower temperatures will affect charging efficiency and even cause sulfation. Higher temperatures can easily cause parameter drift of charger components, or even thermal runaway and battery swelling.
(4) In daily use, pure electric cars should not be driven to the maximum mileage as much as possible, the best mileage is 1/3 ~2/3 of the longest mileage.
(5) When the battery overheating alarm occurs during the driving of the electric vehicle, stop the car to cool the battery, and continue driving after the overheating warning light goes out; if the battery overheating alarm or battery fault

warning light is often on, it should be removed by professionals in time. Non-professionals are strictly prohibited from disassembling the battery pack by themselves to avoid personal injury.

(6) The charger used with the car must have sufficient accuracy and stable voltage and current performance, and pass the company's technical testing and certification. It is strictly forbidden to use chargers with poor quality, low price, and poor aging resistance, otherwise it will damage the power battery.
(7) When the pure electric vehicle is not in use, the battery should be fully charged for storage, and charged once a month to avoid sulfation of the power battery plate due to long-term power shortage.
(8) When the battery catches fire, use dry sand or a nitrogen fire extinguisher to extinguish the fire.

7.3 Early Warning of Multiple Faults in the Power Battery System

As we all know, abnormal fluctuations in various state parameters such as voltage, temperature, and SOC of the power battery system can both trigger various battery faults and may be the main performance characteristics of some battery faults. The common abnormal types of these parameters can be summarized as overvoltage, undervoltage, too fast voltage change, poor battery voltage consistency, too high temperature, too fast temperature rise, too large temperature difference, and SOC diving (i.e., sudden change). Table 7.1 lists the potential battery fault types related to the above battery abnormalities. In the actual operation of the vehicle, timely and accurate state prediction is of vital practical significance for preventing various battery faults.

In the above text, we saw LSTM's long and short-term prediction capabilities for various state parameters of the actual vehicle power battery system. Therefore, this section mainly explores the early warning ability of various power battery system faults based on the existing "threshold-change rate" safety threshold of the NMMP-NEV platform, using the LSTM long and short-term prediction results.

7.3.1 Voltage Faults

Using the actual vehicle operation data as the research object, the LSTM (many-to-one) structure is used. This model structure can avoid the mutual interference between multiple output parameters and has better voltage prediction and abnormal warning capabilities compared to "one-to-one" and "many-to-many" model structures. In order to avoid the impact of inconsistency between cell voltages on the voltage fault warning effect, the total voltage warning model and the cell voltage warning model are established separately. In addition, to reduce the number of models, save

Table 7.1 Various battery abnormalities and potential faults

Abnormal type		Possible battery faults
Total voltage	Overvoltage	The power battery system is overcharged and the charging protection circuit fails
	Undervoltage	The power battery system is over-discharged or the electrode is reversed
Cell voltage	Overvoltage	The power battery system is overcharged and the charging protection circuit fails
	Undervoltage	The power battery system is over-discharged or internally short-circuited
	Voltage changes too quickly	The battery capacity is too low or the connection between the batteries is loose
	Poor consistency	Some individual batteries age too quickly
Temperature	Temperature too high	Some individual batteries are internally short-circuited or there is an internal/external short circuit in the battery pack
	Temperature rises too quickly	The connection between the cells is loose; there is an internal short circuit in the battery or battery pack
	Temperature difference too large	Some cells age too quickly; the connection between the cells is loose; the thermocouple falls off
SOC	SOC plunges	The Hall forward current is too large, the feedback current is too small; the cell voltage is too low, the voltage drops too quickly

computing power and improve warning efficiency, a "head-to-tail series" (Head-To-Tail in Series, HTTS) method is proposed. All cell voltages are connected in series in order, and integrated into a single voltage vector. Training is needed only once to learn the data features of all cell voltages. The trained cell voltage warning model can be shared by all cells, and the prediction effect will not interfere with each other.

We take vehicle J as the research object, and the layout diagram of its power battery system is shown in Fig. 7.2. Among them, 80 battery cells (cell 1 to cell 80) are arranged horizontally, and the other 20 battery cells (cell 81 to cell 100) are arranged vertically.

Since the battery system under study has 100 battery cells, training and building 100 models separately will consume a lot of time, and when performing real-time voltage prediction on an actual running electric vehicle, calling 100 models at once will also seriously reduce the prediction efficiency. To solve this problem, this article connects the data of 100 cell voltages in series with the HTTS method to form a new set of cell voltage data for training, and obtains a universal cell voltage prediction model. This approach offers two advantages: first, it can simplify the final voltage model from 101 models (100 cell voltage models and 1 total voltage model) to two models, namely a total voltage model and a cell voltage model; second, the integrated cell voltage model contains all battery cell voltage data information.

Through the correlation analysis results, it can be known that after excluding all strongly related and weakly related parameters, the parameters that match the battery voltage are the brake pedal stroke value, SOC and vehicle speed. Therefore, we can get two LSTM (4–1) model structures of the total voltage model and the cell voltage model. In order to get the best training and prediction effect, the hyperparameters have been adjusted and optimized in multiple layers. The optimized hyperparameters are: Epoch = 50, BS = 16, TWS = 360, PWS = SWS = 6, and the training dataset is the full year data of the vehicle under study in 2016. The optimization prediction results of various window sizes are shown in Fig. 7.3, and other parameters are not repeated in this part (Fig. 7.4).

For all battery cells of the same material type, size, and factory batch, their OCV, capacity, internal resistance, cycle life and other performance parameters all conform to a normal distribution and a finite degree of dispersion. Therefore, a change in battery consistency means a change in the level of centralized distribution of performance parameters such as open circuit voltage, and this change can be characterized by the standard deviation. According to the theory of normal distribution, the number of single cells whose voltage value differs from the average voltage by $\pm 3\sigma$ should account for more than 99% of the total number of single cells; the better the consistency of the single cell voltage, the smaller the standard deviation, and the more centralized the distribution of the single cell voltage. The NMMP-NEV platform currently uses $\sigma = 0.05$ as the threshold to evaluate the consistency level of battery voltage. When the standard deviation of all battery voltages is within the range of $\sigma > 0.05$, it indicates that the voltage consistency of the current battery system is poor,

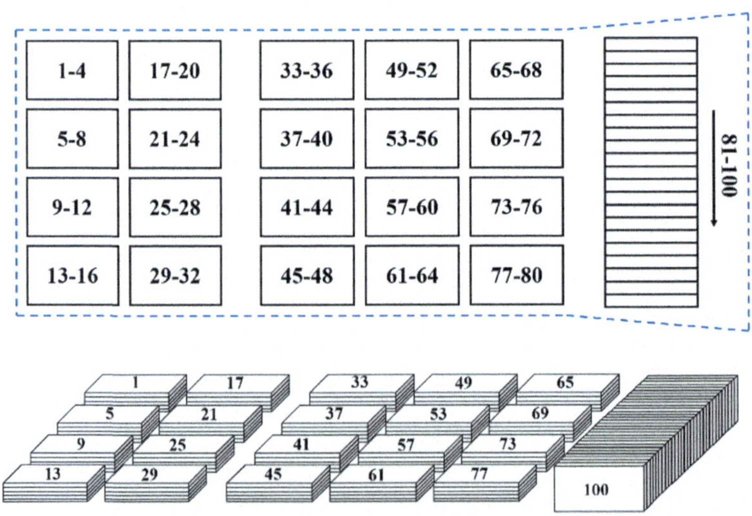

Fig. 7.3 Schematic diagram of the structure layout of the power battery system under study

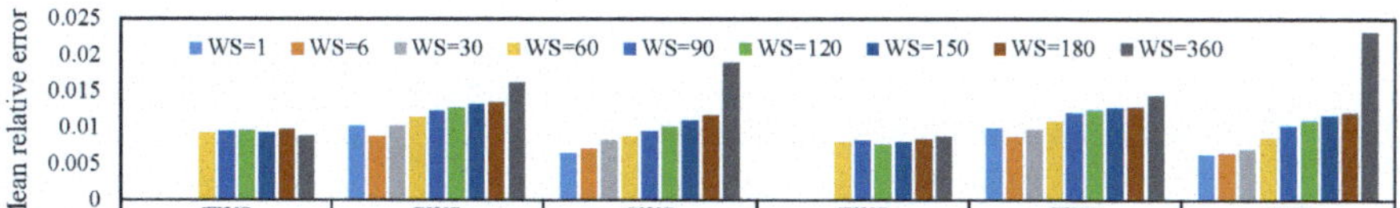

	TWS	PWS	SWS	TWS	PWS	SWS
		Cell voltage			Pack voltage	
WS=1		0.0102	0.0065		0.01	0.0063
WS=6		0.0088	0.0071		0.0089	0.0066
WS=30		0.0102	0.0082		0.0098	0.0071
WS=60	0.0093	0.0115	0.0088	0.0081	0.011	0.0089
WS=90	0.0095	0.0124	0.0095	0.0083	0.0121	0.0104
WS=120	0.0096	0.0128	0.0102	0.0077	0.0125	0.0111
WS=150	0.0093	0.0133	0.0111	0.0081	0.0128	0.0118
WS=180	0.0097	0.0135	0.0118	0.0085	0.0129	0.0122
WS=360	0.0088	0.0162	0.0189	0.0089	0.0145	0.0232

Fig. 7.4 Optimization prediction results of various window sizes

and vice versa. It can also be explained that the battery cells within the range of ± 0.15V from the average voltage need to account for more than 99% of all battery cells.

The single cell voltage, total voltage, rate of voltage increase and standard deviation curve of 100 single cells of vehicle J on the day of the test data are shown in Fig. 7.5. Figure 7.5 shows that the single cell voltage of the day shows high uniformity, but the standard deviation exceeds 0.05 at the moment of charging completion. In addition, it can be seen from the figure that the consistency of the power battery system during the charging phase is much better than during the driving phase, especially during the stable phase in the middle of charging, only at the beginning of charging and near the end of charging, the standard deviation of the single cell voltage is slightly larger. This is because the vehicle is always in a relatively stable operating condition during the charging phase, while during the driving phase, due to the complexity and uncertainty of road conditions, weather and driver behavior, the operating conditions of the battery system are also complex, so the consistency of the single cell voltage is difficult to guarantee.

The prediction results of single cell voltage (taking the voltage of cell 1 as an example) and total voltage before and after hyperparameter optimization are shown in Fig. 7.5, and the average relative error of all single cell voltage predictions is shown in Fig. 7.6. As can be seen from the two figures below, the two LSTM models after hyperparameter optimization have excellent demonstrate high prediction accuracy for both single cell voltage and total voltage, and the predicted values after optimization are closer to the real values; the LSTM single cell voltage prediction model has very small prediction errors for all cells, verifying the effectiveness and reliability of this method.

Table 7.2 lists the voltage abnormality pre-alarm and alarm levels classification and threshold of the studied battery type, where Upack is the total voltage, Ucell is the single cell voltage, △Ucell is the rate of change of single cell voltage, △Upack is the rate of change of total voltage, and σ is the standard deviation of all single cell voltages. Voltage abnormality pre-warning is divided into three levels, where

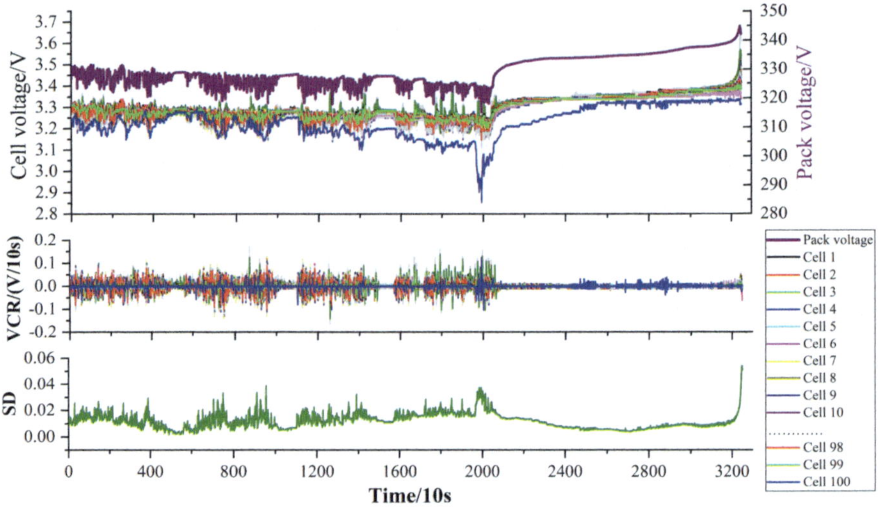

Fig. 7.5 Single cell voltage, total voltage, rate of voltage increase and standard deviation curve

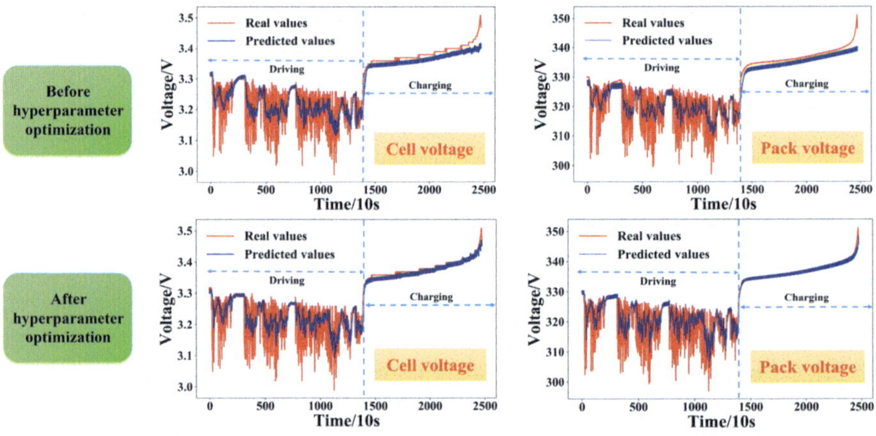

Fig. 7.6 Battery voltage prediction effect before and after hyperparameter optimization

the third level pre-warning state is a safe state, but it will pre-warn the poor battery voltage state; the second level pre-warning indicates that the battery is in a dangerous state, and it needs to stop the vehicle for inspection and log the location of the early warning trigger when conditions permit; when a first-level warning is triggered, the driver needs to immediately park the car in a safe place, and then wait for professionals to come for repair, to avoid accidents that may harm the driver and passengers. It should be emphasized that this threshold method is also universal for other types of batteries, and there are only slight differences in the setting of pre-alarm and alarm levels thresholds due to different battery types and parameters.

Table 7.2 Power battery system voltage abnormal pre/alarm threshold table

Parameter	Pre/alarm type	Classification rules	Level
Total voltage	Overvoltage	$3.6V*n \leq Upack < 3.9V*n$	2
		$Upack \geq 3.9V*n$	1
	Undervoltage	$2.5V*n < Upack \leq 3.0V*n$	3
		$2.0V*n < Upack \leq 2.5V*n$	2
		$Upack \leq 2.0V*n$	1
	Voltage Change Too Fast	$\lvert\Delta Upack\rvert \geq 0.4V*n/10s$	1
Single cell voltage	Overvoltage	$3.6V \leq Ucell < 3.9V$	2
		$Ucell \geq 3.9V$	1
	Undervoltage	$2.5V < Ucell \leq 3.0V$	3
		$2.0V < Ucell \leq 2.5V$	2
		$Ucell \leq 2.0V$	1
	Voltage changes too quickly	$\lvert\Delta Ucell\rvert \geq 0.4V/10s$	1
	Consistency is too poor	$\sigma \geq 0.05$	1

In order to verify the pre-warning effect of the well-trained offline model for various voltage faults, we retrieved and exported the operating data of the vehicle J power battery system on September 15, 2017, when a fault warning occurred, as test data. Figure 7.7 shows the battery voltage curve before and after prediction on this day. The car started operation at 09:27:10, and stopped uploading data after being fully charged at 19:29:30 in the evening, during which it drove and charged once. The 100th cell issued a third-level under-voltage warning of $Ucell < 3.0V$ at 15:56:20 in the afternoon, and the 19th cell issued a second-level over-voltage warning of $Ucell > 3.6V$ at 19:28:10. As can be seen from the figure, since the vehicle started driving that day, the 100th cell has shown different fluctuation characteristics from other batteries, it has deviated from other cells and has always maintained a lower voltage than other cells; the reason may be that the 100th cell has over-discharged or a slight short circuit inside the cell. In addition, at the end of charging, all cell voltages increased rapidly in a short time, but only the 19th cell had over-voltage, which may be due to the failure of the charging protection circuit or the rapid aging of the 19th cell causing its own overcharging.

As shown in Fig. 7.7, because $TWS = 360$ and $SWS = PWS = 6$, the data of the first hour of the test day can only be used as the initial training window during the test, and then predict the voltage for the next minute every minute from an hour later. By locally magnifying the over-voltage and under-voltage areas, we can see that all cell voltages, including the 19th cell voltage, the 100th cell voltage, and the total voltage, can be accurately predicted. Theoretically, the more stable the total voltage or a single cell voltage, the better the prediction effect, so the constructed LSTM model is highly reliable, not only can accurately predict abnormal voltages, but also will not cause false alarms due to poor prediction.

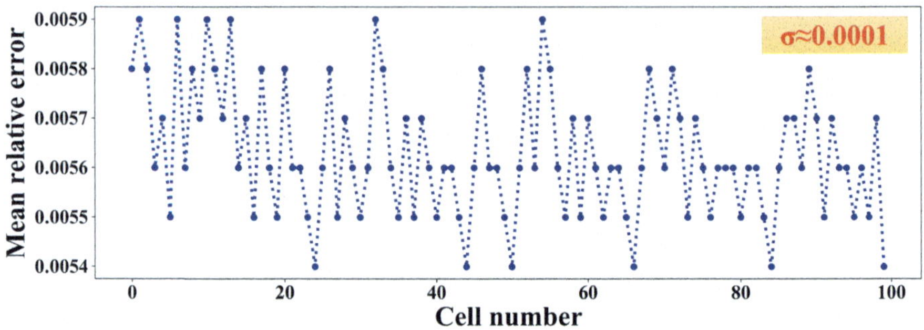

Fig. 7.7 Average relative error of all single cell voltage predictions

Because there was no abnormal voltage change on the test day, the rate of change of the single cell voltage has always been very small ($|\triangle Ucell|<0.4V/10s$). Figure 7.8 shows the curve of the rate of rise of all single cell voltages before and after prediction, the rate of change of voltage before and after prediction is very close, all stably distributed between −0.2 ~0.2, proving that this model can also accurately predict the rate of voltage change (Fig. 7.10).

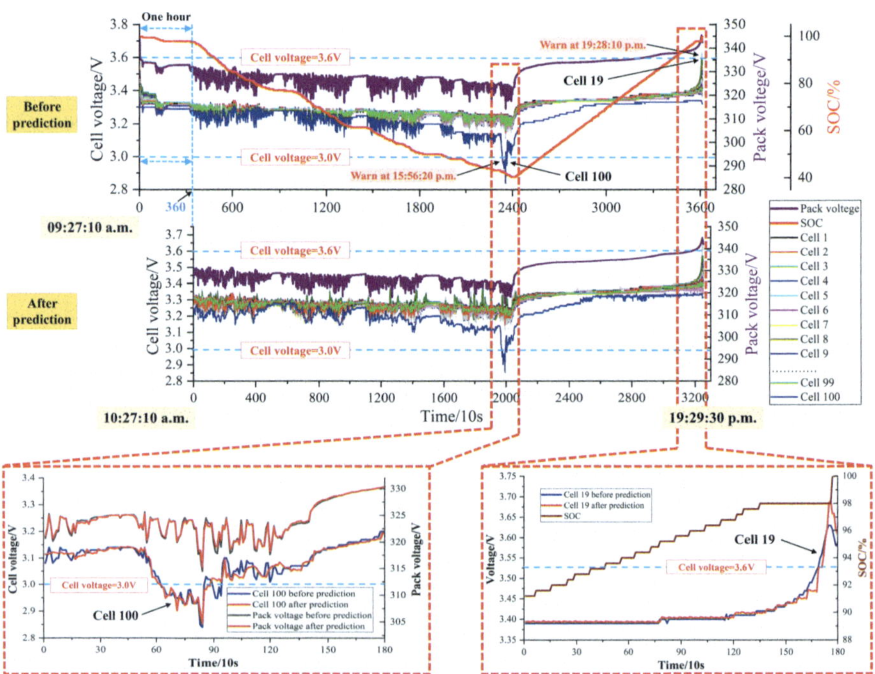

Fig. 7.8 Voltage curve before and after prediction of the studied power battery system

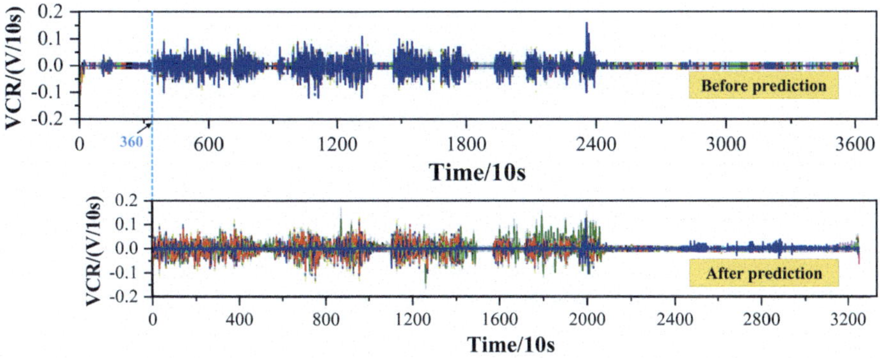

Fig. 7.9 Effect of single cell voltage temperature rise rate prediction

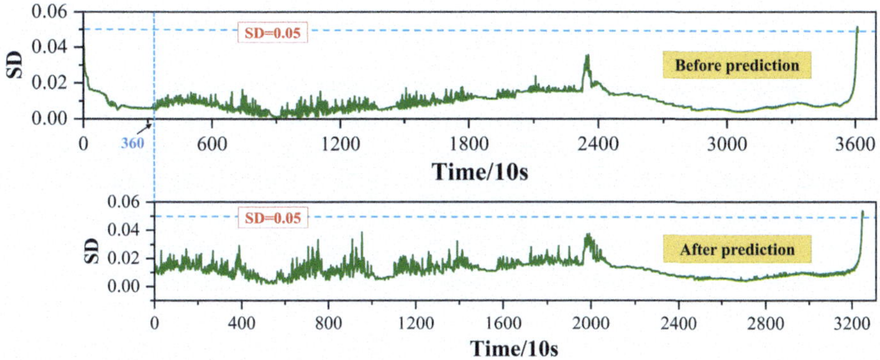

Fig. 7.10 Effect of single cell voltage standard deviation prediction

Figure 7.9 shows the standard deviation curve of single cell voltage before and after prediction. The standard deviation curve of single cell voltage before and after prediction is almost consistent, and the standard deviation exceeding 0.05 at the end of charging can also be accurately predicted.

7.3.2 Temperature Faults

Ambient temperature significantly affects battery temperature. As depicted in the temperature curve of Beijing in 2016 and the temperature curve recorded by Probe 1 on Vehicle J in Fig. 7.11, the size and change pattern of the probe temperature are very similar to the ambient temperature during spring, summer, and fall, when the temperature is above 0 °C, the temperature of the battery box probe changes almost synchronously with the ambient temperature; in the colder winter, the temperature often drops below 0°C, and the battery box usually needs some preheating

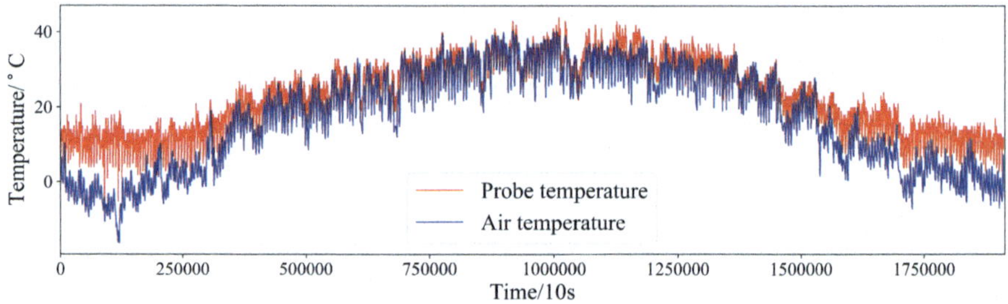

Fig. 7.11 Probe temperature and ambient temperature curve

measures or the temperature rises above 0 °C after the vehicle starts to work normally. According to the analysis results of different LSTM mapping structures and parameter correlations in the above text, all parameters other than the probe temperature itself are irrelevant to it, so the LSTM (1–1) model is more suitable for predicting the probe temperature.

Through hyperparameter optimization, a set of hyperparameters was finally determined: Epoch = 50, BS = 128, TWS = 360, PWS = SWS = 6, the training dataset is the one-year operation data of the studied vehicle in 2016, other parameters and optimization process are not repeated in this part. There are 16 temperature probes distributed inside and outside the battery enclosure. To improve modeling and early warning efficiency, the data of 16 probe temperatures are integrated and trained according to the HTTS method to obtain a general probe temperature prediction model.

Table 7.3 lists the types of temperature abnormality/fault pre/alarm types and corresponding thresholds of the studied power battery system type on the NMMP-NEV platform, where TDmax is the maximum temperature difference, and △Trmax is the maximum temperature rise rate. Temperature abnormality pre/alarm is generally divided into three levels, among which the third level pre-alarm state is a relatively safe state, but it is still necessary to focus on monitoring the temperature probe state that issues the third level pre-alarm; the second level pre-alarm indicates that the power battery system is in a dangerous temperature state, and it is necessary to stop the car to check the trigger pre-alarm position if conditions permit, and resolve the issue promptly; when a first-level alarm is triggered, the driver needs to immediately park the car in a safe place, and then wait for professionals to come for repair, to avoid a thermal runaway accident of the power battery system that may injure the driver and passengers. It needs to be emphasized that this threshold method is also applicable to other types of power battery systems, but due to differences in type and parameters, there are slight differences in the setting of pre/alarm thresholds.

To verify the online early warning effect of the trained offline LSTM temperature early warning model on various temperature abnormalities, we retrieved and exported the operation data of vehicle J on March 5, 2017, when an excessive temperature difference alarm occurred. Figure 7.11 shows the probe temperature and temperature rise curve on this day. The vehicle started running at 7:10 AM", and Probe 1 triggered

Table 7.3 Power battery system temperature abnormality pre/alarm threshold table

Pre/alarm type	Classification rules	Level
Over temperature	40°C ≤ T < 50 °C	3
	50°C ≤ T < 65 °C	2
	T ≥ 65 °C	1
Excessive temperature difference	5 °C ≤ TDmax < 8°C	2
	TDmax ≥ 8°C	1
Rapid temperature rise	ΔTrmax ≥ 5 °C/10s	1

a third-level pre-alarm due to an excessive temperature differential (T > 40 °C) at 09:11:43 in the morning. Since the vehicle started running that day, probe 1 has shown a fluctuation feature that is completely different from other probes, which means that the single cell or module near probe 1 in the battery pack may be experiencing overheating or other faults; fortunately, from the temperature rise rate curve, it can be seen that the temperature rise rate of all probes is still within the safe range of 5 °C/10s.

The abnormality of excessive temperature difference is usually caused by one (or more) temperature probe(s) being too high or too low. The probe temperature and temperature difference curve of the studied vehicle on November 3, 2017, are shown in Fig. 7.12. The vehicle started working at 07:55:47 in the morning, and a second-level warning of excessive temperature difference TDmax > 5 °C occurred at 18:14:55 in the afternoon. As can be seen from the figure, when all probe temperatures rise or fall, probe 16 is always at the maximum temperature position, but probe 11 began diverging from the overall temperature trend in the later charging stage and shows a relatively low temperature, finally triggering the warning of excessive temperature difference.

The one-hour data containing high temperature and excessive temperature difference abnormalities in the above two days (09:00:00–10:00:00 on March 5 and 18:00:00–19:00:00 on November 3) are used as prediction objects, and the data of the previous hour are used as the initial training window for sliding prediction. Figure 7.13 shows the temperature curve before and after the prediction based on the offline LSTM model. Figure 7.13a shows that the model has a very good prediction effect, and the predicted values of temperature and temperature rise rate at the time of high temperature abnormality are very close to the actual values; more importantly, the high temperature abnormality can be predicted timely and accurately, and the temperature rise rate close to the alarm threshold does not cause false alarms. Figure 7.13b shows that the prediction effect of the temperature and temperature difference of the two temperature probes (probe 11 and probe 16) causing the excessive temperature difference alarm is also highly accurate, not only the probe temperature value can be accurately predicted, but also the excessive temperature difference abnormality can be predicted timely and accurately.

Through the stability verification of more validation data, it is found that the LSTM model has very accurate and stable prediction effects for different test data and

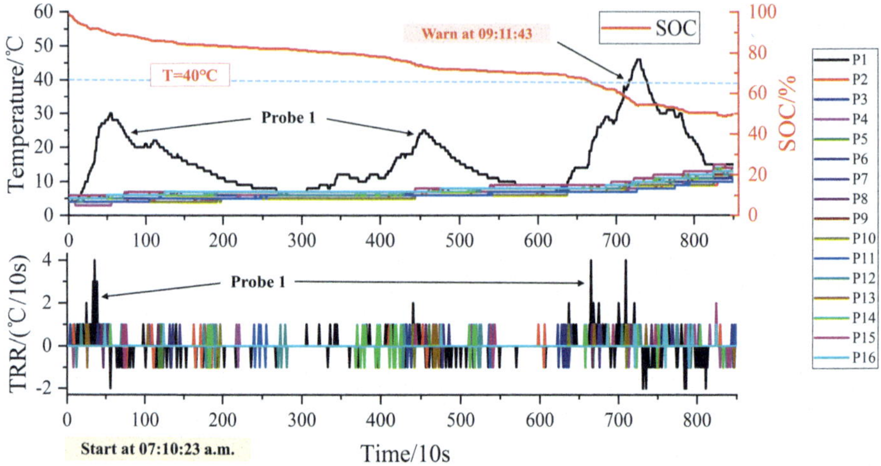

Fig. 7.12 Probe temperature and temperature rise rate curve on March 5, 2017

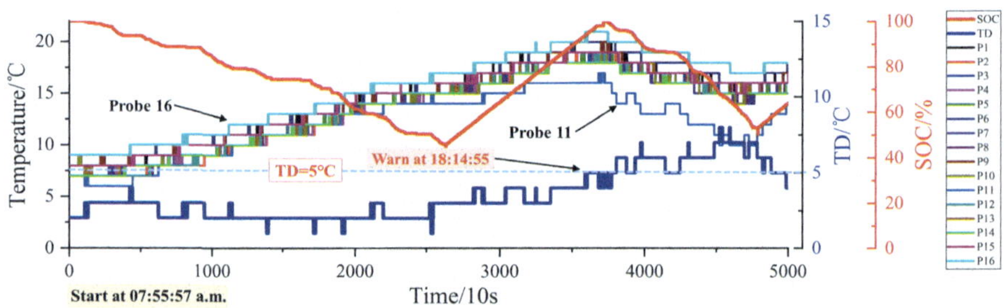

Fig. 7.13 Temperature and temperature difference curve of the probe on November 3, 2017

different types and levels of temperature abnormalities. Since we set the prediction window to PWS = 6, i.e., the next minute, it is possible to give at least one minute advance warning of the above temperature abnormalities. Moreover, our proposed model can flexibly adjust the size of the prediction window PWS as needed, such as extending the prediction window size to 5 min, 10 min or even longer time based on an acceptable prediction accuracy threshold when the training dataset is expanded to a sufficiently large size in the future (for example, the full lifecycle historical data of a car or multiple cars of a certain model). This has extremely important practical significance for temperature abnormalities that are very likely to cause various thermal faults or even thermal runaway accidents of the battery system, and it is very important for drivers and passengers to quickly escape or take emergency measures before danger occurs.

7.3.3 *Multi-fault Early Collaborative Diagnosis Strategy*

In addition to the battery voltage and temperature faults analyzed above, other types of fault characterization parameters of the power battery system can also be modeled and multi-step predicted using the method proposed in this paper, such as predicting whether the SOC will be too high, too low or suddenly dropping in the future during the actual vehicle operation, which can diagnose the battery faults that are likely to occur under these SOC abnormal conditions. Since there has been no SOC diving fault in the training data of the studied vehicle, we take SOC being too low as an example. On January 15, 2017, the power battery system of the research vehicle had a third-level warning of SOC <30%. The data half an hour after the alarm is used as the test data, and the data of the previous hour is used as the initial prediction window for testing. The prediction result of SOC is shown in Fig. 7.14. The results show that the SOC being too low can be predicted timely and accurately.

The continuous accumulation of more different fault characteristic parameters, different models, and different batches of vehicle operation data by the electric vehicle big data monitoring platform will provide us with more abundant training samples

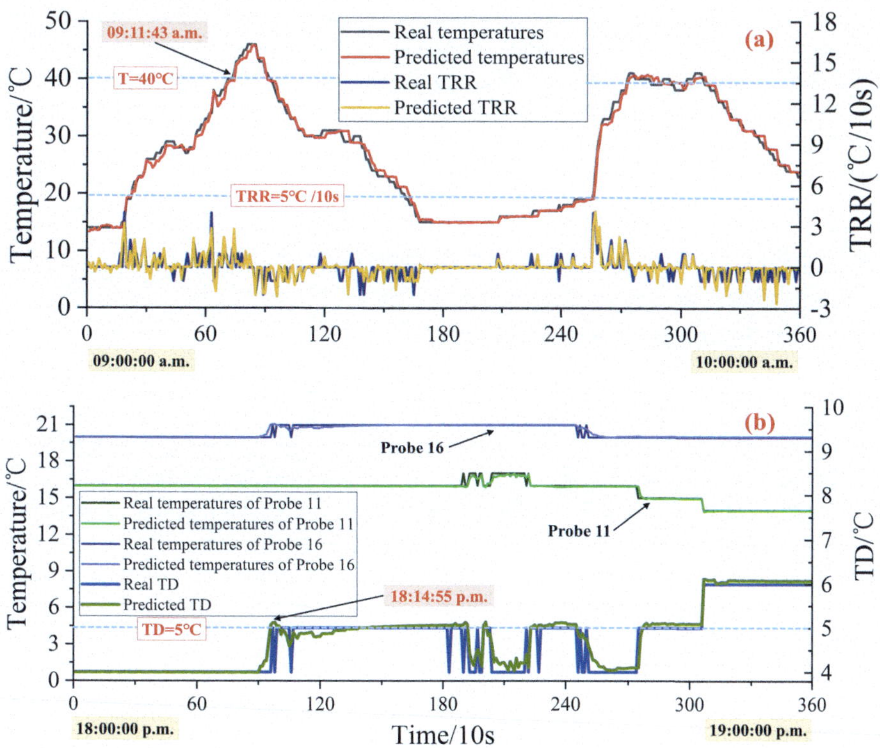

Fig. 7.14 Temperature prediction results: **a** Temperature and temperature rise rate curve before and after prediction; **b** Temperature and temperature difference curve before and after prediction

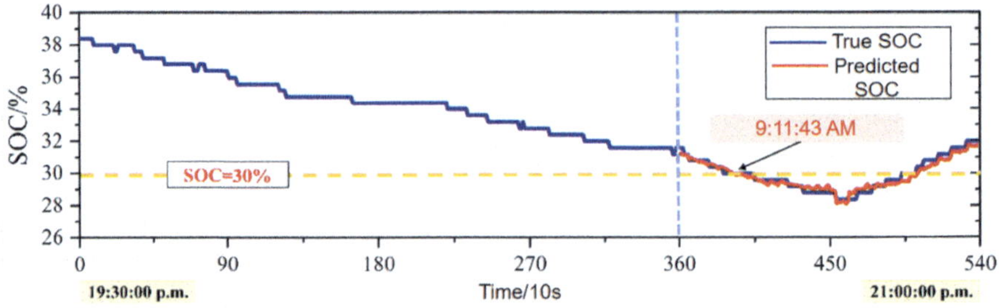

Fig. 7.15 SOC before and after prediction on January 15, 2017

in the future, ensuring consistent and reliable data support for us to establish more comprehensive and accurate prediction models. Through continuous optimization and updating of the model, the method proposed in this research has a broader practical application prospect. We have formulated a multi-stage research plan for the development of multi-fault collaborative warning for the operation of actual vehicle power battery systems, as shown in Fig. 7.15. Through the expansion of data from the full lifecycle of different models, the ultimate goal of this research is to establish a multi-fault collaborative warning model for power battery systems suitable for all stages, all states, and real-world conditions based on accurate multi-parameter collaborative prediction and flexible multi-step prediction window adjustment, to achieve real-time warning of various battery faults during the operation of actual vehicles.

7.4 Safety Control Strategy of Power Battery System

7.4.1 Multi-model Fusion Method

Although the previous text has implemented early warning of various power battery system failures based on the long and short-term prediction capabilities of LSTM, it is still based on conventional threshold and change rate research. In order to achieve early collaborative warning of different types of power battery system failures in "mechanical–electrical-thermal", and to achieve early accurate prediction of safety risks of power battery systems, it is still necessary to combine the proposed MOSE, MMSE, NDWD and other fault diagnosis models, to study the multi-fault collaborative warning strategy of power battery systems based on the "threshold-change rate-model", establish a multi-level warning system, quantitatively evaluate the risk status of power battery systems, and provide a basis for subsequent vehicle safety design and manufacturing, warning system development, and consumer vehicle choice.

From the perspective of algorithms or models, multi-model fusion methods can be divided into linear weighted fusion, cross fusion, waterfall fusion, diverse fusion, prediction fusion, and additive fusion methods.

(1) Linear weighted fusion method

The linear weighted fusion method is a commonly used fusion method from the model perspective, which refers to further weighting and combining the candidate results generated by different recommendation models to generate the final recommendation results. Since the quality of different model results may vary in different application scenarios, the fixed weighting system cannot take advantage of each model, so through mixing or secondary mixing, the results of different models can be weighted and combined in the most reasonable way, often achieving better recommendation effects, as shown in Fig. 7.16. In addition to simple linear models, weighted mixed models often use regression models, Boltzmann machines, gradient boosting decision trees, and other models.

(2) Waterfall fusion method

The waterfall fusion method is often used in recommendation scenarios with a large number of candidate sets. It refers to connecting multiple models in series, and then treating different recommendation models as filters with different granularities to achieve model fusion, as shown in Fig. 7.17. Through waterfall fusion, the filtering results of the previous recommendation model will serve as the candidate input for

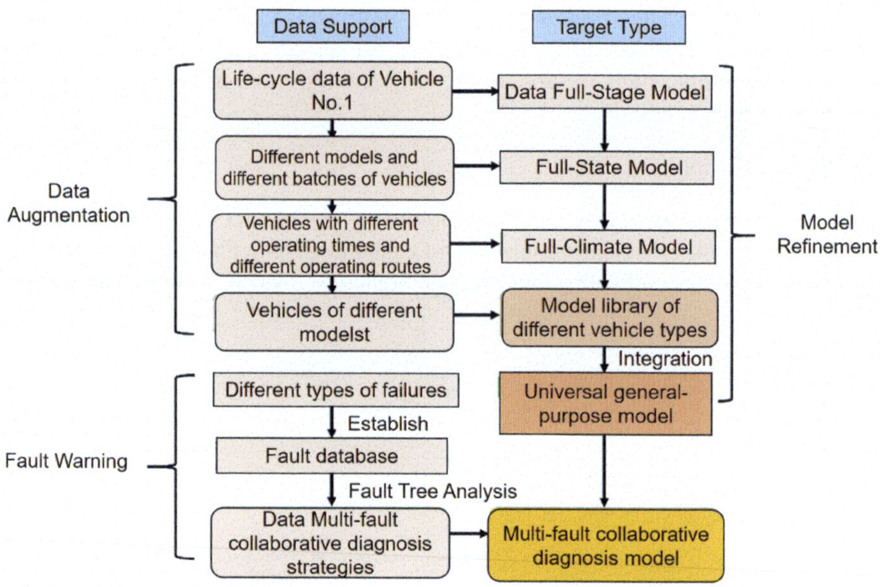

Fig. 7.16 Research plan for multi-fault collaborative diagnosis of actual vehicle power battery system

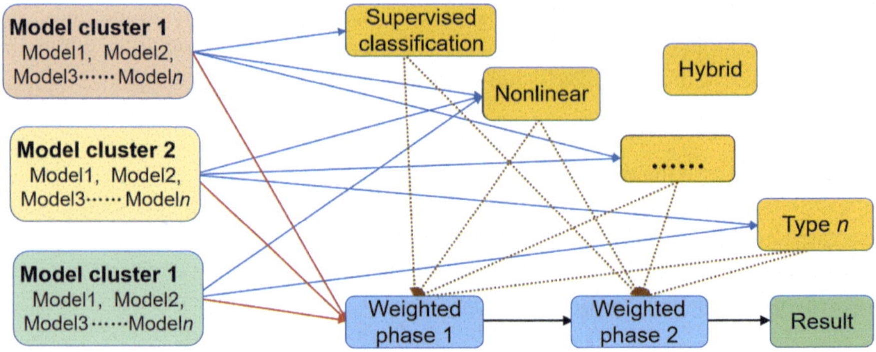

Fig. 7.17 Schematic diagram of weighted mixing technology recommendation

the next recommendation model, so the candidate results will be selected layer by layer in this process, and finally a small number of high-quality result sets will be obtained.

(3) Prediction fusion method

The prediction fusion method can also be regarded as a kind of "prediction algorithm". Its idea is to generate the final prediction results by establishing a second layer prediction model for the prediction results of different models, as shown in Fig. 7.18. The prediction results of different recommendation models can be used as new features, and the feedback data of prediction results can be used as secondary training samples. The second layer prediction model here can use common classification algorithms, such as support vector machines, random forests, maximum entropy, and other methods.

(4) Additive fusion

The additive fusion model is usually a nonlinear tree model composed of many branches. Different branches are nested and added to get the final prediction result. Each branch can extend more and smaller branches. During training, the fitting

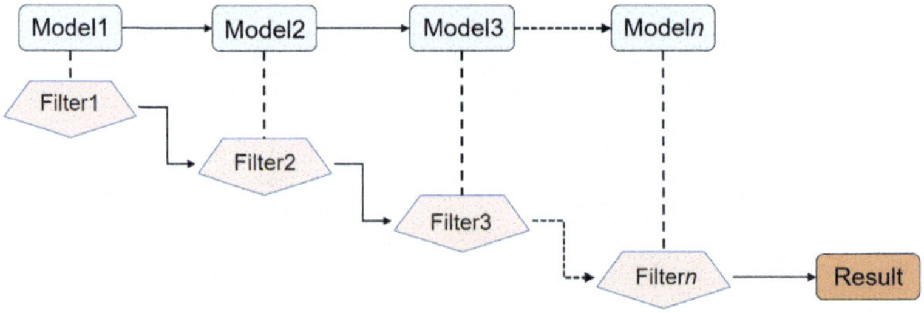

Fig. 7.18 Schematic diagram of waterfall fusion method

target of each branch is the residual between the sum of the prediction results of other branches and the real value. When a given number of branches is reached, the retrained branches will replace the previous branches one by one, and then the final converged prediction result will be obtained through multiple iterations.

(5) Diverse fusion

The difference here refers to the difference between model parameters, features, and samples. Different original data features and data quality have a great impact on the results of the recommendation model, so using data from different sources and extracting different features, input into the recommendation model for training and merging training results. This type of fusion method can effectively solve the common problem of data missing in engineering applications, and the fusion can enhance the model's ability to recognize data features and expand its applicability.

(6) Cross fusion method

The cross fusion method, also often referred to as the Blending method, its idea is to ensure the diversity of results by interweaving the results of different recommendation models in the recommendation results. The cross fusion method is suitable for recommendation scenarios where more results can be displayed at the same time, and is usually used for multiple models with obvious differences.

7.4.2 Safety Control Strategy Based on Multi-model Fusion

With the significant increase in data volume nowadays, data-driven power battery safety control strategies are an inevitable trend. The current single safety control strategy can no longer meet the high safety and reliability requirements of the electric system. As can be seen from the above sections, through the introduction of multi-model fusion methods, we can find that predictive fusion methods and diverse fusion methods are both suitable for the joint prediction of various states of the actual vehicle power battery system, because the data features of different state parameters are different, and the diagnostic effects of different fault diagnosis models are also different. Therefore, based on the LSTM's ability to predict the long and short term status of the power battery system and online SOH estimation, by integrating the MOSE, MMSE, NDWD models and other fault diagnosis models proposed in this paper, we can achieve "mechanical–electrical-thermal" different types of power battery system early warning of faults and real-time SOH assessment, and obtain a high safety, high reliability and high stability power battery system safety control strategy for the full lifecycle operation of the actual vehicle power battery system, as shown in Fig. 7.19, which has important engineering application value (Fig. 7.20).

In order to achieve the safety control management of the actual vehicle power battery system, future plans should start from BMS control and big data monitoring to carry out continuous optimization and expansion research on power battery system safety control strategies. In terms of BMS control, real-time monitoring of various

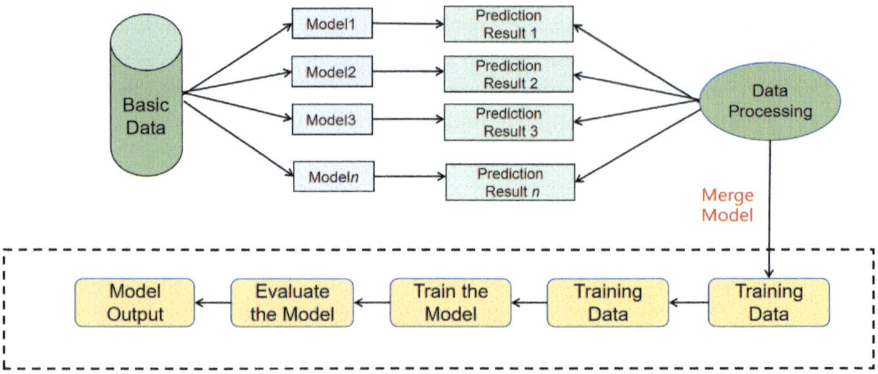

Fig. 7.19 Flowchart of prediction fusion method

states of the power battery system such as voltage, temperature, SOC, SOH, etc., will be carried out, and the LSTM model will be used to predict the long and short term states of the power battery system, providing a data basis for multi-fault early warning of the power battery system. In terms of big data monitoring, in response to the requirements of large capacity, fast response, high computing power, high compatibility and high reliability for thermal runaway warning of actual vehicle power battery system, big data management and analysis technologies such as Hadoop, Spark will be used to build a multi-level analysis system for data collection access, interactive application, storage management, analysis decision, process tracking and big data display on the big data platform, and through thermal runaway accident investigation results and traceability management, reverse trace the cause of the accident, reverse optimize the safety design of the power battery system.

In addition, research on key technologies for collaborative fusion management of different levels of data, models and functions will be carried out, rules for scheduling multi-level tasks such as vehicle information interaction, thermal runaway risk warning and decision management oriented to "vehicle-to-cloud (V2C)" will be formulated, combined with big data platform resource optimization and migration, massive high-load balancing and other technologies, a tiered mechanism for thermal runaway of power battery system and safety evaluation index system combining vehicle end and cloud end will be constructed, a power battery system safety control management system with high-quality prevention and control and decision output capabilities will be established, effectively improving the safety and reliability of the power battery system during actual vehicle operation.

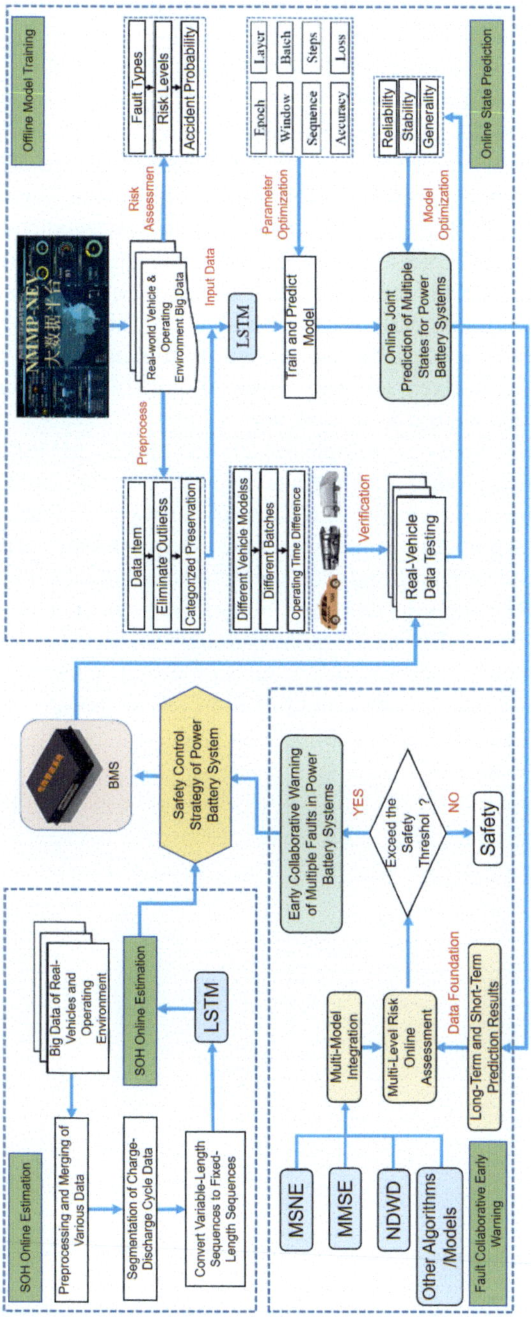

Fig. 7.20 Schematic diagram of safety control strategy for actual vehicle power battery system

7.5 Summary of This Chapter

(1) This chapter systematically summarizes the common safety problems and solutions of power battery systems, such as cell safety, design safety, group safety, etc. Based on the existing "threshold-change rate" alarm rules of the big data monitoring platform, a highly efficient early warning method for multiple faults of the actual vehicle power battery system is proposed using LSTM long and short term prediction results and HTTS method. The trained single cell voltage warning model can be shared by all single cells, and the prediction effect will not interfere with each other. Based on accurate multi-parameter collaborative prediction and flexible multi-step prediction window adjustment, a power battery system multi-fault early warning model with full state and full climate application capabilities is established, providing sufficient time guarantee and technical support for drivers and passengers to quickly escape danger or take emergency measures before danger occurs.

(2) Based on the fusion theory of multiple models, the MOSE, MMSE, NDWD and other fault diagnosis models are integrated to propose a "threshold-change rate-model" multi-fault collaborative diagnosis model for power battery systems, achieving online quantitative assessment of the safety risk status of power battery systems. Combined with long and short term prediction of power battery system status, online SOH estimation and traceability management, a high safety, high reliability and high stability safety control strategy for the full life cycle operation of actual vehicle power battery system is proposed, achieving monitorable and traceable safety status within the full life cycle of power battery system, which has important engineering application value.

The manufacturer's authorised representative in the EU is Springer Nature Customer Service Centre GmbH, Europaplatz 3, 69115 Heidelberg, Germany. If you have any concerns regarding our products, please contact ProductSafety@springernature.com

Printed and bound by CPI Group (UK) Ltd, Croydon, CR0 4YY
21/09/2026
02220009-0005